ENVIRONMENTALSTATS for S-PLUS

Springer
New York
Berlin
Heidelberg
Barcelona
Budapest
Hong Kong
London
Milan
Paris
Singapore
Tokyo

Steven P. Millard

EnvironmentalStats for S-Plus

User's Manual for Windows and UNIX

With 150 figures

Springer

Steven P. Millard
Probability, Statistics & Information
7723 44th Avenue NE
Seattle, WA 98115-5117
USA

Library of Congress Cataloging-in-Publication Data
Millard, Steven P.
ENVIRONMENTALSTATS for S-PLUS : user's manual for Windows and UNIX / Steven P. Millard.
p. cm.
Includes bibliographical references.
ISBN 0-387-98486-0 (softcover : alk. paper)
1. Environmental sciences—Statistical methods—Data processing.
2. S-Plus. I. Title.
GE45.S73M55 1998
363.7′0072—dc21 98-4685

Printed on acid-free paper.

Production managed by Allan Abrams; manufacturing supervised by Thomas King.
Photocomposed copy prepared using the author's Microsoft Word files.
Printed and bound by Maple-Vail Book Manufacturing Group, York, PA.
Printed in the United States of America.

9 8 7 6 5 4 3 2 1

ISBN 0-387-98486-0 Springer-Verlag New York Berlin Heidelberg SPIN 10669767

How to Use This Book

This manual is divided into 11 chapters, plus three appendices. Everyone should read **Chapter 0: A Quick Start**, which describes the module ENVIRONMENTALSTATS for S-PLUS, system and user requirements, how to install the software, how to start the software, and how to get technical support. The last section of the chapter, **A Tutorial**, may be skipped, although you may want to revisit it after learning more about ENVIRONMENTALSTATS for S-PLUS.

Chapter 1 is an introduction to environmental statistics in general and ENVIRONMENTALSTATS for S-PLUS in particular. Chapter 2 contains details about starting and using ENVIRONMENTALSTATS for S-PLUS. Chapters 3-9 contain information about how to use the functions in ENVIRONMENTALSTATS for S-PLUS to design sampling programs and perform graphical and statistical analyses of environmental data. Finally, Chapter 10 discusses Monte Carlo simulation and probabilistic risk assessment, and shows you how to use S-PLUS and ENVIRONMENTALSTATS for S-PLUS functions to perform probabilistic risk assessment.

Appendix A contains an extensive list of references for environmental statistics. Appendix B contains the license agreement for ENVIRONMENTALSTATS for S-PLUS. Appendix C contains an index of key words and terms. In addition to using the index in Appendix C, you are encouraged to use the on-line hypertext help system. The help system contains a utility called **Index** that let's you search for a topic by keywords. Windows users may also use the **Find** utility to search for all occurrences of a word or phrase in the help system.

A Note About Color

It is very easy to produce color plots in S-PLUS and ENVIRONMENTALSTATS for S-PLUS. Many of the built-in plotting functions produce color plots by default. In this manual, however, all of the plots are black-and-white or gray-scale due to the high cost of color printing. Commands for producing color plots are still included in the examples in this manual, but the pictures in the manual will be in black-and-white, whereas the pictures on your computer screen will be in color.

Typographic Conventions

This manual uses the following typographic conventions:

- The *italic font* is used for user-supplied variables within S-PLUS, Windows, and UNIX commands. Sometimes it is also used for emphasis.
- The ***bold italic font*** is used for emphasis.
- The **bold font** is used for Windows and UNIX commands and filenames, as well as for chapter and section headings. Sometimes it is also used for emphasis.
- The **`bold courier font`** is used to display what you type within S-PLUS.
- The `regular courier font` is used to display the names of S-PLUS objects and output from S-PLUS.

Displayed S-PLUS commands are shown with the default S-PLUS prompt > (the greater than sign). Commands that require more than one line of input are displayed with the S-PLUS continuation prompt + (the plus sign).

Cutting and Pasting Examples From the Help Files into S-PLUS

In this manual and the help files, S-PLUS commands are shown with the default S-PLUS prompt (>) and continuation prompt (+), so if you cut S-PLUS commands displayed in the help files and paste them into the S-PLUS command window S-PLUS will indicate syntax errors. Windows users can use the **Command History** tool under the **Tools** menu in the S-PLUS command window to select the lines just pasted into S-PLUS, edit them to remove the S-PLUS prompts, and re-execute them. Alternatively, both Windows and UNIX users can cut the lines from the help files, paste them into a text editor, remove the prompts in the text editor, and then copy the lines from the text editor and paste them into S-PLUS to reproduce the results shown in the help files.

ENVIRONMENTALSTATS for S-PLUS and S-PLUS Version 4.0

ENVIRONMENTALSTATS for S-PLUS Version 1.0 was developed between 1993 and 1997, and runs under S-PLUS for Windows Version 3.3, and S-PLUS for UNIX Version 3.4. ENVIRONMENTALSTATS for S-PLUS Version 1.1 is a slight modification of Version 1.0 and runs under S-PLUS for Windows Version 4.0. This manual was written assuming the user is using one of these version of S-PLUS.

The new version of S-PLUS for Windows, Version 4.0, differs substantially from past versions of S-PLUS in terms of the graphical user interface (GUI). Past versions of

S-PLUS for Windows all relied on the user typing in commands in the command window. S-PLUS 4.0 for Windows includes a point-and-click interface for creating graphs and performing statistical analyses, but the command window still exists, and S-PLUS 4.0 is backwards compatible with functions written in earlier versions of S-PLUS.

You can therefore use ENVIRONMENTALSTATS for S-PLUS with S-PLUS 4.0, but some of the basic S-PLUS commands listed in this manual may be replaced with different commands. For example, in S-PLUS 4.0, the `graphsheet` function replaces the `win.graph` function. (In fact, you don't even need to start a graphics device before plotting; one will open automatically when you click on a plotting icon or call a plotting function in the command window.)

The next major release of ENVIRONMENTALSTATS for S-PLUS for Windows will run under S-Plus 4.0 or later versions, *and* use the built-in GUI capability of S-PLUS 4.0.

Acknowledgments

In the early 1980's, while pursuing a degree in biostatistics, I became aware of a knowledge gap in the field of environmental statistics. There was lots of research going on in the academic field of environmental statistics, but there were lots of poor designs and analyses being carried out in the real-world "field". There were even federal laws mandating incorrect statistical analyses of ground water samples at hazardous waste sites.

One of the first steps to improving the quality of environmental statistics is to improve the quality of communication between statisticians and experts in various environmental fields. Several researchers and practitioners have done this by publishing excellent books and journal articles dealing with general and specific problems in environmental statistics (see the help topic **References for Environmental Statistics**).

The next logical step is to provide the necessary tools to carry out all the great methods and ideas in the literature. In the mid-1980's I sat down with two friends of mine, Dennis Lettenmaier and Jim Hughes, to talk about building a software package for environmental statistics. Dennis and Jim gave me some good ideas, but I never acted on these ideas because I didn't feel like I had the tools to make the kind of software package I wanted to.

The emergence of S-PLUS and help-authoring tools changed all that. S-PLUS, from MathSoft, Inc., is a premiere statistical software package with great graphics that allows users to write their own functions and create pull-down menus. ForeHelp, from ForeFront, Inc., is one of several help-authoring tools that allows anyone who can figure out a word processor to write a hypertext Windows help system and even convert it into an HTML format. The availability of these tools resulted in the creation of ENVIRONMENTALSTATS for S-PLUS.

There are several people who have helped and encouraged me over the past 4 years as I developed ENVIRONMENTALSTATS for S-PLUS. First, I would like to thank Gilbert FitzGerald, former Director of Product Development at MathSoft and now at Sun Micro Systems, for his encouragement and advice. Thanks also to Jim Hughes (University of Washington) for being a statistical consultant's statistical consultant. Thanks to Charlie Roosen and Stephen Kaluzny of MathSoft, and the MathSoft technical support team for answering my numerous questions. Special thanks to the members of the S-news group for constantly giving me insight into the workings of S-PLUS. Also thanks to Tim Cohn and Dennis Helsel (US Geological Survey), Chris Fraley (MathSoft and University of Washington), Jon Hosking (IBM Research Division), Ross Prentice (Fred Hutchinson Cancer Research Center), and Terry Therneau (Mayo Clinic) for their help in answering technical questions. Dick Gilbert has been a constant source of encouragement and advice; thanks! Thanks to the Beta testers for their feedback on early versions.

There would be no ENVIRONMENTALSTATS for S-PLUS if there was no S-PLUS. There would be no S-PLUS if there had not first been S. I am grateful to the researchers of the Bell Laboratories S team at AT&T (now Lucent Technologies) who first created S, including Richard A. Becker, John M. Chambers, Alan R. Wilks, William S. Cleveland, and others. I am also grateful to Doug Martin for leading the creation of S-PLUS, and to MathSoft for constantly enhancing and improving S-PLUS.

I am grateful to John Kimmel and Allan Abrams at Springer-Verlag for their help in transforming this User's Manual into a book.

Finally, and most gratefully, thanks to my wife Stacy and my son Chris for their moral and financial support of this project. I couldn't have done it without you!

ENVIRONMENTALSTATS for S-PLUS is the culmination of a dream I had over ten years ago. I hope it provides you with the tools you need to "get the job done", and I hope you enjoy it as much as I have enjoyed creating it!

Sincerely,

Steven P. Millard, Ph.D.
Statistical Consultant

February, 1998

Table of Contents

0 A Quick Start 1

What is ENVIRONMENTALSTATS for S-PLUS? 1

System Requirements 1

User Pre-requisites—What You Need to Know 1

Installing ENVIRONMENTALSTATS for S-PLUS 2

Installing ENVIRONMENTALSTATS for S-PLUS Under Windows 2

Installing ENVIRONMENTALSTATS for S-PLUS Under UNIX 4

Starting ENVIRONMENTALSTATS for S-PLUS 6

Getting Help for ENVIRONMENTALSTATS for S-PLUS 6

Technical Support 7

A Tutorial 9

1 Introduction 23

What is Environmental Statistics? 23

What is ENVIRONMENTALSTATS for S-PLUS? 24

Intended Audience/Users 25

Summary 26

2 Getting Started 27

Starting EnvironmentalStats for S-Plus 27

Getting Help for ENVIRONMENTALSTATS for S-PLUS 28

Changing Font Sizes in the Help Files 29

Using the Help System 30

Removing ENVIRONMENTALSTATS for S-PLUS From Your S-PLUS Session 31

Customizing EnvironmentalStats for S-Plus 31
Customizing S-Plus 4.0 Under Windows 95 32
Customizing S-Plus under UNIX 35
Summary 37

3 Designing a Sampling Program 39

The Necessity of a Good Sampling Design 39
What is a Population and What is a Sample? 39
Random vs. Judgment Sampling 41
The Basic Process of Designing a Sampling Program 41
Common Mistakes in Environmental Studies 43
Power and Sample Size Calculations 43
Sample Size Calculations for Confidence Intervals 44
Power and Sample Size Calculations for Hypothesis Tests 50
Summary 58

4 Looking At Data 61

Summary Statistics 61
Graphs for Univariate Data 64
Graphs for Bivariate and Multivariate Data 96
Summary 115

5 Probability Distributions 117

Overview 117
What is a Random Variable? 127
What is a Probability Distribution? 130
Probability Density Function (pdf) and Probability Mass Function (pmf) 132
Cumulative Distribution Function (cdf) 136
Quantiles and Percentiles 139
Characteristics of Probability Distributions 140
Generating Random Numbers From Probability Distributions 143
Important Distributions in Environmental Statistics 144
Summary 163

6 Estimating Distribution Parameters and Quantiles 165

Overview 165

Methods for Estimating Distribution Parameters 168
Comparing Different Estimators 175
Confidence Intervals for Distribution Parameters 175
Estimating Parameters for Important Distributions in Environmental Statistics 177
Estimating and Constructing Confidence Intervals for Quantiles 185
Summary 193

7 Prediction and Tolerance Intervals 195

Overview 195
Prediction Intervals 197
Tolerance Intervals 205
Summary 213

8 Hypothesis Tests & Statistical Models 215

Overview 215
Hypothesis Tests 218
Goodness-of-Fit Tests 221
Nonparametric Two-Sample Tests 229
Nonparametric Tests for Trend 234
Summary 241

9 Censored Data 243

Overview 243
Terminology 245
Defining the Detection Limit 247
Commonly Used Data Qualifiers 248
Graphical Assessment of Censored Data 249
Estimating Distribution Parameters 258
Goodness-of-Fit Tests 261
Two-Sample Linear Rank Tests 266
Summary 268

10 Monte Carlo Simulation and Risk Assessment 269

Overview 269
Mathematical Models and Monte Carlo Simulation 271
Uncertainty and Sensitivity Analysis 286

Risk Assessment 295
Guidelines for Conducting and Reporting a Probabilistic Risk Assessment 303
Using EnvironmentalStats for S-Plus for Probabilistic Risk Assessment 305
Summary 341

A References 343

B License Agreement 369

C Index 373

CHAPTER 0 :

A Quick Start

This chapter provides you with all of the information you need to quickly start using ENVIRONMENTALSTATS for S-PLUS. First there is a brief explanation of this module, followed by sections that give information on system requirements, user pre-requisites, instructions for installation, starting the module, and technical support. The last section is a brief tutorial that highlights the major features of ENVIRONMENTALSTATS for S-PLUS. The chapters following this one provide more detailed explanations of ENVIRONMENTALSTATS for S-PLUS.

What Is ENVIRONMENTALSTATS for S-PLUS?

ENVIRONMENTALSTATS for S-PLUS is an S-PLUS module for environmental statistics. It is an add-on module to the statistical software package S-PLUS, and provides a set of powerful yet simple-to-use functions for performing graphical and statistical analyses of environmental data. See **Chapter 1: Introduction** for more information.

System Requirements

Because ENVIRONMENTALSTATS for S-PLUS is an S-PLUS module, it runs under every operating system that S-PLUS runs under, including Windows 3.1, Windows 95, Windows NT, and UNIX. Windows users must have S-PLUS Version 3.3 or later, and UNIX users must have S-PLUS Version 3.4 or later. The module requires approximately 30 MB of disk space.

User Pre-requisites—What You Need to Know

ENVIRONMENTALSTATS for S-PLUS is an S-PLUS module. In order to use ENVIRONMENTALSTATS for S-PLUS, you need to know how to perform basic operations in S-PLUS, such as reading data into S-PLUS, creating basic data objects (e.g., vectors, matrices, and data frames), and using built-in functions. See the S-PLUS documentation for more information on S-PLUS. In addition, you should have a basic knowledge of probability and statistics.

Installing ENVIRONMENTALSTATS for S-PLUS

Installing ENVIRONMENTALSTATS for S-PLUS is a simple process, but the details differ depending on whether you are installing on a Windows or UNIX system. This section contains detailed procedures for both systems. If you have any difficulty with the installation, please contact PSI at **(206) 528-4877**, or send e-mail to the following address: **support@probstatinfo.com**.

Before you install ENVIRONMENTALSTATS for S-PLUS, you should make sure you are using S-PLUS Version 3.3 (Windows) or Version 3.4 (UNIX) or a later version. To determine what version of S-PLUS you are using, start S-PLUS, and then type `version` at the S-PLUS prompt. S-PLUS will print out the version you are using. For example, if you are using S-PLUS Version 4.0 for Windows, you should see the following:

```
> version
Version 4.0 Release 3 for Microsoft Windows : 1997
```

Note: If you just want to install the help files for ENVIRONMENTALSTATS for S-PLUS, you do not need to have S-PLUS installed on your computer.

Installing ENVIRONMENTALSTATS for S-PLUS Under Windows

To install ENVIRONMENTALSTATS for S-PLUS under Windows 3.1 or Windows 95, close all windows applications, and then simply run the **Setup** program included on the ENVIRONMENTALSTATS for S-PLUS distribution disks.

To leave **Setup** before the setup is complete, select **Cancel** from any **Setup** dialog box. When you do this, **Setup** displays an exit dialog box that warns you that you have not completed the setup procedure.

To run the **Setup** program, follow these steps:

1. Start Microsoft Windows 3.1 or Windows 95 if it is not already running. Make sure no Windows applications are running.

2. Insert the disk labeled "**ENVIRONMENTALSTATS for S-PLUS** Disk 1 of 3" ("1 of 4" for Version 1.0) in the appropriate drive, A or B.

3. Choose **Run** from the **File** menu of the Program Manager (Windows 3.1), or simply click on the **Start** button and choose **Run** (Windows 95).

4. In the **Command Line** text field of the **Run** dialog, type **a:\setup** and then press **Enter** or click the **OK** button. (If your disk drive is drive **B:**, type **b:\setup** instead.)

5. At this point **Setup** displays a message that it is initializing itself. Click on **OK** or press **Enter** to continue the installation. Now the installation screen appears and then the ***Welcome!*** dialog box that welcomes you to the Setup program. Click the button labeled **OK** or simply press **Enter**.

6. **Setup** now displays the ***Select Destination Directory*** dialog box. **Setup** will try to find the directory where S-PLUS for Windows is installed. If **Setup** finds the directory, it displays that directory in the **Destination Directory** text box. If **Setup** does not find the directory, the default destination directory is **C:\SPLUSWIN** under Version 1.0 and **C:\Program Files\Splus4** under Version 1.1. To change the destination directory, type the complete path of the directory (including the drive) in the **Destination Directory** text box. Once you are satisfied with the destination directory, click on **OK** or simply press **Enter**.

> **Note:** If you do not have S-PLUS for Windows installed on your computer and you just want to install the help files, you may specify any installation directory that you wish. The actual directory where the files will be installed, however, is
>
> ***SDIR*\MODULE\ENVSTATS**
>
> where ***SDIR*** is the name of the directory you choose in the **Destination Directory** text box.

7. **Setup** now displays the ***Select Components to Install*** dialog box. By default, both the Program Files and Help Files are installed. Only the Program Files are required to use ENVIRONMENTALSTATS for S-PLUS. Also, you may install just the Help Files if you don't have S-PLUS for Windows installed on your computer.

   The components selected for installation are indicated with an X in the associated box. The disk space required to install each component is also indicated. The Program Files require approximately 3 MB, and the Help Files require approximately 10 MB (Version 1.0) or 22 MB (Version 1.1). Click on a box to select or unselect an option as needed. When you have chosen which components to install, click on **OK** or simply press **Enter**. **Setup** installs the selected files to the appropriate sub-directories of the destination directory.

8. **Setup** now displays the ***Add Help Icon to S-PLUS Program Group*** dialog box. **Setup** will add an icon for the ENVIRONMENTALSTATS for S-PLUS Help System to your Program Manager. The default destination program group for the icon is **S-PLUS for Windows** (Version 1.0) or **S-PLUS 4.0** (Version 1.1). This default program group is usually the name of the group that holds your other S-PLUS icons.

   To use a different group name, select from the existing groups displayed in the window, or type in your own new group name in the **S-PLUS Group Name** text box.

   Once you are satisfied with the group name, click on **OK** or simply press **Enter**.

9. **Setup** now displays the ***EnvironmentalStats for S-PLUS Serial Number*** dialog box. Enter your name, company, and ENVIRONMENTALSTATS for S-PLUS serial number in the appropriate text fields. If you do not have a company name, simply type a space in this text box (you must type something). The serial number is located on your ENVIRONMENTALSTATS for S-PLUS floppy disk; it is the 14-character string beginning with "ESWN" (for example, ESWN-011-00256).

   When you have finished entering this information, click on **OK** or press **Enter**. **Setup** will prompt you asking you to verify the information you have typed in. Click on **Yes** or press **Enter** if the information is correct, otherwise click on **No** and you will be able to re-enter the information.

10. Setup now displays the ***Installation Complete!*** dialog box. This dialog box contains information on how to start using ENVIRONMENTALSTATS for S-PLUS and also how to start the help system. Click on **OK** or press **Enter** to finish the installation.

Installing ENVIRONMENTALSTATS for S-PLUS Under UNIX

The installation process for ENVIRONMENTALSTATS for S-PLUS for UNIX consists of three main steps: determining the directory where S-PLUS is installed (here denoted ***SDIR***), copying the contents of the installation CD or tape to a particular sub-directory under this directory, and running ENVIRONMENTALSTATS for S-PLUS for the first time to record your name, company name, and serial number. You need to have write permission to the directory where S-PLUS is installed (i.e., to ***SDIR***).

To install ENVIRONMENTALSTATS for S-PLUS, follow these steps:

1. Determine where S-PLUS is installed. To do this, type **Splus SHOME** at the shell prompt. As noted above, this directory will be denoted ***SDIR***.

2. Verify that you have write permission to the ***SDIR*** directory.

3. Set your current directory to the ***SDIR*** directory.

4. Place the ENVIRONMENTALSTATS for S-PLUS installation CD or tape in the appropriate drive.

5. **Local Drive.** If you are installing from a CD, type the following command at the shell prompt:

    tar xvf /*cdpath*/esfsp10.tar

    where ***cdpath*** denotes the pathname of the CD drive.

    If you are installing from a tape, type the following command at the shell prompt:

    tar xvf /dev/*drivename*

    where ***drivename*** is the name of the local drive. Some common drive names are /dev/rst1, /dev/rst8, and /dev/rmt0.

    Remote Drive. If you are installing from a CD, type the following command at the shell prompt:

    rsh *remotename* -n dd if=/*cdpath*/esfsp10.tar | tar xBf -

    where ***remotename*** is the name of the machine with the CD drive and ***cdpath*** denotes the pathname of the CD drive.

    If you are installing from a tape, type the following command at the shell prompt:

    rsh *remotename* -n dd if=/dev/*drivename* | tar xBf -

    where ***remotename*** is the name of the machine with the tape drive and ***drivename*** is the local name of the tape drive on the remote machine.

6. Now get into the directory ***SDIR*/module/envstats** and type **INSTALL** at the shell prompt.

7. Move to a directory where you normally run S-PLUS. Make sure you still have write permission to the ***SDIR*** directory.

8. Start S-PLUS by typing **Splus** at the shell prompt. Load the ENVIRONMENTALSTATS for S-PLUS module by typing the following command at the S-PLUS prompt:

```
> module(envstats)
```

9. Since this is the first time you have loaded ENVIRONMENTALSTATS for S-PLUS, you will be prompted for information on the user name, company name, and serial number (located on the tape case). Follow the directions for typing in this information.

10. Quit S-PLUS by typing **q()** at the S-PLUS prompt. You may now change the permission settings for the ***SDIR*** directory.

In order to access the help files for ENVIRONMENTALSTATS for S-PLUS, users will need to be able to browse to the file **helpset.HTM** located under the directory ***SDIR*/module/envstats/Html**. If this will not work under your setup, you may wish to move the **Html** directory to another location or server and inform the users of the new location. ***Please ensure, however, that only licensed users of ENVIRONMENTALSTATS for S-PLUS have access to these files.***

Starting ENVIRONMENTALSTATS for S-PLUS

To start ENVIRONMENTALSTATS for S-PLUS, simply start S-PLUS and then type the following command at the S-PLUS prompt:

```
> module(envstats)
```

All of the functions and data sets in ENVIRONMENTALSTATS for S-PLUS are now accessible. See **Chapter 2: Getting Started** for more detailed information on starting and customizing ENVIRONMENTALSTATS for S-PLUS.

Getting Help for ENVIRONMENTALSTATS for S-PLUS

How you access the help system depends upon which operating system you are running.

Getting Help Under Windows

If you are running S-PLUS under Windows, start the help system for ENVIRONMENTALSTATS for S-PLUS by typing

```
> help(module="envstats")
```

or click on the **Help** menu in the S-PLUS command window and select **EnvironmentalStats Help**. You may also access the help files via the **S-PLUS for Windows** or **S-PLUS 4.0** group under the Start Menu, or double-click the **EnvironmentalStats for S-PLUS Help** icon in the **S-PLUS for Windows** or **S-PLUS 4.0** program group.

Getting Help Under UNIX

If you are running S-PLUS under UNIX, in order to view the help files that come with ENVIRONMENTALSTATS for S-PLUS you must use a browser that supports HTML 3.2 or later (e.g., Netscape Navigator 3 or Netscape Communicator 4). To view the help files, simply start your browser, then browse to the file

***SDIR*/module/envstats/Html/helpset.HTM**

where ***SDIR*** denotes the directory in which S-PLUS is installed. If you do not know the directory in which S-PLUS is installed, you can find out by typing the following command at your shell prompt:

Splus SHOME

Note: If your browser cannot find the file **helpset.HTM**, your system administrator may have moved the location of the help files.

You must use your browser to view the help files for ENVIRONMENTALSTATS for S-PLUS. Unlike other S-PLUS modules written for S-PLUS 3.4 for UNIX, you cannot get help for functions or data sets in this module by using the `help` function within S-PLUS. On the other hand, the help for ENVIRONMENTALSTATS for S-PLUS is truly hypertext.

Technical Support

Registered users of ENVIRONMENTALSTATS for S-PLUS on the Software and Support Plan are entitled to one year of technical support from the date of purchase (the first year is included in the purchase price). Only questions specific to ENVIRONMENTALSTATS for S-PLUS will be answered. General questions about the operation of S-PLUS should be directed to technical support at the Data Analysis Products Division of MathSoft, Inc. (see address and phone number below).

Before you contact technical support for ENVIRONMENTALSTATS for S-PLUS, please make sure you have perused this manual and the help system for the information you are seeking. If you are unable to find what you are looking for in this manual or the help files, or you are having technical problems with ENVIRONMENTALSTATS for

S-PLUS, please contact Probability, Statistics & Information at the addresses or phone numbers indicated below. Please have the following information available:

- The platform you are running on (Windows 95, SUN, HP, etc.).
- The version of S-PLUS you are using. Type

```
> version
```

- The version of ENVIRONMENTALSTATS for S-PLUS you are using. After you have attached ENVIRONMENTALSTATS for S-PLUS with the `module` command., type

```
> version.EnvironmentalStats
```

- The license information associated with the copy of ENVIRONMENTALSTATS for S-PLUS that you are using. After you have attached ENVIRONMENTALSTATS for S-PLUS with the `module` command, type

```
> license.EnvironmentalStats
```

Technical Support for ENVIRONMENTALSTATS for S-PLUS

Probability, **S**tatistics & **I**nformation
7723 44th Avenue NE
Seattle, WA 98115-5117
TEL: 206-528-4877
FAX: 206-528-4802
E-Mail: support@ProbStatInfo.com
Web: www.ProbStatInfo.com

Technical Support for S-PLUS

MathSoft, Inc.
Data Analysis Products Division
1700 Westlake Avenue North, Suite 500
Seattle, WA 98109
TEL: 206-283-8802 ext. 235
FAX: 206-283-8691
E-Mail: support@StatSci.com
Web: www.MathSoft.com

A Tutorial

This section is a brief tutorial that highlights some of the major features of ENVIRONMENTALSTATS for S-PLUS. There are several ways to use this section. If you are fairly new to S-PLUS, you may want to briefly skim this section to get an idea of what you can do in ENVIRONMENTALSTATS for S-PLUS, and then come back later after you have read the other chapters of this manual. If you have used S-PLUS for a long time and have just installed ENVIRONMENTALSTATS for S-PLUS, you may want to follow this tutorial in depth right now to get acquainted with some of the features available in this S-PLUS module.

TcCB Data at a Reference and Cleanup Site: Summary Statistics, Graphical Analyses, and Nonparametric Comparisons

The guidance document **Statistical Methods for Evaluating the Attainment of Cleanup Standards, Volume 3: Reference-Based Standards for Soils and Solid Media** (USEPA, 1994b, pp.6.22-6.25) contains measures of 1,2,3,4-Tetrachlorobenzene (TcCB) concentrations (ppb) from soil samples at a reference site and a "cleanup" area. There are 47 observations from the reference site and 77 in the cleanup area.

These data are stored in the data frame `epa.94b.tccb.df` (see the help file **Datasets: USEPA (1994b)**). There is one observation coded as "ND" in this data set as presented in the guidance document. Here, we'll assume this observation is less than the smallest observed value, which is 0.09 ppb. For the purposes of this tutorial, we'll set this one censored observation to the assumed detection limit of 0.09.

```
> epa.94b.tccb.df
     TcCB.orig   TcCB Censored      Area
  1       0.22   0.22        F Reference
  2       0.23   0.23        F Reference
  3       0.26   0.26        F Reference
  .
122      18.40  18.40        F Cleanup
123      51.97  51.97        F Cleanup
124     168.64 168.64        F Cleanup
```

Example 0.1: Computing Summary Statistics for the Two Sites

The ENVIRONMENTALSTATS for S-PLUS help file **Summary Statistics** lists functions for computing summary statistics that are available in ENVIRONMENTALSTATS for S-PLUS but not built into S-PLUS. These include functions to compute the geometric mean, standard deviation, interquartile range, skew, kurtosis, and coefficient of variation, as well as a function called `full.summary` that computes all of these summary statistics and others as well.

```
> attach(epa.94b.tccb.df)

> full.summary(split(TcCB, Area))
                          Cleanup Reference
              Sample Size 77          47
                # Missing  0           0
                     Mean  3.915       0.5985
                   Median  0.43        0.54
        10% Trimmed Mean   0.6846      0.5728
           Geometric Mean  0.5784      0.5382
                     Skew  7.566       0.8729
                 Kurtosis 61.6         2.993
                      Min  0.09        0.22
                      Max 168.6        1.33
                    Range 168.5        1.11
             1st Quartile  0.23        0.39
             3rd Quartile  1.1         0.75
       Standard Deviation 20.02        0.2836
      Interquartile Range  0.87        0.36
Median Absolute Deviation  0.3558      0.2669
 Coefficient of Variation  5.112       0.4739
```

These summary statistics indicate that the observations for the cleanup area are extremely skewed to the right. The medians for the two areas are about the same, but the mean for the cleanup area is much larger, indicating a few or more “outlying” observations with large values. This may be indicative of residual contamination that was missed during the cleanup process.

Example 0.2: Looking at the Data

You can use the built-in S-PLUS functions `hist` and `boxplot` to compare the observations in the two areas. Here are commands to produce histograms:

```
> TcCB.cleanup <- TcCB[Area == "Cleanup"]

> TcCB.ref <- TcCB[Area == "Reference"]

> par(mfrow = c(2,1))

> hist(log(TcCB.ref), xlim = range(log(TcCB)),
+    xlab = "log [ TcCB (ppb) ]",
+    ylab = "Number of Observations",
+    main = "Histogram for Reference Area")

> hist(log(TcCB.cleanup), xlim = range(log(TcCB)),
+    nclass = 25,
+    xlab = "log [ TcCB (ppb) ]",
+    ylab = "Number of Observations",
+    main = "Histogram for Cleanup Area")
```

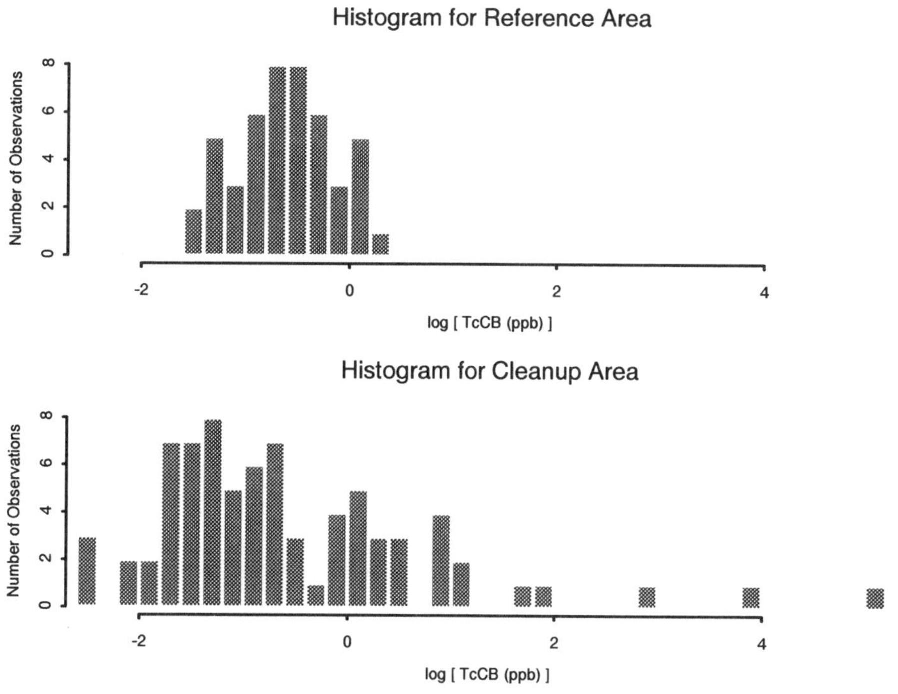

Figure 0.1: Histograms comparing TcCB data at the cleanup and reference areas.

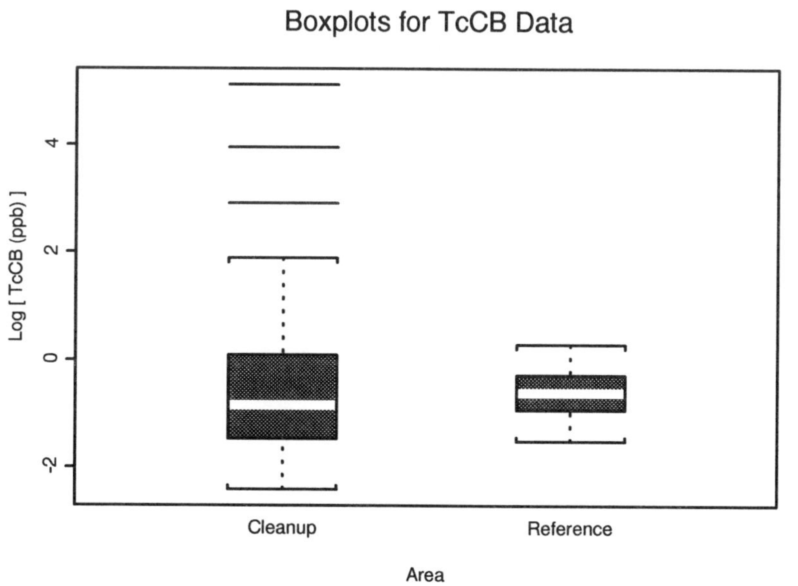

Figure 0.2: Boxplots comparing TcCB data at the cleanup and reference areas.

Here are commands to produce boxplots:

```
> boxplot(split(log(TcCB), Area),
+    xlab = "Area", ylab = "Log [ TcCB (ppb) ]",
+    main = "Boxplots for TcCB Data")
```

Both the histograms and boxplots show that most of the observations for the cleanup area are comparable to (or even smaller than) the observations for the reference area, but, as we found out from looking at the summary statistics for these data, there are a few very large "outliers" in the cleanup area. This may indicate a few "hot spots" in the cleanup area that were missed during the remediation process.

Example 0.3: Plotting the Empirical Cumulative Distribution Functions

You can use `ecdfplot` to plot the empirical cumulative distribution function (ecdf) of the observations for either the reference or cleanup area (or both). The function `cdf.compare` let's you compare an ecdf to a theoretical cdf, or to another ecdf.

First, let's plot the empirical cdf of the reference area data by itself.

```
> ecdfplot(TcCB.ref, xlab = "TcCB (ppb)",
+    main = "Empirical CDF for Reference Area")
```

Figure 0.3 shows that the data are right-skewed.

Now let's create another plot comparing this empirical cdf with the cdf of a lognormal distribution.

```
> cdf.compare(TcCB.ref, dist = "lnorm")
```

Figure 0.4 shows that these data may probably be adequately fit by a lognormal distribution.

Now let's compare the empirical cdf's of the reference and cleanup areas.

```
> cdf.compare(log(TcCB.ref), log(TcCB.cleanup))
```

The results are shown in **Figure 0.5**. As we saw with both the histograms (**Figure 0.1**) and boxplots (**Figure 0.2**), the cleanup area has quite a few extreme values compared to the reference area.

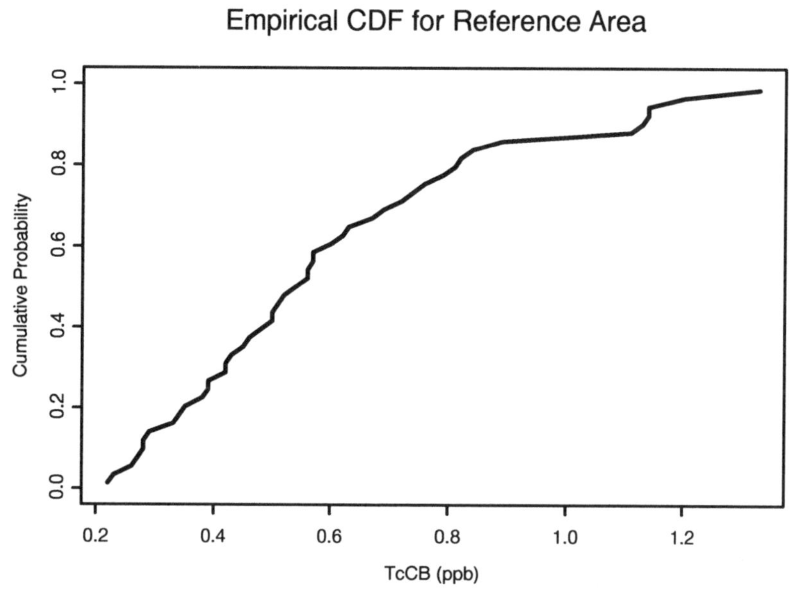

Figure 0.3: Empirical cdf plot of TcCB data for the reference area.

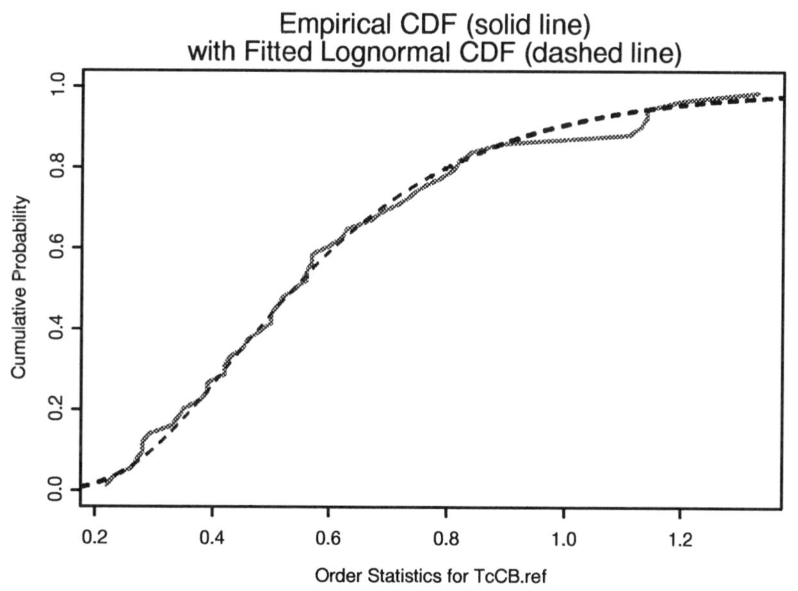

Figure 0.4: Empirical cdf of reference area TcCB data compared to a lognormal cdf.

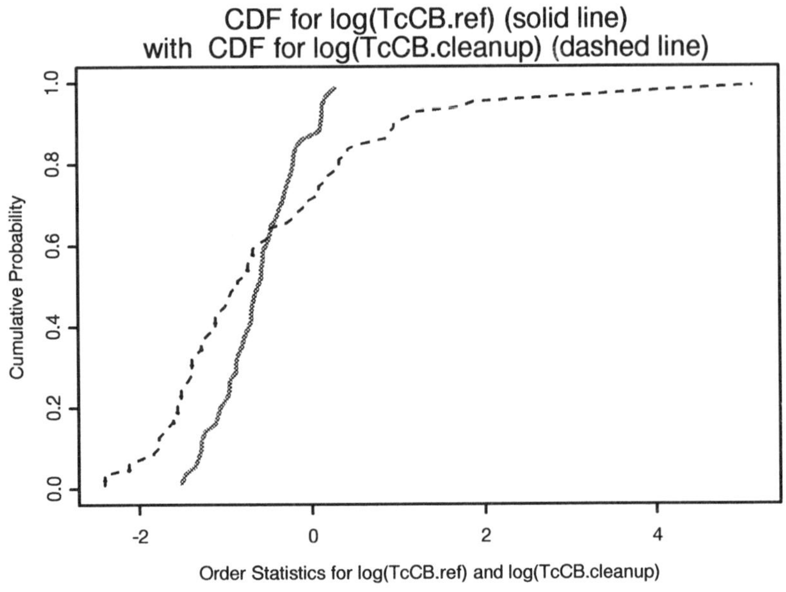

Figure 0.5: Quantile plots comparing TcCB data at the cleanup and reference areas.

Example 0.4: Assessing Goodness-of-Fit with Quantile-Quantile (Q-Q) Plots

The `qqplot` function has been modified in ENVIRONMENTALSTATS for S-PLUS to let you specify a theoretical distribution, what kind of line to add to the plot (if any), and whether to estimate the parameters of the theoretical distribution. Also, both standard and Tukey mean-difference Q-Q plots can be produced.

Let's create a Q-Q plot for the reference area TcCB data assuming they come from a lognormal distribution, and also look at the corresponding Tukey mean-difference Q-Q plot.

```
> qqplot(TcCB.ref, dist = "lnorm",
+    estimate.params = T, add.line = T,
+    qq.line.type = "0-1")

> qqplot(TcCB.ref, dist = "lnorm",
+    estimate.params = T, plot.type = "Tukey",
+    add.line = T)
```

The results are shown in **Figure 0.6** and **Figure 0.7**. As we saw in **Figure 0.4**, the lognormal model appears to be a fairly good fit to these data.

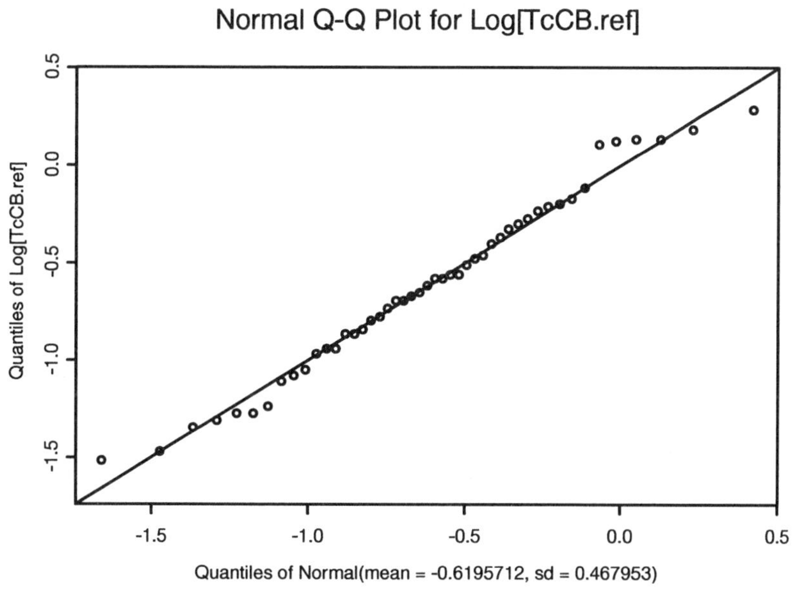

Figure 0.6: Normal Q-Q plot based on estimated parameters with 0-1 line.

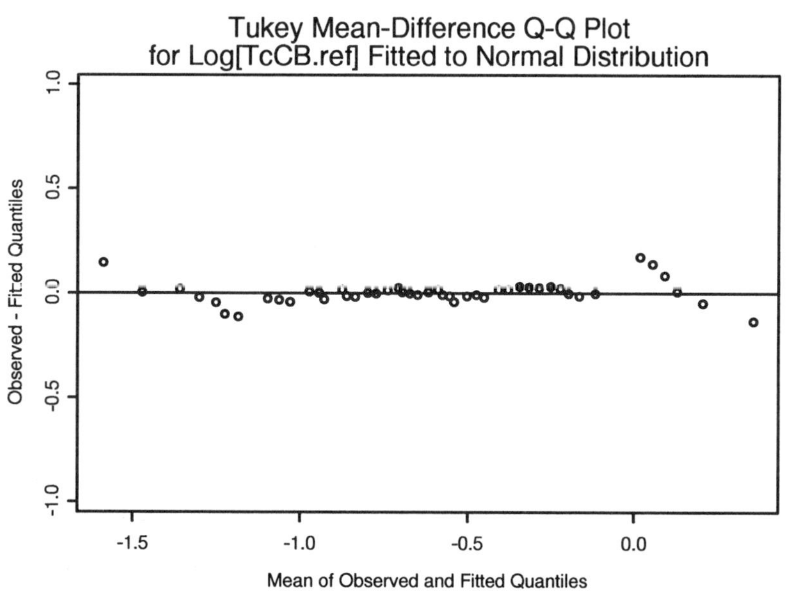

Figure 0.7: Tukey m-d plot for log-transformed reference area TcCB data.

Example 0.5: Estimating Distribution Parameters

EnvironmentalStats for S-Plus contains several function for estimating distribution parameters. The functions `elnorm` and `elnorm.alt` let you estimate the parameters of a lognormal distribution, given a set of observations. The function `elnorm` estimates the mean and standard deviation of the log-transformed distribution. The function `elnorm.alt` (alternative parameterization) estimates the mean and coefficient of variation of the original distribution. Both of these functions allow you to compute confidence intervals as well.

Here are the results of these two functions using the reference area TcCB data.

```
> elnorm(TcCB.ref, ci = T)

Results of Distribution Parameter Estimation
--------------------------------------------

Assumed Distribution:            Lognormal

Estimated Parameter(s):          meanlog = -0.6195712
                                 sdlog   =  0.467953

Estimation Method:               mvue

Data:                            TcCB.ref

Sample Size:                     47

Confidence Interval for:         meanlog

Confidence Interval Method:      Exact

Confidence Interval Type:        two-sided

Confidence Level:                95%

Confidence Interval:             LCL = -0.7569673
                                 UCL = -0.4821751
```

```
> elnorm.alt(TcCB.ref, ci = T)

Results of Distribution Parameter Estimation
--------------------------------------------

Assumed Distribution:            Lognormal

Estimated Parameter(s):          mean = 0.5989072
                                 cv   = 0.4899539

Estimation Method:               mvue

Data:                            TcCB.ref

Sample Size:                     47

Confidence Interval for:         mean

Confidence Interval Method:      Land

Confidence Interval Type:        two-sided

Confidence Level:                95%

Confidence Interval:             LCL = 0.5243786
                                 UCL = 0.7016992
```

Example 0.6: Testing for Goodness-of-Fit

ENVIRONMENTALSTATS for S-PLUS contains several new functions not available in S-PLUS for testing goodness of fit. Here, we'll use the Shapiro-Wilk test to test the goodness-of-fit of the reference area TcCB data to a lognormal distribution.

```
> tccb.ref.gof <- sw.gof(log(TcCB.ref))

> tccb.ref.gof

Results of Goodness-of-Fit Test
-------------------------------

Test Method:                     Shapiro-Wilk GOF

Hypothesized Distribution:       Normal

Estimated Parameter(s):          mean = -0.6195712
                                 sd   =  0.467953

Estimation Method:               mvue

Data:                            log(TcCB.ref)
```

```
Sample Size:                     47

Test Statistic:                  W = 0.9789915

Test Statistic Parameter:        n = 47

P-value:                         0.5512532

Alternative Hypothesis:          True cdf does not equal the
                                 Normal Distribution.
```

ENVIRONMENTALSTATS for S-PLUS also contains a plotting method for the results of goodness-of-fit tests, as well as a function called `plot.gof.summary`, which produces four summary plots on one page.

```
> plot.gof.summary(tccb.ref.gof)
```

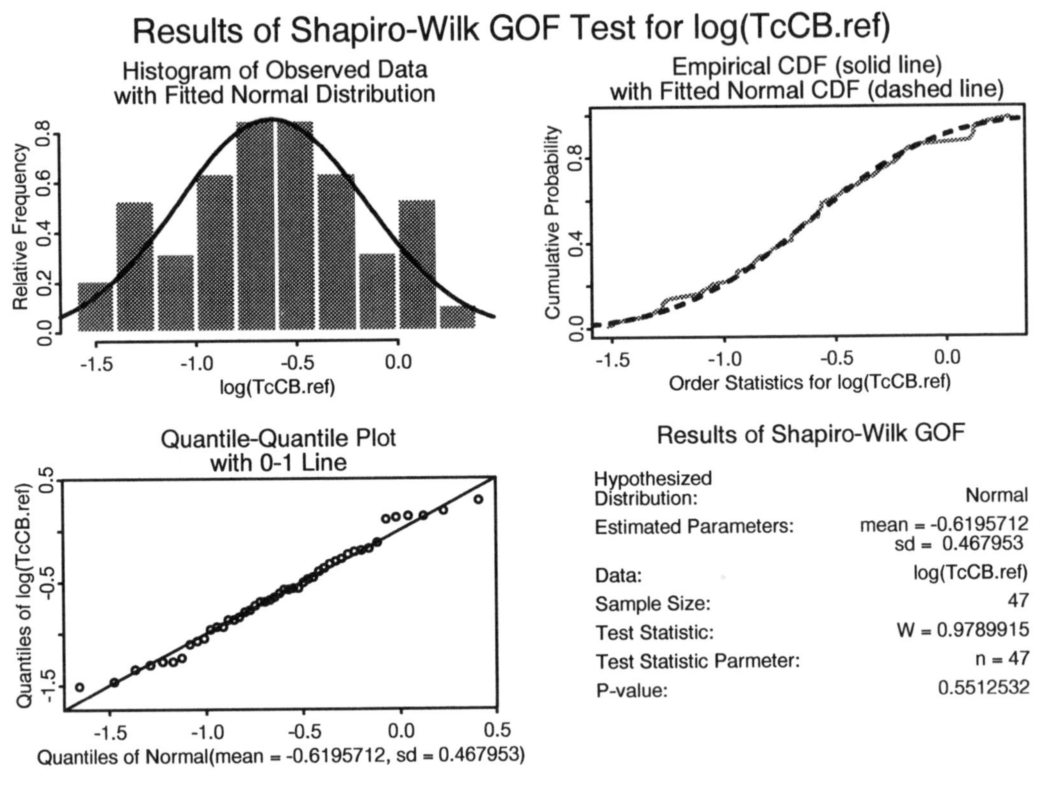

Figure 0.8: Summary plots of normal goodness-of-fit test for log-transformed reference area TcCB data.

Example 0.7: Estimating Quantiles and Computing Confidence Limits

ENVIRONMENTALSTATS for S-PLUS contains several functions for estimating quantiles and optionally constructing confidence limits for the quantiles. Let's estimate the 90th percentile of the distribution of the reference area TcCB, assuming the true distribution is a lognormal distribution, and compute a 95% confidence interval for this 90th percentile.

```
> eqlnorm(TcCB.ref, p = 0.9, ci = T)

Results of Distribution Parameter Estimation
--------------------------------------------

Assumed Distribution:            Lognormal

Estimated Parameter(s):          meanlog = -0.6195712
                                 sdlog   =  0.467953

Estimation Method:               mvue

Estimated Quantile(s):           90'th %ile = 0.9803307

Quantile Estimation Method:      qmle

Data:                            TcCB.ref

Sample Size:                     47

Confidence Interval for:         90'th %ile

Confidence Interval Method:      Exact

Confidence Interval Type:        two-sided

Confidence Level:                95%

Confidence Interval:             LCL = 0.8358791
                                 UCL = 1.215498
```

Example 0.8: Comparing the TcCB Concentrations at the Two Areas Using Nonparametric Tests

ENVIRONMENTALSTATS for S-PLUS contains functions for performing general two-sample linear rank tests (to test for a shift in location) and a special quantile test that tests for a shift in the tail of one of the distributions. Here, we'll perform the usual Wilcoxon Rank Sum test and the quantile test.

```
> two.sample.linear.rank.test(TcCB.cleanup,
+   TcCB.ref, alternative = "greater")

Results of Hypothesis Test
--------------------------

Null Hypothesis:            Fx(t) = Fy(t)

Alternative Hypothesis:     Fx(t) > Fy(t) for at least one t

Test Name:                  Two-Sample Linear Rank Test:
                            Wilcoxon Rank Sum Test
                            Based on Normal Approximation

Data:                       x = TcCB.cleanup
                            y = TcCB.ref

Sample Sizes:               nx = 77
                            ny = 47

Test Statistic:             z = 1.171872

P-value:                    0.1206242

> quantile.test(TcCB.ref, TcCB.cleanup, target.r = 9)

Results of Hypothesis Test
--------------------------

Null Hypothesis:            e = 0

Alternative Hypothesis:     0 < e <= 1, where
                            Fy(t) = (1-e)*Fx(t) + e*Fz(t),
                            Fz(t) <= Fx(t) for all t,
                            and Fx != Fz

Test Name:                  Quantile Test

Data:                       x = TcCB.ref
                            y = TcCB.cleanup

Sample Sizes:               nx = 47
                            ny = 77

Test Statistics:            k (# y obs of r largest) = 9
                            r                        = 9
```

```
Test Statistic Parameters: m            = 47
                           n            = 77
                           quantile.ub  =  0.928

P-value:                   0.01136926
```

The Wilcoxon Rank Sum Test is not significant at the 0.10 level, while the quantile test is significant at the 0.011 level. The quantile test picked up the portion of large outlying values in the cleanup area TcCB data (see **Figure 0.**1 and **Figure 0.2**).

CHAPTER 1 : Introduction

Welcome to ENVIRONMENTALSTATS for S-PLUS! This user's manual provides step-by-step guidance to using this software. ENVIRONMENTALSTATS for S-PLUS is an S-PLUS module for environmental statistics. This chapter gives brief descriptions of environmental statistics, ENVIRONMENTALSTATS for S-PLUS, and the intended users of this software and user's manual.

What is Environmental Statistics?

Environmental statistics is simply the application of statistical methods to problems concerning the environment. Examples of activities that require the use of environmental statistics include:

- Monitoring air or surface water quality.
- Monitoring groundwater quality near a hazardous or solid waste site.
- Using risk assessment to determine whether a potentially contaminated area needs to be cleaned up, and, if so, how much.
- Assessing whether a previously contaminated area has been cleaned up according to some specified criterion.
- Using hydrological data to predict the occurrences of floods

In addition, there are several specialized branches of ecology that use environmental statistics such as animal population dynamics and general ecological modeling.

Environmental statistics is a special field of statistics. Probability and statistics deal with situations in which the outcome is not certain. They are built upon the concepts of a ***population*** and a ***sample*** from the population.

Probability deals with predicting the characteristics of the sample, given that you know the characteristics of the population (e.g., what is the probability of picking an ace out of a deck of 52 well-shuffled standard playing cards?). ***Statistics*** deals with inferring the characteristics of the population, given information from one or more samples from the population (e.g., after 100 times of randomly choosing a card from a deck of 20 unknown playing cards and then replacing the card in the deck, no ace has appeared; therefore the deck probably does not contain any aces).

The field of environmental statistics is relatively young and employs several statistical methods that have been developed in other fields of statistics, such as sampling design, exploratory data analysis, basic estimation and hypothesis testing, quality control, multiple comparisons, survival analysis, and Monte Carlo simulation. Nonetheless, special problems have motivated innovative research, and both traditional and new journals now report on statistical methods that have been developed in the context of environmental monitoring. (See the help topic **References: Environmental Statistics—General References** for a list of general text books on environmental statistics.)

In addition, environmental legislation such as the Clean Water Act, the Clean Air Act, the Comprehensive Emergency Response, Compensation, and Liability Act (CERCLA), and the Resource and Recovery Act (RCRA) have spawned environmental regulations and agency guidance documents that mandate or suggest various statistical methods for environmental monitoring (see the help topic **References: Guidance Documents and Regulations**).

What is EnvironmentalStats for S-Plus?

EnvironmentalStats for S-Plus, created by Dr. Steven P. Millard of Probability, Statistics & Information (PSI), is a new ***comprehensive*** software package for environmental statistics. An add-on module to the statistical software package S-Plus (from MathSoft, Inc.), it provides a set of powerful yet simple-to-use functions for performing graphical and statistical analyses of environmental data. EnvironmentalStats for S-Plus brings the major environmental statistical methods found in the literature and regulatory guidance documents into one statistical package, along with an extensive hypertext help system that explains *what* these methods do, *how* to use these methods, and *where* to find them in the environmental statistics literature. Also included are numerous built-in data sets from regulatory guidance documents and the environmental statistics literature.

EnvironmentalStats for S-Plus, combined with S+SpatialStats and S+GISLink or S-Plus for ArcView (from MathSoft, Inc.) provide the environmental scientist, researcher, and technician with all the tools needed to "get the job done"!

Because ENVIRONMENTALSTATS for S-PLUS is an S-PLUS module, you automatically have access to all the features and functions of S-PLUS, including powerful graphics, standard hypothesis tests, and the flexibility of a programming language. In addition, specific features of ENVIRONMENTALSTATS for S-PLUS include functions to:

- Plot probability and cumulative distribution functions.
- Create probability (Q-Q) plots for a variety of distributions.
- Estimate distribution parameters and quantiles.
- Compute confidence intervals.
- Perform, print, and plot the results of goodness-of-fit tests.
- Compute prediction and tolerance intervals.
- Estimate and test for trend.
- Easily create power and sample size plots.
- Analyze Type I left-censored (less-than-detection limit) data.
- Perform probabilistic risk assessment.

Built-in data sets let you reproduce examples in selected environmental literature and guidance documents. In addition, the hypertext help system includes numerous references, abstracts, and a glossary of statistical terms.

Intended Audience/Users

ENVIRONMENTALSTATS for S-PLUS and this user's manual are intended for anyone who has to make sense of environmental data, including hydrologists, soil scientists, atmospheric scientists, geochemists, environmental engineers and consultants, hazardous and solid waste site managers, and regulatory agency analysts and enforcement officers. Some parts of ENVIRONMENTALSTATS for S-PLUS incorporate statistical methods that have appeared in the environmental literature but are not commonly found in any statistical software package. Some parts are specifically aimed at users who are required to collect and analyze environmental monitoring data in order to comply with federal and state Superfund, RCRA, CERCLA, and Subtitle D regulations for environmental monitoring at hazardous and solid waste sites. All of the functions in ENVIRONMENTALSTATS for S-PLUS, however, are useful to anyone who needs to analyze environmental data.

ENVIRONMENTALSTATS for S-PLUS is an S-PLUS module. In order to use it, you need to know how to perform basic operations in S-PLUS, such as reading data into S-PLUS, creating basic data objects (e.g., vectors, matrices, and data frames), and using built-in functions. See the S-PLUS documentation for more information on S-PLUS. In addition, you need to have a basic knowledge of probability and statistics. This user's manual is not meant to be a text book in environmental statistics. See the references listed Appendix A and in the help topic **References: Environmental Statistics—General References** for a list of useful books and articles on this subject.

Summary

- Environmental statistics is the application of statistics to environmental problems.
- ENVIRONMENTALSTATS for S-PLUS is an S-PLUS module for environmental statistics. It includes several functions for creating graphs and performing statistical analyses that are commonly used in environmental statistics.
- To use ENVIRONMENTALSTATS for S-PLUS you should be familiar with basic features of S-PLUS and have an elementary knowledge of probability and statistics.

CHAPTER 2 :

Getting Started

Chapter 0: A Quick Start provides you with the minimum information you need to install and start using ENVIRONMENTALSTATS for S-PLUS. Chapter 0 also gives information on how to obtain technical support for ENVIRONMENTALSTATS for S-PLUS. This current chapter provides you with more details on starting and customizing the module, and using the help system.

Starting EnvironmentalStats for S-Plus

To start ENVIRONMENTALSTATS for S-PLUS, simply start S-PLUS and then type the following command at the S-PLUS prompt:

```
> module(envstats)
```

This command attaches the library of ENVIRONMENTALSTATS for S-PLUS functions to the second position in your search list. (Type `search()` at the S-PLUS prompt to see the listing of directories on your search list.) You can now access any of the functions in the ENVIRONMENTALSTATS for S-PLUS module.

Note: Some of the functions in the ENVIRONMENTALSTATS for S-PLUS module mask built-in S-PLUS functions. The masked functions are modified versions of the built-in functions and have been created to support the other functions in ENVIRONMENTALSTATS for S-PLUS, but the modifications should not affect normal use of S-PLUS. If you experience unexpected behavior of built-in S-PLUS functions after attaching ENVIRONMENTALSTATS for S-PLUS, try unloading the module (see **Removing ENVIRONMENTALSTATS for S-PLUS From Your S-PLUS Session** below). All of the functions in ENVIRONMENTALSTATS for S-PLUS, whether they mask built-in functions or not, are described in the help system.

Getting Help for ENVIRONMENTALSTATS for S-PLUS

How you access the help system for ENVIRONMENTALSTATS for S-PLUS depends upon which operating system you are running.

Getting Help Under Windows

If you are running S-PLUS under Windows, you may start the help system for ENVIRONMENTALSTATS for S-PLUS within S-PLUS by typing

```
> help(module="envstats")
```

at the S-PLUS prompt, or by clicking on the **Help** menu in the S-PLUS command window and selecting **EnvironmentalStats Help**. You may want to customize your use of the help system so that it always starts whenever you start S-PLUS (see the section **Customizing ENVIRONMENTALSTATS for S-PLUS** later in this chapter). To do this, simply add the above line to your `.First` function.

Under S-PLUS 4.0, you can also start the help system for ENVIRONMENTALSTATS for S-PLUS by using the `?` operator. For example, typing

```
> ?pdfplot
```

will bring up the help file for the `pdfplot` function.

You may also access the help files for ENVIRONMENTALSTATS for S-PLUS outside of S-PLUS via the **S-PLUS for Windows** or **S-PLUS 4.0** shortcut under the **Programs** menu under the **Start** menu (Windows 95), or by double-clicking the **EnvironmentalStats for S-PLUS Help** icon (a yellow question mark) in the **S-PLUS for Windows** or **S-PLUS 4.0** program group (Windows 3.1 and Windows 95).

Getting Help Under UNIX

The help system that comes with ENVIRONMENTALSTATS for S-PLUS for UNIX platforms is based on HyperText Markup Language (HTML), version 3.2. This help system was created with a help-authoring tool called ForeHelp, and the original help system was based on Windows help. Using ForeHelp, the help system based on Windows help was then translated into HTML files.

In order to view the help files that come with ENVIRONMENTALSTATS for S-PLUS you must use a browser that supports HTML 3.2 or later (e.g., Netscape Navigator 3 or Netscape Communicator 4). To view the help files, simply start your browser, then browse to the file

***SDIR*/module/envstats/Html/helpset.HTM**

where ***SDIR*** denotes the directory in which S-PLUS is installed. If you do not know the directory in which S-PLUS is installed, you can find out by typing the following command at your shell prompt:

Splus SHOME

Note: If your browser cannot find the file **helpset.HTM**, your system administrator may have moved the location of the help files.

You must use your browser to view the help files for ENVIRONMENTALSTATS for S-PLUS. Unlike other S-PLUS modules written for S-PLUS 3.4 for UNIX, you cannot get help for functions or data sets in this module by using the `help` function within S-PLUS. On the other hand, the help for ENVIRONMENTALSTATS for S-PLUS is truly hypertext.

Changing Font Sizes in the Help Files

The help files were written with larger font sizes than are conventionally used, based on the author's frustration with squinting at help files (for other applications) for long periods of time. Depending on the fonts you have available on your computer, you may want to change the font sizes in the help windows for ENVIRONMENTALSTATS for S-PLUS. The directions for changing font sizes are given below.

Changing Font Sizes Under Windows 95

If you are using Windows 95, you can change the font sizes for the help windows by doing the following. Place the mouse cursor inside the help window, and click on the right mouse button. A menu will appear. Move the mouse cursor down to the **Font** item, then select **Small**, **Normal**, or **Large**. This setting affects only the help system you are currently using.

You may also change the setting of font sizes for all of your windows by going to the Start button, selecting Control Panel, double-clicking on the Display icon, and clicking on the Appearance tab. See the help for Windows 95 for more information.

Changing Font Sizes Under UNIX

See the documentation that comes with your browser for instructions on how to do this.

Using the Help System

The help topic **Functions and Datasets** listed in the contents shows you that you can access the help files for functions by category or alphabetically, and you can access data sets by source or alphabetically. The help file **Functions By Category** lists of all of the functions in ENVIRONMENTALSTATS for S-PLUS by what they do or the kind of data they handle (e.g., **Censored Data**, **Counting: Combinations, Factorials, Permutations**, etc.) The help file **Datasets by Source** lists all of the built-in data sets in ENVIRONMENTALSTATS for S-PLUS by the source of the data sets.

The help system for ENVIRONMENTALSTATS for S-PLUS is a separate application from the help system for S-PLUS. Words that are underlined and highlighted in color (green or blue or whatever color your help system or browser uses for jumps) are jumps to other help windows *within* the ENVIRONMENTALSTATS for S-PLUS help system. In each help file, if a word appears under the paragraph heading SEE ALSO but is not underlined and highlighted, you must use the S-PLUS help system to look up this word.

For example, in the help system for ENVIRONMENTALSTATS for S-PLUS, look up the topic `pdfplot`. Within that help file, under the paragraph heading SEE ALSO, you will see:

SEE ALSO

```
.Distribution.frame, Probability Distributions and
  Random Numbers, cdfplot, ecdfplot, qqplot,
  qqplot.gestalt, plot, plot.default, par, title.
```

You can access the help files for `.Distribution.frame`, `Probability Distributions and Random Numbers`, `cdfplot`, `ecdfplot`, `qqplot`, and `qqplot.gestalt` directly by simply clicking on these words. To look at the help file for the functions `plot`, `plot.default`, `par`, and `title`, however, you must use the S-PLUS help system.

Special Buttons for Help Under UNIX

In order to mimic the operation of a Windows help system, the help files under UNIX include a set of buttons at the top of each help window to aid in using the help system. These buttons are labeled Contents, Index, Search, Back, Forward, and Exit. Clicking on one of these buttons produces the action shown below.

Button Name	Action
Contents	Go to Table of Contents page.
Index	Go to the Index page, which lists all of the keywords in alphabetical order and provides jumps to the appropriate help files.
Search	List all of the files used in the help system.
Back	Go to the page (topic) that was previously displayed.
Forward	Go to the page (topic) that was displayed after the current page.
Exit	Exit the help system.

Table 2.1: Buttons for Help Under UNIX.

Removing ENVIRONMENTALSTATS for S-PLUS From Your S-PLUS Session

To remove the ENVIRONMENTALSTATS for S-PLUS module from your S-PLUS session, simply type the following command at the S-PLUS prompt:

```
> module(envstats, unload=T)
```

This command removes the library of ENVIRONMENTALSTATS for S-PLUS functions from your search list.

Customizing ENVIRONMENTALSTATS for S-PLUS

If you plan to use ENVIRONMENTALSTATS for S-PLUS extensively, you may want to customize your S-PLUS startup routine to automatically attach the ENVIRONMENTALSTATS for S-PLUS module each time you start S-PLUS. You can do this by adding the line `module(envstats)` to your `.First` function. If you do not have a `.First` function, you can create one by simply typing the following command:

```
> .First <- function() {module(envstats)}
```

If you are running under Windows and you want to automatically attach the ENVIRONMENTALSTATS for S-PLUS module each time you start S-PLUS *and* start the ENVIRONMENTALSTATS for S-PLUS help system, make sure your `.First` function includes the following lines:

```
module(envstats)
help(module="envstats")
```

If you are going to use EnvironmentalStats for S-Plus for several different projects, it is a good idea to use separate directories for each project. In addition, if you are running S-Plus under Windows, it is often a good idea to create separate icons (menu items or shortcuts) for each particular project. The following sections explain how to do this, depending on the operating system you are using.

Customizing S-Plus 4.0 Under Windows 95

To keep separate directories (folders) for each project, create a directory (folder) for the project and two subdirectories (subfolders) under that directory (folder), one called **_Data** and the other called **_Prefs**. Then, whenever you want to work on that project, you may either start up S-Plus as usual and attach the **_Data** directory for that project to your search list using the `attach` function (see the S-Plus help file for `attach`), or you may start up S-Plus using a special menu item (icon) you created for that project. The second method is usually preferred over the first. The following sections describe how to create folders for new projects, how to create special menu items for new projects, and how to use S-Plus and EnvironmentalStats for S-Plus for new projects.

Creating a New Folder for a Specific Project

Suppose you have a folder on your **C:** drive called **PROJECTS** where you organize all of your current projects. If you are working on a project called WQ (stands for water quality), you would perform the following steps to create a new folder for this project:

1. Start the **Windows Explorer**. You can do this by pointing to the **Start** button, clicking on the right mouse button, and selecting **Explore**, or by pointing to the **Start** button, clicking on the left mouse button, and selecting **Windows Explorer** under **Programs**.

2. Under the **All Folders** (left-hand) column, select the drive and directory in which you want to create the new project folder. For this example, you would select the **C:** drive and then the **PROJECTS** folder.

3. From the **File** menu, choose **New** and then **Folder**. A new folder appears in the right-hand column with a highlighted text box. The default name is **New Folder**. Type in the name you want for the new folder. In our example, we would type **WQ**.

4. Click on the folder icon or simply press **Enter**.

5. Now double-click on the newly-created folder. It now appears in the left-hand (All Folders) column and the right-hand column is labeled **Contents of 'WQ'** (or whatever you named the new folder). The right-hand column should be empty.

6. From the **File** menu, choose **New** and then **Folder**. A new folder appears in the right-hand column with a highlighted text box. The default name is **New Folder**.

7. Type in **_Data** in the text box.

8. Click on the folder icon or simply press **Enter**.

9. Click on the folder **WQ** in the left-hand (**All Folders**) column.

10. From the **File** menu, choose **New** and then **Folder**. A new folder appears in the right-hand column with a highlighted text box. The default name is **New Folder**.

11. Type in **_Prefs** in the text box.

12. Click on the folder icon or simply press **Enter**.

You can now start S-PLUS and attach the **_Data** directory in your new project directory using the `attach` function. For our example, you would start S-PLUS and then type:

```
> attach("C:\\PROJECTS\\WQ\\_Data")
```

You can also attach this new **_Data** directory to your search list by using the Object Browser. See Chapter 6 of the **S-PLUS 4 User's Guide** for more information.

Using Separate Icons (Menu Items) for Separate Projects

Rather than using the `attach` function, it is often easier to create a new icon (and thus menu item) that will automatically start S-PLUS so that S-PLUS uses the **_Data** directory in this new project directory as the S-PLUS working directory. Also, you may want to use different preferences for different projects. The next two sections describe how to do this.

Creating a New Menu Item for a Specific Project

1. Start S-PLUS using the default icon or menu item. Go to the **Options** menu, choose **General Settings...**, click on the **Startup** tab, and make sure the box labeled "Set S_PROJ to Working Directory" is ***not*** checked. Save these settings then exit S-PLUS.

2. In the **Windows Explorer**, select the **Windows** folder (usually this is on the **C:** drive), then select the **Start Menu** folder (within the **Windows** folder), then select the **Programs** folder (within the **Start Menu** folder), then select the **S-PLUS 4.0** folder (within the **Programs** folder).

3. Within the **S-PLUS 4.0** folder, select the file named **S-PLUS 4.0**. This file is a shortcut file.

4. From the **Edit** menu, choose **Copy**.

5. From the **Edit** menu, choose **Paste**. A new shortcut file appears with the name **Copy of S-PLUS 4.0**.

6. Click on the name of the new shortcut file once to highlight it, and then a second time to make the text box appear. Now type a new name for the shortcut file. In our example, you would probably type something like **WQ** or **WQ project**.

7. Click on the file icon or simply press **Enter**.

8. From the **File** menu, choose **Properties** so that the properties dialog box pops up. Click on the **Shortcut** tab.

9. In the **Target**: text box you will see something like the following:

   "C:\Program Files\splus4\cmd\SPLUS.EXE"

   In this text box, put a space after the above line, and then set the **S_PROJ** environment variable to the directory that you want S-PLUS to start in. That is, for our example you would add the following text:

   S_PROJ="C:\PROJECTS\WQ"

   Do ***not*** specify the **_Data** directory as part of the value of **S_PROJ**!

> **Note:** In earlier versions of S-PLUS, the working directory was specified in the text box labeled **Start in:**. The value of the string in this text box is now ignored unless within S-PLUS you go to the **Options** menu, choose **General Settings...**, click on the **Startup** tab, and check the box labeled "Set S_PROJ to Working Directory". This method is not recommended because it makes it easy to get confused which **_Data** and which **_Prefs** directories are being used.

10. Click on **OK** or simply press **Enter**.

You may customize the values of several other S-PLUS environment variables for this icon as well. See Chapter 15 of the **S-PLUS 4 Programmer's Guide** for more information.

Running S-PLUS and ENVIRONMENTALSTATS for S-PLUS for a Specific Project

Once you have created a new directory for a project and created a new icon (menu item) for that project, simply choose that menu item when you find it on the **Programs** menu in order to start S-PLUS running. S-PLUS will use that project directory's **_Data** directory as its S-PLUS working directory and use the directory's **_Pref** directory to store your preferences for the project..

The *first* time you start S-PLUS with this new menu item, you will have to attach the ENVIRONMENTALSTATS for S-PLUS module directly using the `module` function:

```
> module(envstats)
```

Now create a `.First` function that includes the line `module(envstats)`, for example:

```
> .First <- function() {
+    module(envstats)
+    help(module="envstats")}
```

From now on, when you start S-PLUS using this new menu item, the ENVIRONMENTALSTATS for S-PLUS module will be automatically attached to your search list.

Customizing S-PLUS under UNIX

To keep separate directories for each project, create a directory for the project and a subdirectory called **.Data** under that directory. Then, whenever you want to work on that project, get into that directory and start S-PLUS. Within S-PLUS, make sure that your `.First` function includes the line `module(envstats)`.

For example, suppose you have a directory called **projects** where you organize all of your current projects. If you are working on a project called **WQ** (stands for water quality), you would type the following commands from your login directory:

```
% cd projects

% mkdir WQ

% cd WQ

% mkdir .Data
```

or to do this in two lines, type

```
% cd projects

% mkdir WQ WQ/.Data
```

> **Note:** the percent symbol (%) represents the shell prompt in this example; do not type it when issuing the commands.

Now to work on the **WQ** project, simply get into the **WQ** directory and start S-PLUS:

```
% cd

% cd projects/WQ

% Splus -e
```

The *first* time you start S-PLUS in this directory, you will have to attach the ENVIRONMENTALSTATS for S-PLUS module directly using the `module` function:

```
> module(envstats)
```

Now create a `.First` function that includes the line `module(envstats)`, for example:

```
> .First <- function() {
+    module(envstats)}
```

From now on, when you start S-PLUS in the **WQ** directory, the ENVIRONMENTALSTATS for S-PLUS module will be automatically attached to your search list.

Summary

- To attach the ENVIRONMENTALSTATS for S-PLUS library of functions to your search list, type the following command at the S-PLUS prompt:

```
> module(envstats)
```

- Windows users can start the help system for ENVIRONMENTALSTATS for S-PLUS within S-PLUS by typing

```
> help(module="envstats")
```

  or by clicking on the **Help** menu and selecting **EnvironmentalStats**. Version 4.0 users can also use the `?` operator. Windows users can also start the help system outside of S-PLUS via the **S-PLUS for Windows** or **S-PLUS 4.0** shortcuts, or by double-clicking the **ENVIRONMENTALSTATS for S-PLUS Help** icon in the **S-PLUS for Windows** or **S-PLUS 4.0** program group.

- UNIX users can start the help system by browsing to the file

  ***SDIR*/module/envstats/Html/helpset.HTM**

  where ***SDIR*** denotes the directory where S-PLUS is installed.

- You can remove the ENVIRONMENTALSTATS for S-PLUS library of functions from your search list by typing

```
> module(envstats, unload=T)
```

- It is a good idea to keep separate directories for separate projects. You can automatically attach the ENVIRONMENTALSTATS for S-PLUS functions to your search list by creating a `.First` function that includes the line `module(envstats)`. Windows users can also include the line `help(module="envstats")` to automatically start the help system as well.

CHAPTER 3 :

Designing a Sampling Program

The first and most important step of any environmental study is to design the sampling program. This chapter discusses the basics of designing a sampling program, and shows you how to use ENVIRONMENTALSTATS for S-PLUS to help you determine required sample sizes.

The Necessity of a Good Sampling Design

A study is only as good as the data upon which it is based. No amount of advanced, cutting-edge statistical theory and techniques can rescue a study that has produced poor quality data, not enough data, or data irrelevant to the question it was meant to answer.

From the very start of an environmental study, there must be a constant dialog between the data producers (field and lab personnel, data coders, etc.), the data users (scientists and statisticians), and the ultimate decision maker (the person for whom the study was instigated in the first place). All persons involved in the study must have a clear understanding of the study objectives and the limitations associated with the chosen physical sampling and analytical (measurement) techniques before anyone can make any sense of the resulting data.

As an example, **Table 3.1** shows a list of qualifiers commonly used by analytical laboratories to report results. All persons involved in the study should understand the meaning and source of any potential data qualifiers, and the sampling plan must include details about how to handle data tagged with any of these qualifiers.

What is a Population and What is a Sample?

In everyday language, the word "population" refers to all the people or organisms contained within a specific country, area, region, etc. When we talk about the population of the United States, we usually mean something like "the total number of people who currently reside in the U.S."

Qualifier	Meaning
J	Result is of limited use due to discrepancies in holding times, blank analyses, duplicate analyses, spike analyses, or laboratory contamination problems.
L	(Formerly known as "trace"). Result is of limited use because it is between the instrument/method detection limit and the contract detection/quantitation limit.
N	Result probably acceptable but is just outside the calibration range or the recovery is just outside the specification range.
R	Results are unusable due to discrepancies in analytical technique/protocol, improper calibration, outside calibration range, outside specified recovery windows, or blunder.
U	Reading was below instrument detection limit (inorganics), or method detection limit (organics).

Table 3.1: Commonly used analytical laboratory data qualifiers.

In the field of statistics, however, the term ***population*** is defined operationally by the question we ask: it is the entire collection of measurements about which we want to make a statement (Zar, 1996, p.15; Berthoux and Brown, 1994, p.7; Gilbert, 1987, Chapter 2).

For example, if the question is "What does the concentration of dissolved oxygen look like in this stream?", the question must be further refined until a suitable population can be defined: "What is the average concentration of dissolved oxygen in a particular section of a stream over a particular three-day period?" In this case, the population is the set of all possible measurements of dissolved oxygen in that section of the stream within that time period. The section of the stream, the time period, the method of taking water samples, and the method of measuring dissolved oxygen all define the population.

A ***sample*** is defined as some subset of a population (Zar, 1996, p.16; Berthoux and Brown, 1994, p.7; Gilbert, 1987, Chapter 2). If the sample contains all the elements of the population, it is called a ***census***. Usually, a population is too large to take a census, so a portion of the population is sampled. The statistical definition of the word sample (a selection of individual population members) should not be confused with the more common meaning of a ***physical sample*** of soil (e.g., 10g of soil), water (e.g., 5ml of water), air (e.g., 20 cc of air), etc.

Random vs. Judgment Sampling

Judgment sampling involves subjective selection of the population units by an individual or group of individuals (Gilbert, 1987, Chapter 3). **The uncertainty inherent in the results of a judgment sample cannot be quantified and statistical methods cannot be applied to judgment samples.** Judgment sampling does not refer to using prior information and the knowledge of experts to define the population and plan the study. Gilbert (1987, p.19) also describes "haphazard" sampling, which is a kind of judgment sampling with the attitude that "any sample will do" and can lead to "convenience" sampling, in which samples are taken in convenient places at convenient times.

Probability sampling or ***random sampling*** involves using a random mechanism to select samples from the population (Gilbert, 1987, Chapter 3). **All statistical methods used to quantify uncertainty assume some form of random sampling has been used to obtain a sample.** At the simplest level, a simple random sample is used in which each member of the population has an equal chance of being chosen, and the selection of any member of the population does not influence the selection of any other member. Other probability sampling methods include stratified random sampling, composite sampling, two-stage sampling, ranked set sampling, systematic sampling, double sampling, and sequential sampling (see the help file **Glossary: Sample and Sampling Methods** for more information).

The Basic Process of Designing a Sampling Program

A sampling design or sampling plan is a set of instructions to use to scientifically investigate some question or problem and come up with a quantifiable answer. A sampling design consists of the following steps (cf., Gilbert, 1987, pp.3-4, Keith, 1991, p.8):

1. **Define the Question(s) and Attainment Objective(s).** Define the question(s) or problem(s) to be addressed. All questions must be clearly stated, in exact terms that will allow you to come up with an answer by measuring something. This entails gathering lots of information on the specific site, including what chemicals, cleanup standards, and population parameters are of concern. Almost always, each question should be stated in terms of one or several hypothesis tests. For each hypothesis test, the significance level and power must be determined.

2. **Define the Population(s).** Determine what population(s) is (are) defined by the question in Step 1.

3. **Define the Physical Samples.** Determine how to physically sample the population(s) defined in Step 2.

4. **Determine the Random Sampling Method.** Determine the method of random sampling (e.g., simple random sampling, stratified random sampling, cluster sampling, etc.). (See the help file **Glossary: Sample and Sampling Methods** for more information.)

5. **Define an Analysis Plan.** Determine how to analyze the resulting data, including what estimates will be computed, what hypothesis tests will be executed, and how the results of the analysis will be reported.

6. **Develop a QA/QC Plan.** Develop a quality assurance/quality control (QA/QC) plan to insure the integrity of the final results. A good QA/QC plan covers instructions for sample collection in the field, handling, laboratory analysis and reporting, data coding, statistical analyses, and reports. Embedded in these instructions is the ***chain of custody procedures*** for documenting who has custody of the samples and the current conditions of the samples from the point of collection in the field to the analysis at the laboratory. Chain of custody procedures are used to insure that samples are not lost, tampered with, or improperly stored or handled. See Keith (1991) for more information.

7. **Estimate the Required Sample Size(s).** Use information from previous studies and/or conduct a pilot study to quantify the amount of variability that is typical in samples. Use this information to estimate the required sample size(s), based on either estimating a parameter with a specified amount of precision (e.g., maximum length of the confidence interval), or detecting a specified difference with a specified significance level and power (see the section **Power and Sample Size Calculations** later in this chapter). Sample size requirements should take into account that a certain proportion of the samples will be unusable due to loss, mislabeling, mishandling, or some other factor that keeps the samples from meeting the specified QA/QC standards.

> **Note**: Even if information on sample variability is available from previous studies, it is almost always advisable to conduct a pilot study in order to "fine tune" the QA/QC sampling plan and the overall sampling design.

8. **Determine Whether the Study Can Be Completed.** Assess whether enough funds and staff are available to answer the question, given the required sample size(s) and method(s) of sampling. If not, you need to redefine the problem, or change the required precision, significance level, or power.

9. **Create a Written Protocol.** Develop a written protocol for implementing the sampling design and QA/QC programs.

See Gilbert (1987), Keith (1991), USEPA (1989a), USEPA (1992b), USEPA (1994a), USEPA (1994b), and USEPA (1996) for more information on sampling design.

Common Mistakes in Environmental Studies

The most common mistakes that occur in environmental studies are:

- **Using Judgment Sampling to Obtain Samples.** When judgment sampling is used to obtain samples, there is no way to quantify the precision and bias of any type of estimate computed from these samples.

- **Failing to Randomize Over Potentially Influential Factors.** There are an enormous number of factors that can influence the final measure associated with a single sampling unit, including the person doing the sampling, the device used to collect the sample, the weather and field conditions when the sample was collected, the method used to analyze the sample, the laboratory to which the sample was sent, etc. A good sampling design controls for as many potentially influencing factors as possible, and randomizes over the factors that can't be controlled. For example, if there are four persons who collect data in the field, and two laboratories are used to analyze the results, you would not send all the samples collected by persons 1 and 2 to laboratory 1 and all the samples collected by persons 3 and 4 to laboratory 2, but rather send samples collected by each person to each of the laboratories.

- **Collecting Too Few Samples to Have a High Degree of Confidence in the Results.** The ultimate goal of an environmental study is to answer one or more basic questions. These questions should be stated in terms of hypotheses that can be tested using statistical procedures. In this case, you can determine the probability of rejecting the null hypothesis when in fact it is true (a Type I error), and the probability of not rejecting the null hypothesis when if fact it is false (a Type II error). Usually, the Type I error is set in advance, and the probability of correctly rejecting the null hypothesis when in fact it is false (the power) is calculated for various sample sizes. Too often, this step of determining power and sample size is neglected, resulting in a study from which no conclusions can be drawn with any large degree of confidence.

Following the steps outlined in the previous section will keep you from committing these common mistakes.

Power and Sample Size Calculations

ENVIRONMENTALSTATS for S-PLUS contains several functions to assist you in determining how many samples you need for a given degree of confidence in the results of a sampling program (see the help file **Power and Sample Size Calculations**). These functions are based on the ideas of confidence intervals and hypothesis tests. If you are unfamiliar with these concepts, please see **Chapter 6: Estimating Population Parameters and Quantiles** and **Chapter 8: Hypothesis Tests**.

A very important point to remember is that no matter what you come up with for estimates of required sample sizes, it is always a good idea to assume you will lose some percentage of your observations (say 10%) due to sample loss, sample contamination, data coding errors, misplaced forms, etc.

Sample Size Calculations for Confidence Intervals

Table 3.2 lists the functions available in ENVIRONMENTALSTATS for S-PLUS for computing required sample sizes and half-widths associated with a confidence interval. There are functions to compute the half-width of the confidence interval given the user-specified sample size(s), functions to compute the required sample size(s) given the user-specified half-width, and functions to plot the relationship between sample size and half-width.

Function	Description
`ci.norm.half.width`	Compute the half-width of a confidence interval for the mean of a normal distribution or the difference between two means.
`ci.norm.n`	Compute the sample size necessary to achieve a specified half-width of a confidence interval for the mean of a normal distribution or the difference between two means.
`plot.ci.norm.design`	Create plots for a sampling design based on a confidence interval for the mean of a normal distribution or the difference between two means.
`ci.binom.half.width`	Compute the half-width of a confidence interval for a binomial proportion or the difference between two proportions.
`ci.binom.n`	Compute the sample size necessary to achieve a specified half-width of a confidence interval for a binomial proportion or the difference between two proportions.
`plot.ci.binom.design`	Create plots for a sampling design based on a confidence interval for a binomial proportion or the difference between two proportions.

Table 3.2: Sample size functions for confidence intervals.

Confidence Intervals for the Mean of a Normal Distribution or the Difference Between Two Means

The help files for `ci.norm.half.width` and `ci.norm.n` explain how to compute a confidence interval for a mean of a normal distribution or the difference between two means, as well as the relationship between the sample size(s), estimated standard deviation, confidence level, and half-width. The function `ci.norm.half.width` computes the half-width associated with the confidence interval, given the sample size(s), estimated standard deviation, and confidence level. The function `ci.norm.n` computes the sample size(s) required to achieve a specified half-width, given the estimated standard deviation and confidence level. The function `plot.ci.norm.design` plots the relationships between sample size(s), half-width, estimated standard deviation, and confidence level.

Example 3.1: Computing Sample Size for a Specified Half-Width of a Confidence Interval for Mean Aldicarb Concentration

The guidance document **Statistical Analysis of Ground-Water Monitoring Data at RCRA Facilities** (USEPA, 1989b) contains an example on pages 6-3 to 6-5 that computes confidence intervals for the mean aldicarb concentration (ppb) at three difference compliance wells. The data in this example are stored in the data frame `epa.89b.aldicarb1.df` (see the help file **Datasets: USEPA (1989b)**). The observed means are 23.1, 24.6, and 4.5, respectively. Each mean is based on a sample size of four observations taken once per month from January to April.

```
> epa.89b.aldicarb1.df
   Aldicarb Month Well
 1     19.9     1    1
 2     29.6     2    1
 3     18.7     3    1
 4     24.2     4    1
 .
 9      5.6     1    3
10      3.3     2    3
11      2.3     3    3
12      6.9     4    3

> attach(epa.89b.aldicarb1.df)

> sapply(split(Aldicarb, Well), mean)
    1     2     3
 23.1 24.65 4.525
```

Consider just the first compliance well. If you are willing to make the assumption that the distribution of aldicarb at this compliance well is normally distributed, that the distribution does not change over time, and that there is no serial correlation between monthly samples (very important assumptions), you can use the function `ci.norm.n` to determine the required sample size in order to achieve a half-width that is 10% of the observed mean (23.1) for a 98% confidence interval for this mean:

```
> objects(2)
[1] "Aldicarb" "Month"    "Well"

> mu.hat <- mean(Aldicarb[Well==1])

> mu.hat
[1] 23.1

> sigma.hat <- sd(Aldicarb[Well==1])

> sigma.hat
[1] 4.93491

> ci.norm.n(half.width = 0.1 * mu.hat,
+    sigma.hat = sigma.hat, conf = 0.98)
[1] 28
```

The required sample size is 28, so more than two years of monitoring are required assuming observations are taken once per month. This sample size calculation assumes the estimated standard deviation based on 28 observations is essentially the same as the one we computed based on the four observations. In practice, it is a good idea to use a conservative (slightly larger) estimate of standard deviation based on expert judgement or previous studies. This practice will yield slightly larger required sample sizes.

More generally, you can use the function `plot.ci.norm.design` to look at the relationship between half-width and sample size for a 98% confidence interval for the mean aldicarb level at the first compliance well:

```
> plot.ci.norm.design(sigma.hat = sigma.hat,
+    conf = 0.98, digits = 2,
+    range.x.var = c(2, 25))
```

The result of calling `plot.ci.norm.design` is shown in **Figure 3.1**. This plot shows that with only 4 observations, the half-width of the confidence interval is about 10 ppb. The half-width decreases dramatically as the sample size increases up to 10. With 25 observations, the half-width is about 2.5 ppb.

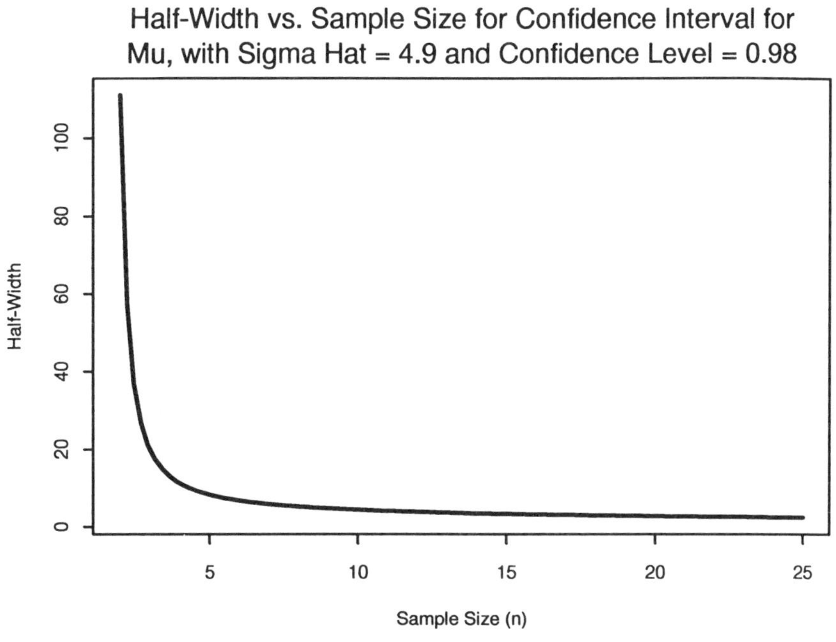

Figure 3.1: Result of calling `plot.ci.norm.design`.

Confidence Intervals for Binomial Proportions

The help files for `ci.binom.half.width` and `ci.binom.n` explain how to compute a confidence interval for a binomial proportion or the difference between two proportions, as well as the relationship between the sample size(s), estimated proportion(s), confidence level, and half-width. The function `ci.binom.half.width` computes the half-width associated with the confidence interval, given the sample size(s), estimated proportion(s), and confidence level. The function `ci.binom.n` computes the sample size(s) required to achieve a specified half-width, given the estimated proportion(s) and confidence level. The function `plot.ci.binom.design` plots the relationships between sample size(s), half-width, estimated proportion(s), and confidence level.

Example 3.2: Computing Sample Size for a Specified Half-Width of a Confidence Interval for the Difference Between Percent Detects at Two Different Wells

The guidance document **Statistical Analysis of Ground-Water Monitoring Data at RCRA Facilities** (USEPA, 1989b) contains an example on pages 8-5 to 8-7 comparing the proportion of detected values of cadmium (µg/L) at a background well vs. a compliance well. The estimated proportion of detects is 8/24 (33%) at the background well and 24/64 (37.5%) at the compliance well. The data in this example are stored in the data frame `epa.89b.cadmium.df` (see the help file **Datasets: USEPA (1989b)**).

```
> epa.89b.cadmium.df
   Cadmium.orig Cadmium Censored  Well.type
 1          0.1   0.100        F Background
 2         0.12   0.120        F Background
 3          BDL   0.000        T Background
 4         0.26   0.260        F Background
 .
85          BDL   0.000        T Compliance
86          BDL   0.000        T Compliance
87          BDL   0.000        T Compliance
88          BDL   0.000        T Compliance

> attach(epa.89b.cadmium.df)

> #  Compute percent detected at each well type

> sapply(split(!Censored, Well.type), mean)
 Background Compliance
  0.3333333      0.375
```

You can use the function `ci.binom.n` to determine the required sample size in order to achieve a half-width of 0.1 (10 percentage points) for the 95% confidence interval for the difference between the proportion of detects of cadmium at the background and compliance wells:

```
> objects(2)
[1] "Cadmium"       "Cadmium.orig" "Censored"
[4] "Well.type"

> p.hat.back <-
+    mean(!Censored[Well.type=="Background"])

> p.hat.back
[1] 0.3333333
> p.hat.comp <-
+    mean(!Censored[Well.type=="Compliance"])

> p.hat.comp
[1] 0.375

> ci.binom.n(half.width = 0.1,
+    p.hat.or.p1.hat = p.hat.back,
+    p2.hat = p.hat.comp)
[1] 195
```

Thus, assuming equal sample sizes, the required sample size at each well is 195, for a total of 390 observations!

More generally, you can use the function plot.ci.binom.design to look at the relationship between half-width and sample size for a 95% confidence interval for the difference between the proportion of detects at the background and compliance wells.

```
> plot.ci.binom.design(p.hat.or.p1.hat = p.hat.back,
+    p2.hat = p.hat.comp, digits=3)
Warning messages:
  The sample sizes 'n1' and 'n2' are too small,
  relative to the given values of 'p1.hat' and
  'p2.hat', for the normal approximation to work well
  for the following element indices:
   1 2 3 4 5 6 7 8 9 10 11 12 13
   in: ci.binom.half.width(n.or.n1 = x, n2 = n2,
  p.hat.or.p1.hat = p.hat.or.p1.hat,  ....
```

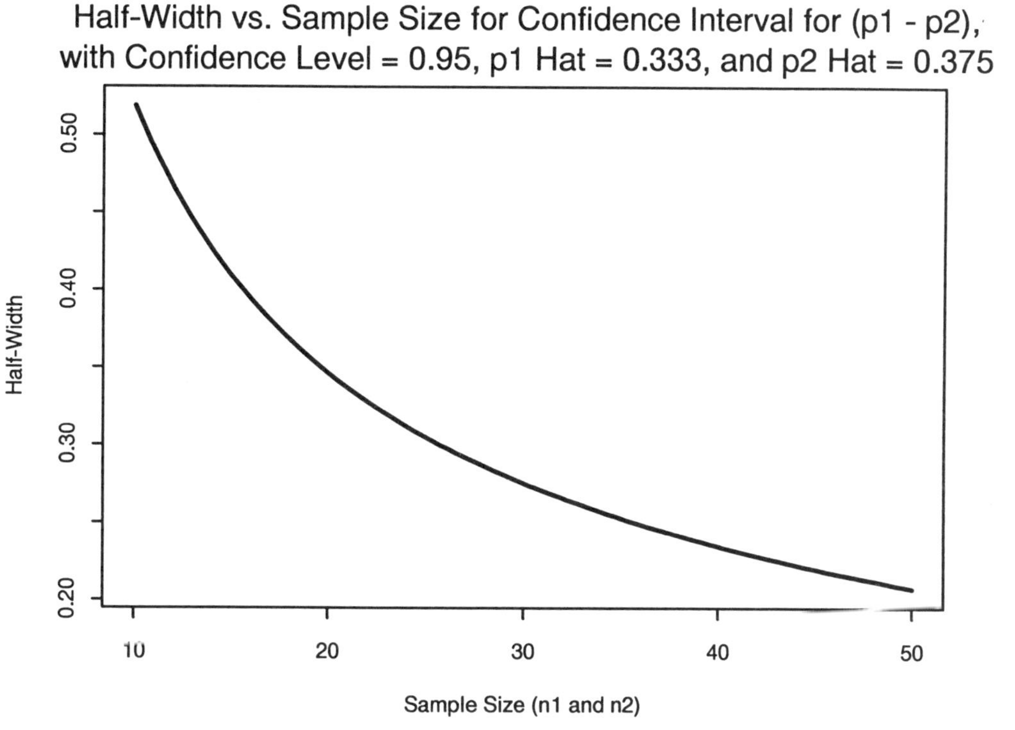

Figure 3.2: Result of calling plot.ci.binom.design.

The result of calling plot.ci.binom.design is shown in **Figure 3.2**. This plot shows that even with 40 observations at each well, the half-width of the confidence interval for the difference between the two proportions is about 0.25 (25 percentage points).

Power and Sample Size Calculations for Hypothesis Tests

Table 3.3 lists the functions available in ENVIRONMENTALSTATS for S-PLUS for computing power, required sample sizes, or the minimal detectable difference associated with a hypothesis test. There are also functions to plot the relationships between these quantities.

Function	Description
`t.test.n`	Compute the sample size necessary to achieve a specified power for a one- or two-sample t-test.
`t.test.power`	Compute the power of a one- or two-sample t-test.
`t.test.scaled.mdd`	Compute the scaled minimal detectable difference associated with a one- or two-sample t-test.
`plot.t.test.design`	Create plots for a sampling design based on a one- or two-sample t-test.
`aov.n`	Compute the sample sizes necessary to achieve a specified power for a one-way fixed-effects analysis of variance test.
`aov.power`	Compute the power of a one-way fixed-effects analysis of variance.
`prop.test.n`	Compute the sample size necessary to achieve a specified power for a one- or two-sample proportion test.
`prop.test.power`	Compute the power of a one- or two-sample proportion test.
`prop.test.mdd`	Compute the minimal detectable difference associated with a one- or two-sample proportion test.
`plot.prop.test.design`	Create plots involving sample size, power, difference, and significance level for a one- or two-sample proportion test.

Table 3.3: Sample size functions for hypothesis tests.

Testing a Mean or a Difference Between Means for a Normal Distribution

The help files for `t.test.power`, `t.test.n`, and `t.test.scaled.mdd` explain how to use Student's t-test to perform a hypothesis test for the mean of a normal distribution or the difference between two means, as well as the relationship between the sample size(s), scaled minimal detectable difference (scaled mdd), type I error level (α-level), and power. The function `t.test.power` computes the power associated with the t-test, given the sample size(s), scaled mdd, and α-level. The function `t.test.n` computes the sample size(s) required to achieve a specified power, given the scaled mdd and α-level. The function `t.test.scaled.mdd` computes the scaled minimal detectable difference associated with user-specified values of power, sample size(s), and α-level. The function `plot.t.test.design` plots the relationships between sample size(s), power, scaled mdd, and α-level.

Example 3.3: Computing the Required Sample Size to Detect a Mean Aldicarb Level Greater Than an MCL for a Specified Power

Consider again the aldicarb data of **Example 3.1** stored in the data frame `epa.89b.aldicarb1.df` (see the help file **Datasets: USEPA (1989b)**). This data frame contains aldicarb levels at three wells measured once per month over a period of four months.

```
> attach(epa.89b.aldicarb1.df)

> objects(2)
[1] "Aldicarb" "Month"    "Well"

> Well
 [1] 1 1 1 1 2 2 2 2 3 3 3 3

> sapply(split(Aldicarb, Well), mean)
    1     2     3
 23.1 24.65 4.525

> sapply(split(Aldicarb, Well), sd)
       1        2        3
 4.93491 2.282542 2.101388
```

The maximum (allowed) concentration limit (MCL) in this example is 7 ppb. You can use the function `t.test.n` to determine the required sample size to detect a mean aldicarb level greater than the MCL of 7 ppb at the third compliance well with a power of 95%, assuming the true mean is 10 or 14. We'll use the estimated standard deviation from the first four months of data (i.e., 2.1) to estimate the true population standard deviation, set α=0.01, and assume an upper one-sided alternative (third compliance well mean larger than 7):

```
> sigma <- sd(Aldicarb[Well==3])

> sigma
[1] 2.101388

> t.test.n(delta.over.sigma = (c(10, 14) - 7)/sigma,
+     alpha = 0.01, sample.type="one",
+     alternative="g")
[1] 11  5
```

Thus, if the true mean at the third compliance well is 10 ppb, you need at least 11 observations to have 95% power of detecting a mean level greater than the MCL of 7 ppb. If the true mean is 14 ppb, you only need 5 observations to have 95% power. Both of these computations assume the true standard deviation is equal to the standard deviation estimated from the four observations shown in the example. In practice, it is a good idea to use a conservative (slightly larger) estimate of standard deviation based on expert judgment or previous studies. This practice will yield slightly larger required sample sizes.

More generally, you can use the function `plot.t.test.design` to look at the relationship between power and sample size for an upper one-sided test of the mean aldicarb level at the third compliance well, assuming an α-level of 0.01 and a true mean level of 10 ppb:

```
> delta.over.sigma <- (10 - 7) / sigma

> delta.over.sigma
[1] 1.427628

> plot.t.test.design(range.x.var=c(2, 20),
+     delta.over.sigma = delta.over.sigma,
+     alpha = 0.01, sample.type="one",
+     alternative="greater", digits = 2)
```

The result of calling `plot.t.test.design` is shown in **Figure 3.3**. This plot shows that the power is negligible for a sample size less than 5, and that the power increases dramatically as the sample size increases from 5 to 10.

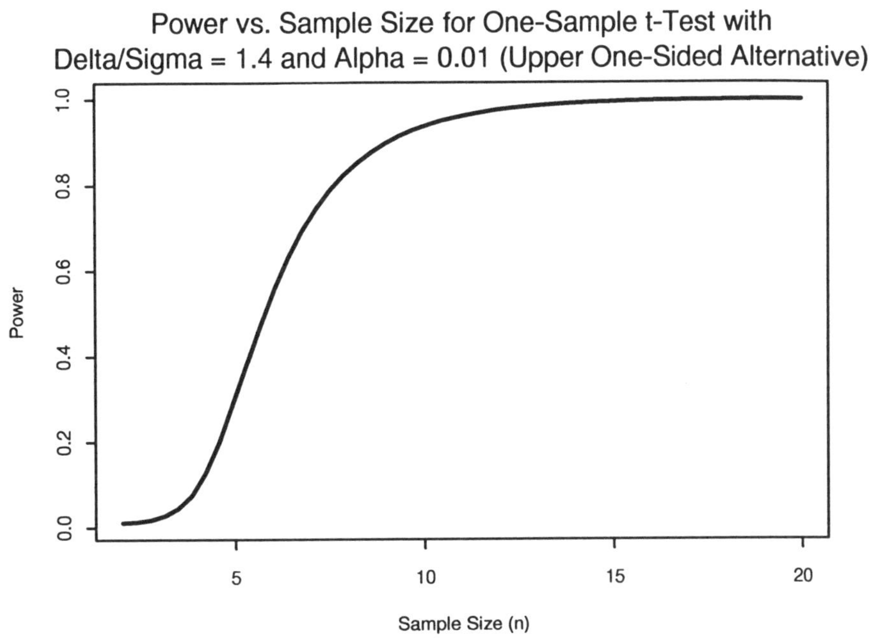

Figure 3.3: Result of calling `plot.t.test.design`.

Testing the Differences Between Several Means for a Normal Distribution (One-Way ANOVA)

The help files for `aov.power` and `aov.n` explain how to perform a hypothesis test to determine whether at least one of several means of normal populations differs from the rest of the means, using the standard F-test for a one-way analysis of variance (ANOVA) procedure. These help files also explain the relationship between sample sizes, the actual differences between the means, the α-level, and the power.

The function `aov.power` computes the power associated with the one-way ANOVA, given the sample sizes, true mean values, and α-level. The function `aov.n` computes the sample sizes required to achieve a specified power, given the true mean values and α-level.

Example 3.4: Computing the Power of a One-Way ANOVA to Compare Lead Concentrations

The guidance document **Statistical Analysis of Ground-Water Monitoring Data at RCRA Facilities** (USEPA, 1989b) contains an example on pages 5-11 to 5-14 that shows log-transformed concentrations of lead (µg/L) at two background wells and four compliance wells, where observations were taken once per month over four months. These data are stored in the data frame `epa.89b.loglead.df` (see the help file **Datasets: USEPA (1989b)**).

```
> epa.89b.loglead.df
    LogLead Month Well  Well.type
 1 4.060443     1    1 Background
 2 3.990834     2    1 Background
 3 3.401197     3    1 Background
 .
22 5.210032     2    6 Compliance
23 5.289781     3    6 Compliance
24 5.080161     4    6 Compliance

> attach(epa.89b.loglead.df)

> round(sapply(split(LogLead, Well), mean), 1)
   1 2   3   4   5   6
 3.8 4 4.5 4.2 4.8 5.3
```

Assume the true mean levels at each well are 3.9, 3.9, 4.5, 4.5, 4.5, and 5, respectively. Compute the power of a one-way ANOVA to test for mean differences between wells. Use α=0.05, and assume the true standard deviation is equal to the one estimated from the data in this example.

First, look at the distribution of the data by creating some boxplots (see **Chapter 4: Looking At Data** for more information on boxplots).

```
> boxplot(split(LogLead, Well), xlab = "Well Number",
+   ylab="Log [ Lead (ug/L) ]",
+   main="Lead Concentrations at Six Wells")
```

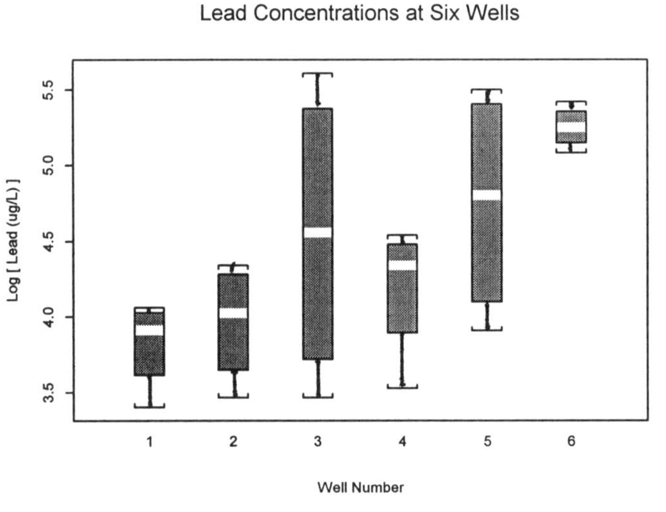

Figure 3.4: Boxplots for the lead data.

The distribution of data at each well is shown in **Figure 3.4**. Note that the usual assumption of equal variances for each well appears suspect in this example.

Now perform the one-way ANOVA to get the estimated variance at each well:

```
> aov.list <- aov(LogLead ~ Well,
+    data=epa.89b.loglead.df)

> summary(aov.list)
          Df Sum of Sq  Mean Sq  F Value      Pr(F)
Well       5  5.744727 1.148945 3.346924 0.02598835
Residuals 18  6.179112 0.343284
```

Thus, the estimated variance for each well is 0.34, based on the residual mean square error.

Now call the function `aov.power`, using the estimated standard deviation:

```
> aov.power(n.vec = rep(4, 6),
+    mu.vec = c(3.9,3.9,4.5,4.5,4.5,5),
+    sigma=sqrt(0.34))
[1] 0.5523148
```

So there is slightly more than a 50% chance of detecting a difference between the six monitoring wells using these assumed values for the true means and standard deviation. This small level of power is not surprising given the small sample sizes at each well.

Let's see how the power increases if we increase the sample size at each well from 4 to 8.

```
> aov.power(n.vec = rep(8, 6),
+    mu.vec = c(3.9,3.9,4.5,4.5,4.5,5),
+    sigma=sqrt(0.34))
[1] 0.9318802
```

So doubling the sample size from 4 to 8 at each well increases the power of detecting a difference between the wells from about 55% to about 93%.

Testing a Binomial Proportion or a Difference Between Binomial Proportions

The help files for `prop.test.power`, `prop.test.n`, and `prop.test.mdd` explain how to use exact and approximate tests to perform a hypothesis test for a binomial proportion or the difference between two proportions, as well as the relationship between the sample size(s), minimal detectable difference (mdd), type I error level (α-level), and power. The function `prop.test.power` computes the power associated with the test, given the sample size(s), mdd, and α-level. The function `prop.test.n` computes the sample size(s) required to achieve a specified power, given the mdd and α-level. The function `prop.test.mdd` computes the minimal detectable difference associated with user-specified values of power, sample size(s), and α-level. The function `plot.prop.test.design` plots the relationships between sample size(s), power, mdd, and α-level.

Example 3.5: Computing the Power to Detect a Difference in the Proportion of Detected Values at Two Different Wells

Consider again the cadmium data of **Example 3.2** stored in the data frame `epa.89b.cadmium.df` (see the help file **Datasets: USEPA (1989b)**). The estimated proportion of detects is 37.5% (24/64) at the compliance well and 33% (8/24) at the background well. You can use the function `prop.test.power` to determine how adding another 20 observations to the background well to increase the sample size from 24 to 44 will affect the power of detecting a difference in the proportion of detects of cadmium between the background and compliance wells.

Set the compliance well to "group 1" and set the background well to "group 2". Assume the true probability of a "detect" is 0.4 at the compliance well and 1/3 at the background well, use a 5% significance level, and use the upper one-sided alternative (probability of a "detect" at the compliance well is greater than the probability of a "detect" at the background well).

```
> attach(epa.89b.cadmium.df)

> objects(2)
[1] "Cadmium"       "Cadmium.orig" "Censored"
[4] "Well.type"

> p.hat.back <-
+    mean(!Censored[Well.type=="Background"])

> p.hat.back
[1] 0.3333333

> p.hat.comp <-
+    mean(!Censored[Well.type=="Compliance"])

> p.hat.comp
[1] 0.375

> n.back <- sum(Well.type == "Background")

> n.back
[1] 24

> n.comp <- sum(Well.type == "Compliance")
```

```
> n.comp
[1] 64

> prop.test.power(n.or.n1 = n.comp,
+    p.or.p1 = 0.4,
+    n2 = c(n.back, 44), p0.or.p2 = 1/3,
+    alt="greater", sample.type="two")
[1] 0.08953013 0.12421135
```

So as the sample size at the background well increases from 24 to 44, the power increases from 9% to 12%, but it is relatively very small for both sample sizes!

More generally, you can use the function `plot.prop.test.design` to look at the relationship between power and sample size or power and minimal detectable difference for an upper one-sided test comparing the proportion of detects at the compliance well with the proportion of detects at the background well. Here, we'll determine the relationship between power and minimal detectable difference to detect a difference in the proportion of detects of cadmium between the background and compliance wells.

As before, set the compliance well to "group 1" and the background well to "group 2". Assume the true probability of a "detect" at the background well is 1/3, use a 5% significance level, use the given sample sizes of 64 observations at the compliance well and 24 observations at the background well, and use the upper one-sided alternative (probability of a "detect" at the compliance well is greater than the probability of a "detect" at the background well).

```
> plot.prop.test.design(x.var = "delta",
+    y.var = "power",
+    range.x.var = c(0.05, 0.65),
+    n.or.n1 = n.comp, n2 = n.back,
+    p0.or.p2 = p.hat.back,
+    sample.type = "two.sample",
+    alternative = "greater", warn = F,
+    digits = 2, title.cex = 0.85)
```

The result of calling `plot.prop.test.design` is shown in **Figure 3.5**. Note that the minimal detectable difference (p1-p2) increases from 0.32 to 0.37 to 0.40 as the required power increases from 80% to 90% to 95%. Thus, in order to detect a difference in probability of detection between the compliance and background wells, the probability of detection at the compliance well must be 0.65, 0.70, or 0.74 (depending on the required power).

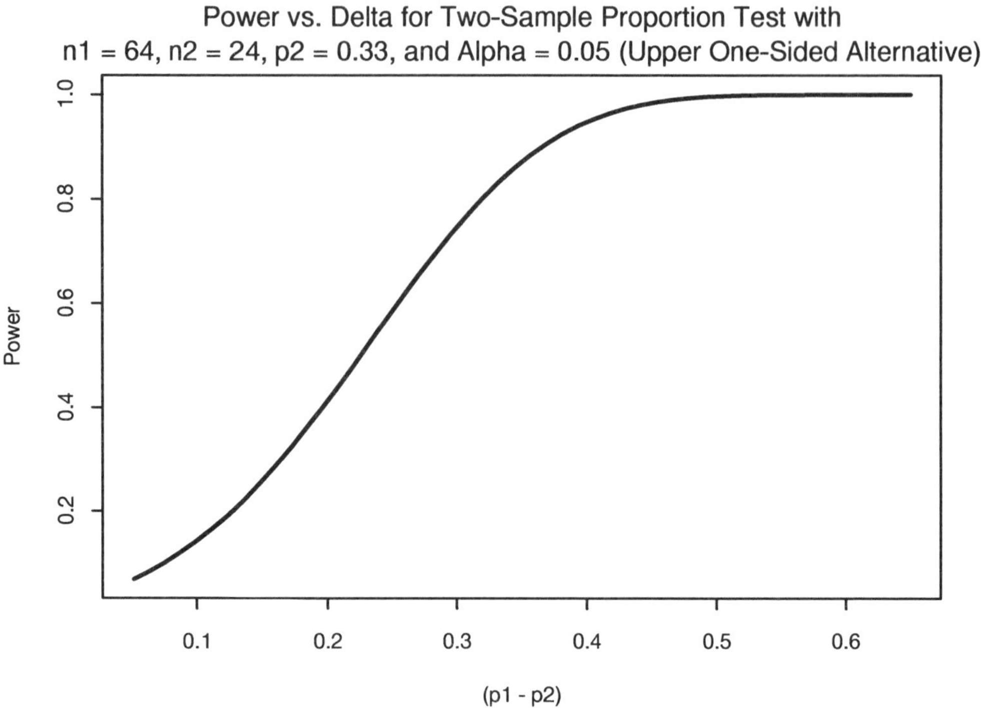

Figure 3.5: Result of calling `plot.prop.test.design`.

Summary

- A study is only as good as the data upon which it is based. No amount of advanced, cutting-edge statistical theory and techniques can rescue a study that has produced poor quality data, not enough data, or data irrelevant to the question it was meant to answer.

- From the very start of an environmental study, there must be a constant dialog between the data producers, the data users, and the ultimate decision maker.

- A ***sample*** is defined as some subset of a population. The statistical definition of the word sample (a selection of individual population members) should not be confused with the more common meaning of a ***physical sample*** of soil (e.g., 10g of soil), water (e.g., 5ml of water), air (e.g., 20 cc of air), etc.

- ***Judgment sampling*** involves subjective selection of the population units by an individual or group of individuals. The uncertainty inherent in the results of a judgment sample cannot be quantified and statistical methods cannot be applied to judgment samples.

- ***Probability sampling*** or ***random sampling*** involves using a random mechanism to select samples from the population. All statistical methods used to quantify uncertainty assume some form of random sampling has been used to obtain a sample.

- The general process of designing an environmental sampling program includes defining the attainment objectives, defining the population, defining the physical sample, determining the method of random sampling, defining the analysis plan, developing a QA/QC plan, estimating the required sample sizes, determining whether enough resources are available to carry out the study, and creating a written protocol to follow during each phase of the study.

- Common mistakes in environmental studies include using judgment sampling to select population units, failing to randomize over potentially confounding factors, and collecting too few samples to have a high degree of confidence in the results.

- You can use the ENVIRONMENTALSTATS for S-PLUS functions listed in **Table 3.2** and **Table 3.3** (and the help file **Power and Sample Size Calculations**) to estimate required sample sizes for an environmental study.

CHAPTER 4 :

Looking At Data

Once you have a collection of observations from your environmental study, you should thoroughly examine the data in as many ways as is possible. One of the main strengths of S-PLUS is the plethora of functions for graphically exploring your data. When the first widely available commercial statistical software packages came out in the 1960's, the emphasis was on statistical summaries of data, such as means, standard deviations, and measures of skew and kurtosis. It is still true that "a picture is worth a thousand words," and no amount of summary statistics can replace a good graph to explain your data. Helsel and Hirsch (1992, Chapters 1, 2, and 16) and USEPA (1996) give a good overview of statistical and graphical methods for exploring environmental data. Cleveland (1993, 1994) and Chambers et al. (1983) are excellent general references for methods of graphing data. This chapter discusses some of the built-in functions in S-PLUS for looking at your data, as well as some new and modified functions available in ENVIRONMENTALSTATS for S-PLUS.

Summary Statistics

The two S-PLUS help files **Mathematical Operations** and **Robust/Resistant Techniques** list several functions available in S-PLUS for computing summary statistics. Functions for computing measures of location include `mean`, `median`, and `location.m` (a robust method of estimating location). The function `var` computes the traditional estimator of variance, while the functions `mad`, `scale.a`, and `scale.tau` are robust measures of spread. S-PLUS also contains the functions `quantile` to compute empirical quantiles and `summary` to compute a simple summary of the data.

Example 4.1: Summary Statistics for TcCB Concentrations in Two Areas

The guidance document **Statistical Methods for Evaluating the Attainment of Cleanup Standards, Volume 3: Reference-Based Standards for Soils and Solid Media** (USEPA, 1994b, pp.6.22-6.25) contains measures of 1,2,3,4-Tetrachlorobenzene (TcCB) concentrations (ppb) from soil samples at a reference site and a "cleanup" area. There are 47 observations from the reference site and 77 in the cleanup area.

These data are stored in the data frame epa.94b.tccb.df (see the help file **Datasets: USEPA (1994b)**). There is one observation coded as "ND" in this data set as presented in the guidance document. Here, we'll assume this observation is less than the smallest observed value, which is 0.09 ppb. For the purposes of this example, we'll set this one censored observation to the assumed detection limit of 0.09.

```
> epa.94b.tccb.df
      TcCB.orig   TcCB Censored      Area
  1        0.22   0.22        F Reference
  2        0.23   0.23        F Reference
  3        0.26   0.26        F Reference
  4        0.27   0.27        F Reference
  .
121        6.61   6.61        F Cleanup
122       18.40  18.40        F Cleanup
123       51.97  51.97        F Cleanup
124      168.64 168.64        F Cleanup
```

We can use built-in S-PLUS functions to look at some summary statistics for both the reference and the cleanup area. We can do this one of two ways, by either creating separate variables to hold the TcCB data for each area, or by using the sapply function:

```
> attach(epa.94b.tccb.df)

> objects(2)
[1] "Area"       "Censored"   "TcCB"        "TcCB.orig"

> TcCB.ref <- TcCB[Area == "Reference"]

> TcCB.clean <- TcCB[Area == "Cleanup"]

> mean(TcCB.ref)
[1] 0.5985106

> mean(TcCB.cleanup)
[1] 3.915195

> sapply(split(TcCB, Area), mean)
  Cleanup Reference
 3.915195 0.5985106

> sapply(split(TcCB, Area), median)
 Cleanup Reference
    0.43      0.54
```

```
> sapply(split(TcCB, Area), var)
  Cleanup  Reference
 400.6243 0.08045208
```

By looking at the mean, median, and variance for each area, it is apparent that the cleanup area must have at least one very large, perhaps "outlying", value, since the medians are fairly close but the means and variances are very different. This is in fact the case, as the `summary` function shows:

```
> summary(TcCB.ref)
  N Min. 1st Qu. Median   Mean 3rd Qu. Max.
 47 0.22    0.39   0.54 0.5985    0.75 1.33

> summary(TcCB.cleanup)
  N Min. 1st Qu. Median  Mean 3rd Qu.  Max.
 77 0.09    0.23   0.43 3.915     1.1 168.6
```

The EnvironmentalStats for S-Plus help file **Summary Statistics** lists other functions for computing summary statistics that are available in EnvironmentalStats for S-Plus but not built into S-Plus. These include functions to compute the geometric mean, standard deviation, interquartile range, skew, kurtosis, and coefficient of variation, as well as a function called `full.summary` that computes all of these summary statistics and others as well.

```
> full.summary(split(TcCB, Area))
                             Cleanup Reference
                Sample Size  77        47
                  # Missing   0         0
                       Mean   3.915     0.5985
                     Median   0.43      0.54
            10% Trimmed Mean   0.6846    0.5728
             Geometric Mean   0.5784    0.5382
                       Skew   7.566     0.8729
                   Kurtosis  61.6       2.993
                        Min   0.09      0.22
                        Max 168.6       1.33
                      Range 168.5       1.11
               1st Quartile   0.23      0.39
               3rd Quartile   1.1       0.75
         Standard Deviation  20.02      0.2836
        Interquartile Range   0.87      0.36
  Median Absolute Deviation   0.3558    0.2669
   Coefficient of Variation   5.112     0.4739
```

Probability-Weighted Moments and L-Moments

ENVIRONMENTALSTATS for S-PLUS contains two functions called `pw.moment` and `l.moment` for computing probability-weighted moments and *L*-moments, respectively. Also, the functions `skew` and `kurtosis` include an argument called `method` that lets you specify the method of computing these quantities: either the traditional method based on moments, or else a method based on *L*-moments.

Probability-weighted moments were introduced by Greenwood et al. (1979) in the hydrology literature as a tool to derive estimates of distribution parameters for distributions that can be (perhaps only be) expressed in inverse form (i.e., instead of characterizing the distribution by the formula for its cumulative distribution function (cdf), the distribution is characterized by the formula for the pth quantile).

Hosking (1990) introduced the concept of *L*-moments, which are a generalization of probability-weighted moments. *L*-moments have several advantages over conventional moments, including:

- *L*-moments can characterize a wider range of distributions because they always exist as long as the distribution has a finite mean.
- *L*-moments are estimated by linear combinations of order statistics, so estimators based on *L*-moments are more robust to the presence of outliers than estimators based on conventional moments.
- Based on research experience, *L*-moment estimators are less biased and approximate their asymptotic distribution more closely in finite samples than estimators based on conventional moments.
- *L*-moment estimators are sometimes more efficient (have a smaller root-mean-square-error) than even maximum likelihood estimators for small samples.

See the help files **Abstract: Greenwood et al. (1979)** and **Abstract: Hosking (1990)**, and the help files for `pw.moment` and `l.moment` for more information on probability-weighted moments and *L*-moments.

Graphs for Univariate Data

S-PLUS includes several functions for plotting univariate data, including functions to create bar charts, dot charts, pie charts, histograms, boxplots, quantile plots (also called empirical cumulative distribution function plots), quantile-quantile (Q-Q) plots (also called probability plots), and autocorrelation plots. ENVIRONMENTALSTATS for S-PLUS includes new and modified versions of some of these functions (see the help file **Plotting Probability Distributions**).

S-PLUS is a constantly evolving software package. With the addition of the Trellis library of functions developed at AT&T Bell Laboratories in Murray Hill, N.J. (now Lucent Technologies), there are duplicate functions for producing a number of standard graphs. **Table 4.1** displays the names of the original S-PLUS functions for producing certain univariate data plots and the analogous Trellis functions. Trellis functions can be used in multi-panel displays (see the section **Graphs for Bivariate and Multivariate Data** later in this chapter). In this chapter, we will show you how to use both types of functions.

Plot Type	S-PLUS Function	Trellis Function
Bar Chart	`barplot`	`barchart`
Dot Chart	`dotchart`	`dotplot`
Pie Chart	`pie`	`piechart`
Strip Plot		`stripplot`
Histogram	`hist`	`histogram`
Density Plot	`density, lines`	`densityplot`
Boxplot	`boxplot`	`bwplot`
Empirical CDF Plot	`ecdfplot`*, `cdf.compare`**	
Probability or Q-Q Plot	`qqnorm, qqplot`**	`qqmath, qq`

* This function is only available in ENVIRONMENTALSTATS for S-PLUS.
** These functions have been modified in ENVIRONMENTALSTATS for S-PLUS.

Table 4.1: A list of S-PLUS and analogous Trellis univariate plotting functions.

Starting a Graphics Device

If you are using S-PLUS 3.3 or 3.4, then before you can create any kind of plot in S-PLUS, you need to start some kind of graphics device (either a window graphics device or a hard-copy graphics device). Before the Trellis functions were added to S-PLUS, you started a window graphics device with the `win.graph` function under S-Plus for Windows or the `motif` function under S-Plus for UNIX. In the current version of S-PLUS, the best way to start a window graphics device is to use the `trellis.device` function.

If you are running S-PLUS for Windows 3.3, type

```
> trellis.device(win.graph)
```

and if you are running S-PLUS for UNIX 3.4, type

```
> trellis.device(motif)
```

The default color scheme for a trellis device is "Trellis". If you prefer a different color scheme (e.g., "Windows Standard"), you may change it (see the S-PLUS documentation for information on how to change the color scheme of a graphics device).

If you are using S-PLUS 4.0 for Windows, a window graphics device (a graph sheet) is automatically started when you call a plotting function if no graphics devices are currently active.

Bar Charts and Dot Charts

The S-PLUS function `barplot` and the Trellis function `barchart` create bar charts. The S-PLUS function `dotchart` and the Trellis function `dotplot` create dot charts. Bar charts and dot charts are useful for displaying data sets in which each observation has a category or name associated with it.

Example 4.2: Comparing Lead Concentrations at Six Wells

The data frame `epa.89b.lead.df` contains concentrations of lead (μg/L) in ground water samples from two background wells and four compliance wells, where observations were taken once per month over four months (see the help file **Datasets: USEPA (1989b)**). Let's compute the average lead concentration at each well, then display the results in a bar chart (**Figures 4.1** and **4.3**) and dot chart (**Figures 4.2** and **4.4**).

```
> epa.89b.lead.df
    Lead Month Well  Well.type
 1  58.0     1    1 Background
 2  54.1     2    1 Background

23 198.3     3    6 Compliance
24 160.8     4    6 Compliance

> attach(epa.89b.lead.df)

> well.means <- sapply(split(Lead, Well), mean)

> well.means
     1      2       3      4     5       6
 47.05 55.725 132.225 70.425 142.5 192.025
```

```
> barplot(well.means, ylim = c(0, 200),
+    names = format(1:6), xlab = "Well Number",
+    ylab = "Average Lead Concentration (ug/L)",
+    main = "Bar Chart for Lead Data")
```

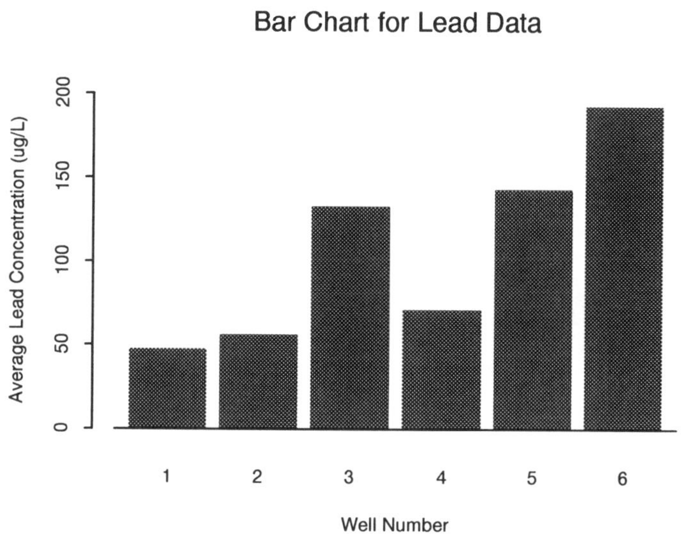

Figure 4.1: Bar chart for lead data.

```
> dotchart(well.means, labels = paste("Well", 1:6),
+    xlim = c(0, 200), main = "Dot Chart for Lead Data",
+    xlab = "Average Lead Concentration (ug/L)")
```

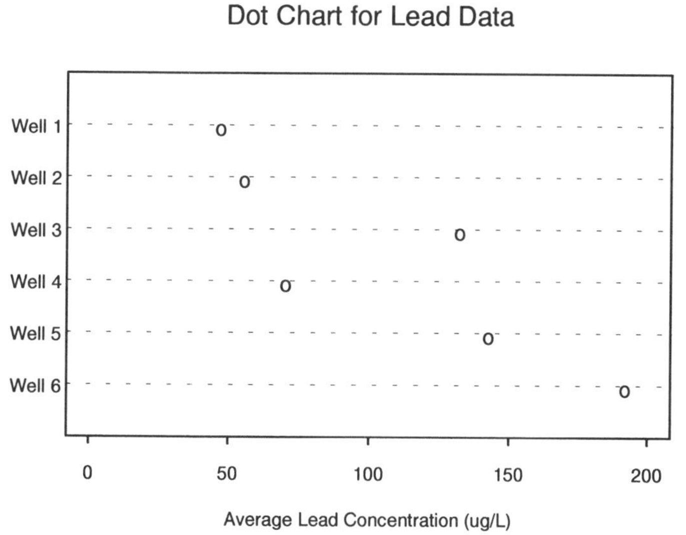

Figure 4.2: Dot chart for lead data.

```
> barchart(names(well.means) ~ well.means,
+    xlim = c(0, 200), ylab = "Well Number",
+    xlab = "Average Lead Concentration (ug/L)",
+    main = list("Bar Chart for Lead Data", cex = 1.5))
```

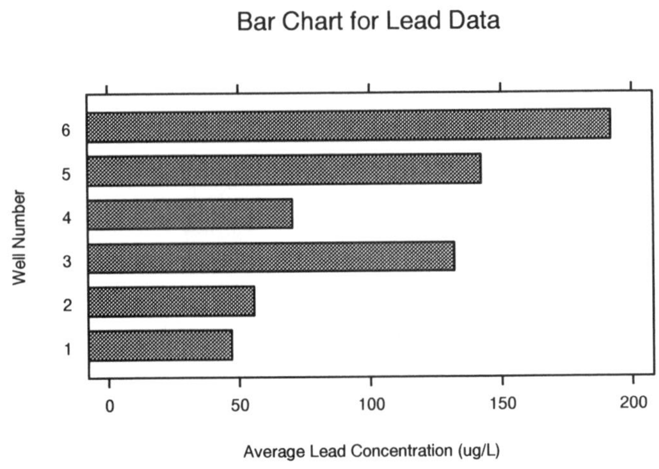

Figure 4.3: Bar chart created with the Trellis function `barchart`.

```
> dotplot(names(well.means) ~ well.means,
+    xlim = c(0, 200), ylab = "Well Number",
+    xlab = "Average Lead Concentration (ug/L)",
+    main = list("Dot Chart for Lead Data", cex = 1.5))
```

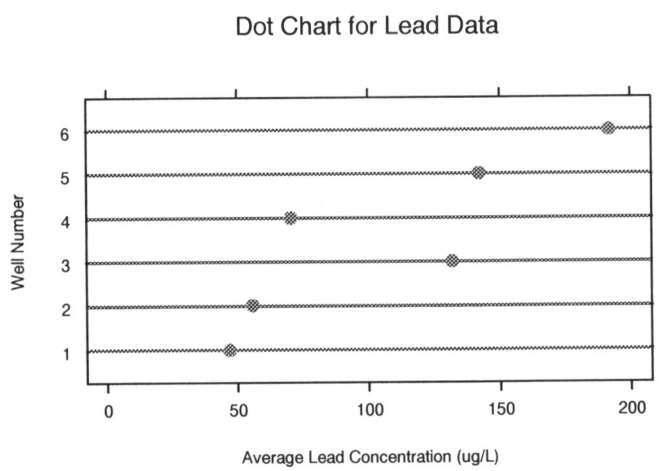

Figure 4.4: Dot chart created with the Trellis function `dotplot`.

Pie Charts

Pie charts are a commonly used graphical tool to display percentages. Experiments in graphical perception, however, have shown that they convey information far less reliably than bar charts or dot charts (Cleveland, 1994, pp.262-264). The S-PLUS function `pie` and the Trellis function `piechart` both produce pie charts.

Example 4.3: Displaying the Percent of "Below-Detection-Limit" Values

The data frame `epa.92c.arsenic2.df` contains arsenic concentrations (ppb) in ground water sampled from three background wells and one compliance well, where observations were taken once per month over six months (see the help file **Datasets: USEPA (1992c)**). You can use a pie chart to show the percentage of non-missing observations that were reported as "below the detection limit" of 5 ppb.

> **Note:** Observations reported as "below the detection limit" are sometimes called ***censored observations***. Observations that are not censored are sometimes called ***complete observations***. See the help file **Glossary: Censored Data** for more information.

The result of calling `pie` is shown in **Figure 4.5** and the result of calling `piechart` is shown in **Figure 4.6**. The pie charts show us that the percent of censored observations is almost the same as the percent of complete observations.

```
> epa.92c.arsenic2.df
   Arsenic.orig Arsenic Censored Month Well  Well.type
 1           <5     5.0        T     1    1 Background
 2           <5     5.0        T     2    1 Background
 3           8      8.0        F     3    1 Background
 .
22                   NA        F     4    4 Compliance
23           8      8.0        F     5    4 Compliance
24           14    14.0        F     6    4 Compliance

> attach(epa.92c.arsenic2.df)

> #  Compute the number of complete and
> #  censored observations.

> bdl.vec <- table(Censored[!is.na(Arsenic)])

> bdl.vec
 FALSE TRUE
    11    9

> bdl.vec <- as.vector(bdl.vec)

> names(bdl.vec) <- c("Complete", "Censored")
```

```
> pie(bdl.vec, names = c("Complete", "Censored"), col = 1,
+    rotate = F, main = "Pie Chart for Arsenic Data")
```

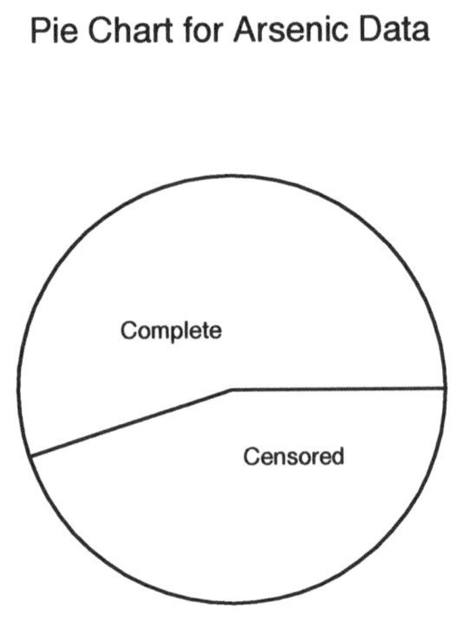

Figure 4.5: Pie chart for the arsenic data.

```
> piechart(names(bdl.vec) ~ bdl.vec)
```

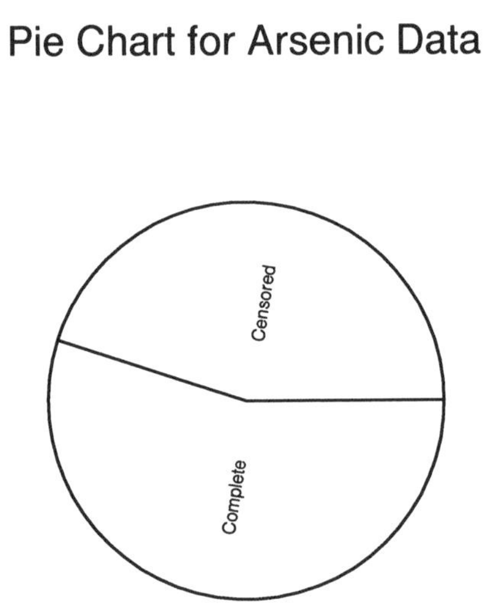

Figure 4.6: Pie chart for arsenic data created with the Trellis function `piechart`.

Strip Plots

A strip plot is a one-dimensional scatterplot. It is simply a plot of each observation showing the value of the observation. Strip plots are used to look at the distribution of one data set, or to compare the distributions of two or more data sets. Use the Trellis function `stripplot` to create strip plots.

Example 4.4: Comparing the Distribution of TcCB in the Cleanup and Reference Area

Figure 4.7 shows a strip plot of the log-transformed TcCB data for the reference and cleanup areas. It was created with these commands:

```
> stripplot(Area ~ log(TcCB), data = epa.94b.tccb.df,
+    jitter = T, main = list(cex = 1.5,
+    "Strip Plot for Log-Transformed TcCB Data"))
```

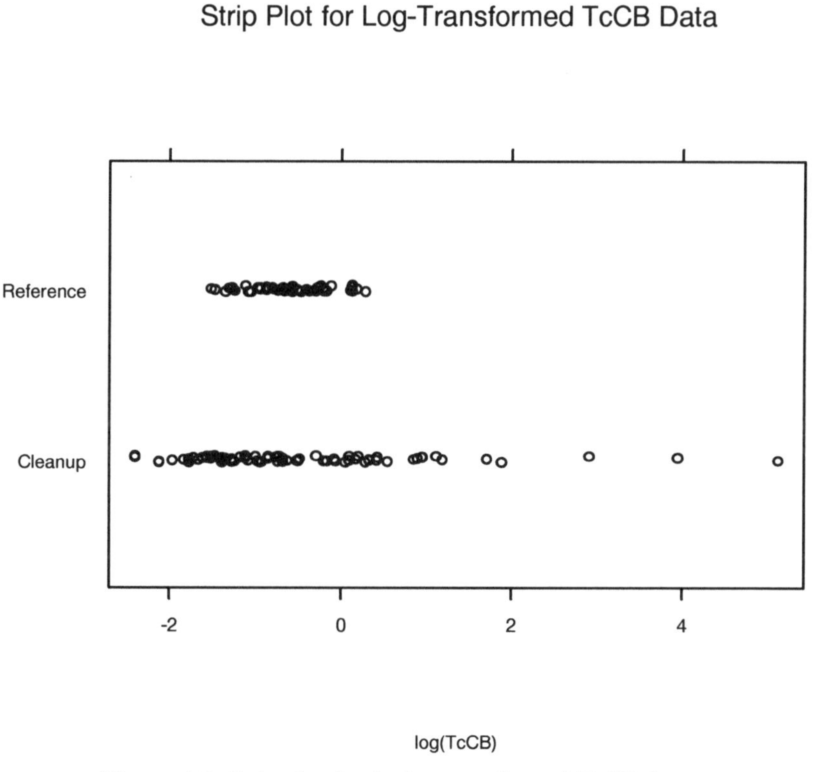

Figure 4.7: Strip plot for the log-transformed TcCB data.

The strip plot shows that most of the observations for the cleanup area are comparable to (or even smaller than) the observations for the reference area, but, as we found out from looking at the summary statistics for these data, there are a few very large "outliers" in the cleanup area. This may indicate a few "hot spots" in the cleanup area that were missed during the remediation process.

Histograms

The S-PLUS function `hist` and the Trellis function `histogram` produce histograms. A histogram is a useful way of displaying the distribution of a data set, but as Helsel and Hirsch (1992, p.19) point out, the appearance of a histogram depends upon how you decide to divide the data into classes (bins).

Example 4.5: Histograms for the TcCB Data

Figure 4.8 shows a histogram of the TcCB data for the reference area. This histogram is very typical of environmental data: it is skewed to the right.

```
> attach(epa.94b.tccb.df)

> objects(2)
[1] "Area"      "Censored"  "TcCB"      "TcCB.orig"

> TcCB.ref <- TcCB[Area == "Reference"]

> hist(TcCB.ref, xlab = "TcCB (ppb)",
+   ylab = "Number of Observations",
+   main = "Histogram for Reference Area TcCB Data")
```

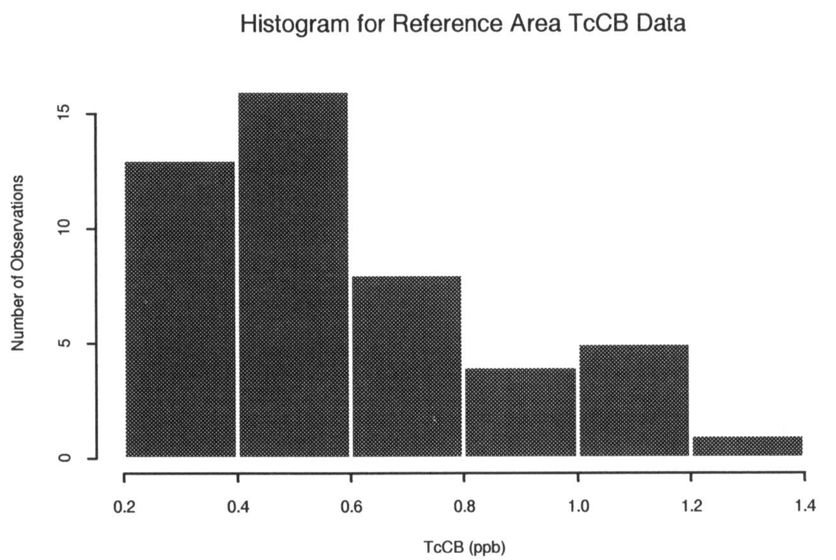

Figure 4.8: Histogram for reference area TcCB data.

Figure 4.9 shows the histograms for the log-transformed reference area and log-transformed cleanup area TcCB data.

```
> TcCB.cleanup <- TcCB[Area == "Cleanup"]

> par(mfrow = c(2,1))

> hist(log(TcCB.ref), xlim = range(log(TcCB)),
+    xlab = "log [ TcCB (ppb) ]",
+    ylab = "Number of Observations",
+    main = "Histogram for Reference Area")

> hist(log(TcCB.cleanup), xlim = range(log(TcCB)),
+    nclass = 25,
+    xlab = "log [ TcCB (ppb) ]",
+    ylab = "Number of Observations",
+    main = "Histogram for Cleanup Area")
```

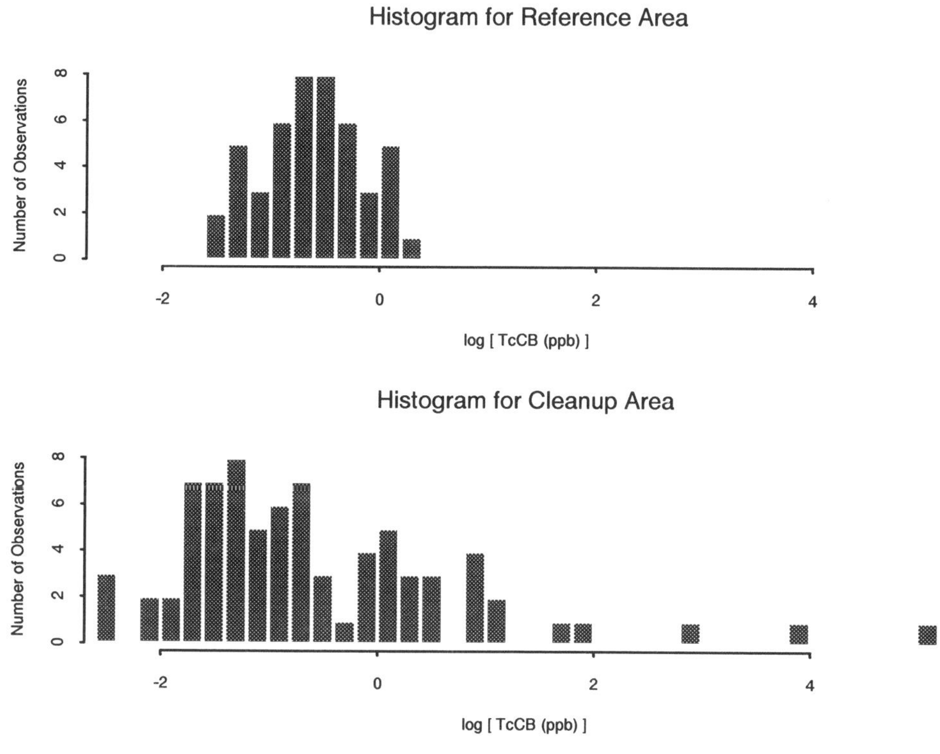

Figure 4.9: Histograms comparing TcCB data at the cleanup and reference areas.

Compare the histograms in **Figure 4.9** to the strip plot in **Figure 4.7**. They convey essentially the same information, but the strip plot shows every single observation, while the histograms summarize the distribution.

Density Plots

If we could take larger and larger sample sizes from the population and make the bins of our histogram narrower and narrower, we would eventually end up with a picture that resembles the true probability density of the underlying population (see the ENVIRONMENTALSTATS for S-PLUS help file **Glossary: Probability Density Function (pdf) and Probability Mass Function (pmf)** for more information).

A density plot is an estimate of the true underlying probability density, based on a sample of observations. The basic idea is that for each *x*-value, a window (local neighborhood) is constructed about that value, the number of observations within the window are counted, and these counts are then scaled to produce an estimate of the probability density at that *x*-value. The Trellis function `densityplot` produces density plots.

Example 4.6: Estimated Density of the Reference Area TcCB Data

Figure 4.10 shows the estimated density based on the reference area TcCB data. Compare this figure with the histogram shown in **Figure 4.8**.

```
> densityplot(~ TcCB, data = epa.94b.tccb.df,
+    subset = Area == "Reference", width = 0.2,
+    xlab = "TcCB (ppb)",
+    main = list(cex = 1.5,
+    "Density Plot for the Reference Area TcCB Data"))
```

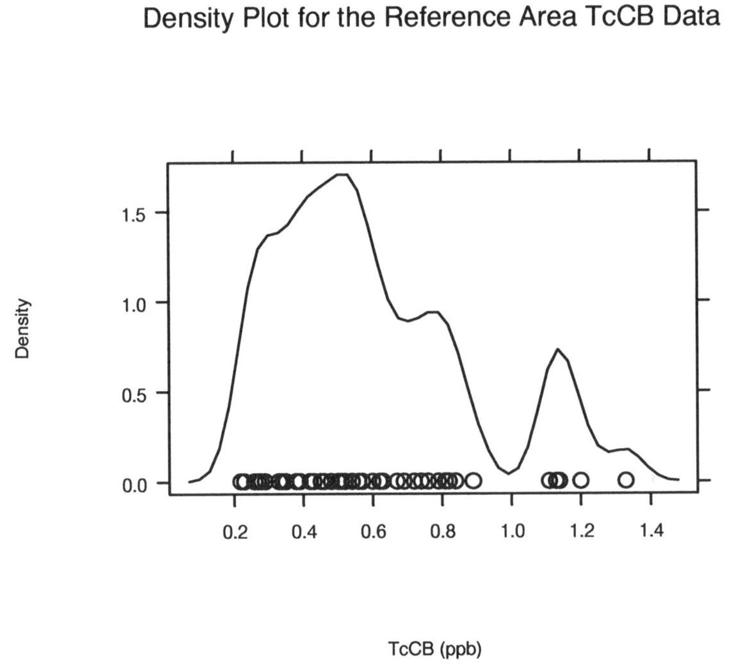

Figure 4.10: Density plot for the reference area TcCB data.

Boxplots

You can use the S-PLUS function `boxplot` or the Trellis function `bwplot` to create a boxplot for one or more data sets. A boxplot (sometimes called a box-and-whisker plot) is a simple graphical display invented by John Tukey that summarizes the distribution of a set of observations. A standard boxplot consists of

- A "box" defined by the 25th and 75th percentiles.
- A line across the box at the median (50th percentile).
- A line (whisker) drawn from the 25th percentile to the smallest observation within one "step" of the 25th percentile (the lower adjacent value).
- A line (whisker) drawn from the 75th percentile to the largest observation within one "step" of the 75th percentile (the upper adjacent value).
- Observations over one step away from the box ends plotted with a symbol (e.g., asterisk) or line.

Figure 4.11 below illustrates a boxplot using the log-transformed TcCB soil data for the cleanup area.

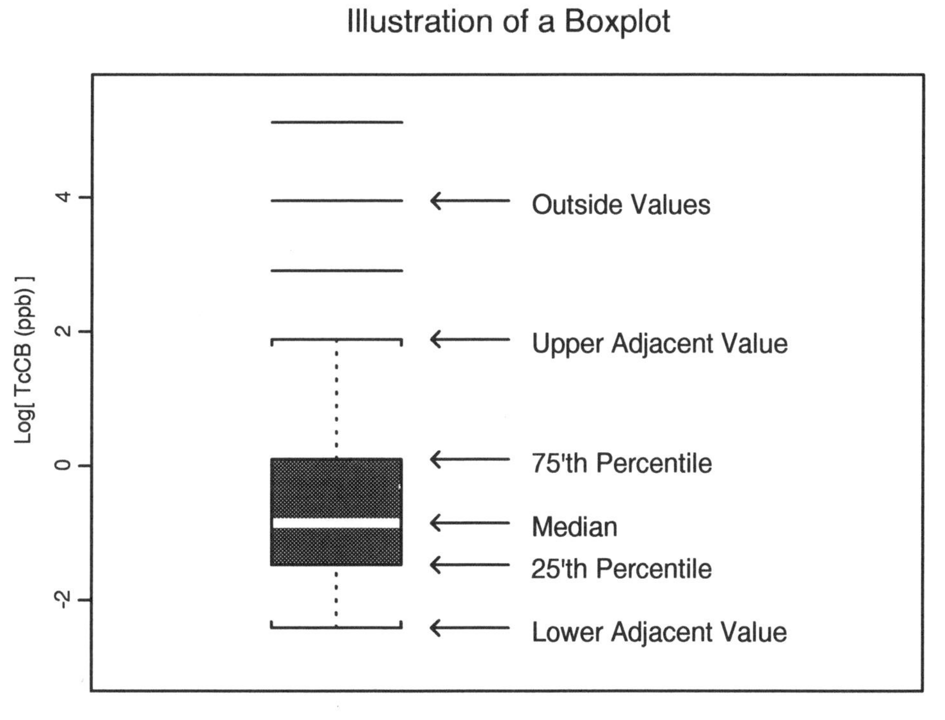

Figure 4.11: A boxplot based on the TcCB data for the cleanup area.

A "step" is conventionally defined as 1.5 times the interquartile range. (The interquartile range is the difference between the 75th and 25th percentiles.) Given this algorithm for constructing boxplots, you would expect to see at least one "outside value" only about 0.7% of the time if you were always looking at data from a normal distribution.

A boxplot quickly allows you to see:

- The center of the data (the median).
- The spread (variability) of the data.
- The skewness of the data (by looking at the relative lengths of the box halves).
- The presence of any "unusual" (outside) values.

Outside values are not necessarily outliers or bad observations! Indeed, since much environmental data is right-skewed, you may see lots of boxplots with outside values when you plot variables on their original scales.

Example 4.7: Boxplot vs. Histogram

Figure 4.12 compares the histogram and boxplot for the log-transformed cleanup area TcCB data. Note how much easier it is to identify summary statistics with the boxplot and still obtain a visual feel for the data. See Cleveland (1993, pp.25-27), Helsel and Hirsch (1992, pp.24-26), and the S-PLUS help file for `boxplot` for more information on boxplots.

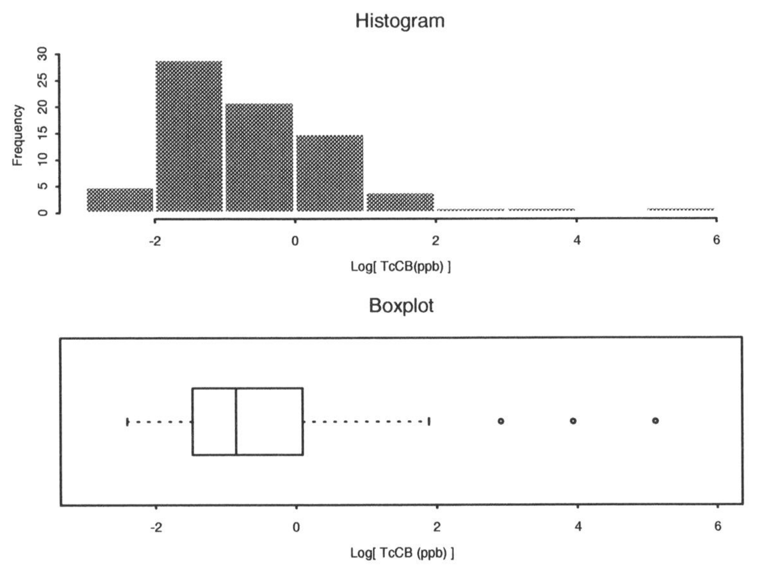

Figure 4.12: Comparison of a histogram and boxplot.

Example 4.8: Using Boxplots to Compare Cleanup and Reference Areas

Figure 4.13 (made with `boxplot`) and **Figure 4.14** (made with `bwplot`) show the two boxplots produced by dividing the log-transformed TcCB data into the reference and cleanup areas. Compare these figures with the histograms in **Figure 4.9**.

```
> boxplot(split(log(TcCB), Area), xlab = "Area",
+   ylab = "Log [ TcCB (ppb) ]",
+   main = "Boxplots for TcCB Data")
```

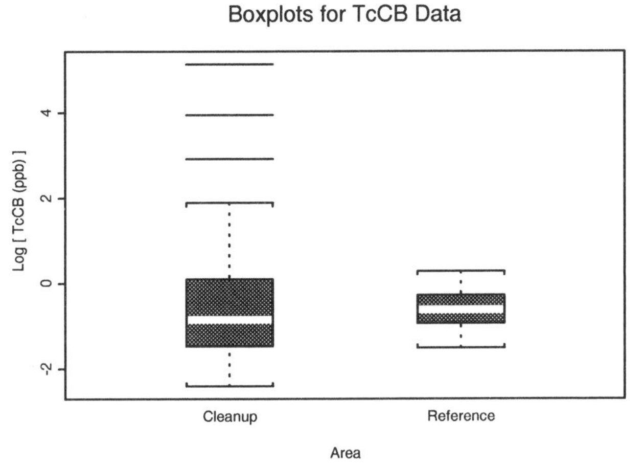

Figure 4.13: Boxplots comparing TcCB data at the cleanup and reference areas.

```
> bwplot(Area ~ log(TcCB), data = epa.94b.tccb.df,
+   main = list("Boxplots for TcCB Data", cex = 1.5))
```

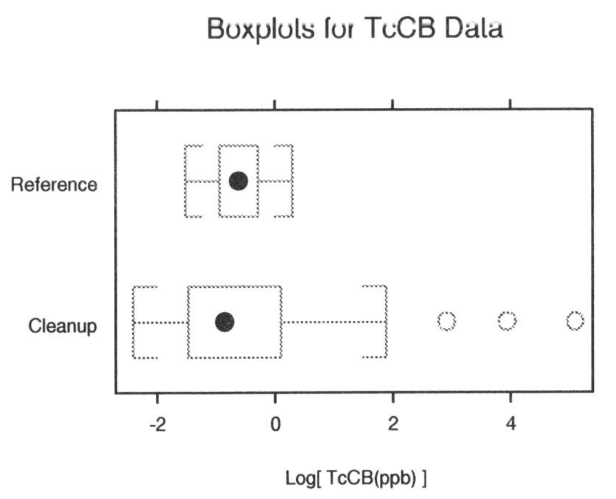

Figure 4.14: Boxplots for TcCB data produced by the function `bwplot`.

Empirical Cumulative Distribution Function (ECDF) Plots or Quantile Plots

The empirical cumulative distribution function (ecdf) plot, also called a quantile plot (Helsel and Hirsch, 1992, pp.21-24) graphically shows all of the quantiles (or percentiles) of the data set. You can use the ENVIRONMENTALSTATS for S-PLUS function `ecdfplot` to create empirical cdf (quantile) plots (see the help file **Plotting Probability Distributions**).

The function `cdf.compare` (built into S-PLUS but modified in ENVIRONMENTALSTATS for S-PLUS) allows you to compare the empirical cdf of a data set with the cdf from some specified theoretical distribution or with the empirical cdf of another data set.

> **Note:** See **Chapter 5: Probability Distributions** for more information on theoretical probability distributions that are used to model environmental data.

Example 4.9: The Empirical CDF of the Reference Area TcCB Data

The empirical cdf plot of the TcCB data for the reference area is shown in **Figure 4.15**. Based on this plot, you can easily pick out the median as about 0.55 and the quartiles as about 0.4 and 0.75. You can also see that the empirical cdf quickly rises, then pretty much levels off after about 0.8, so the data are skewed to the right.

Helsel and Hirsch (1992, p.22) note that ecdf plots or quantile plots, unlike histograms, don't require you to figure out how to divide the data into classes, and, unlike boxplots, all of the data are displayed in the graph.

> **Note:** Some authors (e.g., Chambers et al., 1983, pp.11-16; Cleveland, 1993, pp.17-20) reverse the axes on a quantile plot, i.e., the order statistics are on the *y*-axis and the cumulative probabilities are on the *x*-axis.

Example 4.10: Comparing the Reference Area TcCB Data to a Lognormal Distribution

Figure 4.16 shows the empirical cdf for the reference area TcCB data compared to the cdf from a lognormal distribution (the parameters of this distribution were estimated from the data).

> **Note:** In the call to `cdf.compare`, the abbreviation `"lnorm"` indicates a lognormal distribution. See **Chapter 5: Probability Distributions** for a full list of abbreviations used for probability distribution names.

We see that the lognormal distribution appears to fit these data quite well. Usually, however, quantile-quantile (Q-Q) plots, not empirical cdf plots, are used to visually assess the goodness of fit of a theoretical distribution to a data set (see the next subsection).

```
> ecdfplot(TcCB.ref, xlab = "TcCB (ppb)",
+   main = "Empirical CDF for Reference Area")
```

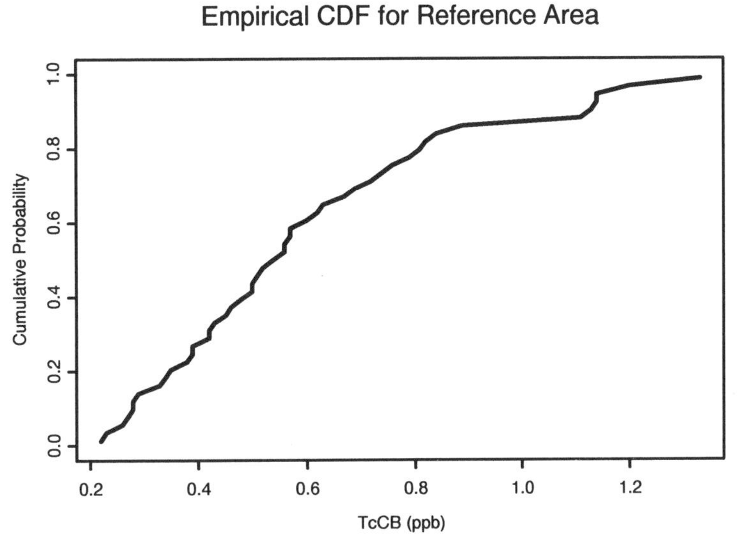

Figure 4.15: Empirical cdf plot of TcCB data for the reference area.

```
> cdf.compare(TcCB.ref, dist = "lnorm")
```

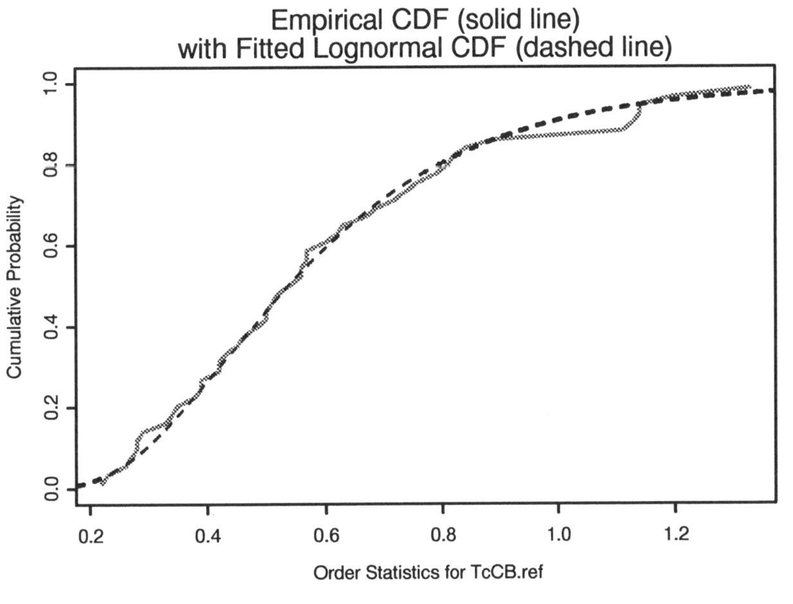

Figure 4.16: Empirical cdf of reference area TcCB data compared to a lognormal cdf.

Example 4.11: Comparing the Cleanup and Reference Area TcCB Data with Empirical CDF Plots

Just as with histograms and boxplots, you can also use empirical cdf plots to compare data sets. **Figure 4.17** compares the ecdf for the reference area with the ecdf for the cleanup area for the log-transformed TcCB data. As we saw with both the histograms (**Figure 4.9**) and boxplots (**Figures 4.13** and **4.14**), the cleanup area has quite a few extreme values compared to the reference area.

See the ENVIRONMENTALSTATS for S-PLUS help files for `ecdfplot`, **Glossary: Empirical Cumulative Distribution Function (ecdf)**, and **Glossary: Quantile Plot** for more information on ecdf (quantile) plots.

```
> cdf.compare(log(TcCB.ref), log(TcCB.cleanup))
```

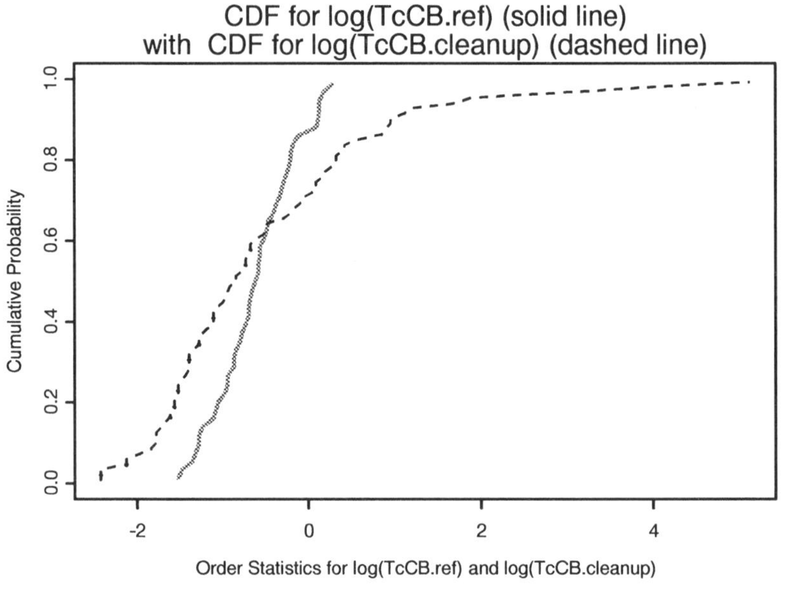

Figure 4.17: Quantile plots comparing TcCB data at the cleanup and reference areas.

Probability Plots or Quantile-Quantile (Q-Q) Plots

A probability plot, also called a quantile-quantile (Q-Q) plot, is a plot of the observed order statistics from a random sample (the empirical quantiles) against their (estimated) mean or median values based on an assumed distribution, or against the empirical quantiles of another set of data (Helsel and Hirsch, 1992, pp.27-35,40-45; Cleveland, 1993, pp.21-25).

Probability plots are used to assess whether data come from a particular distribution, or whether two data sets have the same parent distribution. If the distributions have the same shape (but not necessarily the same location or spread parameters), then the plot will fall roughly on a straight line. If the distributions are exactly the same, then the plot will fall roughly on the straight line $y=x$.

In a Q-Q plot, the extreme points have more variability than points toward the center. A U-shaped Q-Q plot indicates that the underlying distribution for the observations on the y-axis is skewed to the right relative to the underlying distribution for the observations on the x-axis. An upside-down-U-shaped Q-Q plot indicates the y-axis distribution is skewed-left relative to the x-axis distribution. An S-shaped Q-Q plot indicates the y-axis distribution has shorter tails than the x-axis distribution. Conversely, a plot that is bent down on the left and bent up on the right indicates that the y-axis distribution has longer tails than the x-axis distribution. See the help file **Glossary: Probability Plot or Quantile-Quantile (Q-Q) Plot** for more information.

S-PLUS comes with four built-in functions for producing Q-Q plots: `qqnorm`, `qqplot`, `qqmath`, and `qq`. The function `qqnorm` produces a normal (Gaussian) Q-Q plot; that is, the data set is compared to a normal distribution. The function `qqplot` is used to compare the empirical distributions of two data sets. The functions `qqmath` and `qq` are Trellis functions; `qqmath` allows you to compare a data set to quantiles from the normal or any other probability distribution (but is a bit tricky to use for distributions other than the normal), and `qq` is used to compare the empirical distributions of two data sets.

The function `qqplot` has been modified in ENVIRONMENTALSTATS for S-PLUS to easily allow comparing a data set to either a hypothetical distribution or to compare two data sets (see the ENVIRONMENTALSTATS for S-PLUS help file for `qqplot`). The default behavior of `qqplot` is to produce a normal Q-Q plot.

Example 4.12: Normal Q-Q Plots for the Reference Area TcCB Data

Let's look at a normal Q-Q plot of the reference area TcCB data, both on the original scale and a log-transformed scale. **Figure 4.18** and **Figure 4.19** show these two probability plots. The plot based on the original data shows a U-shape, indicating these data are skewed to the right relative to a normal distribution (see **Figure 4.8**). The plot based on the log-transformed data, however, shows the points falling in roughly a straight line, indicating these data may be reasonably modeled with a lognormal distribution (see **Figure 4.9**).

```
> qqplot(TcCB.ref)
```

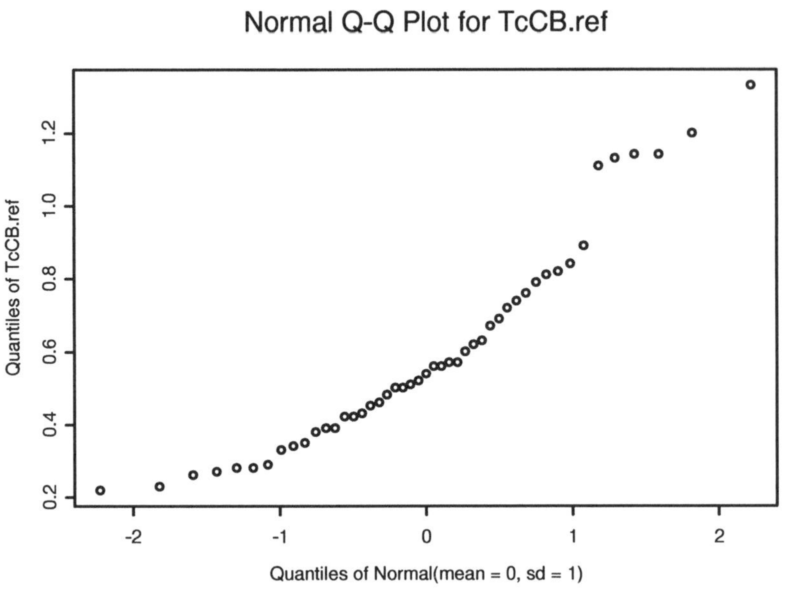

Figure 4.18: Normal probability plot for the reference area TcCB data.

```
> qqplot(log(TcCB.ref))
```

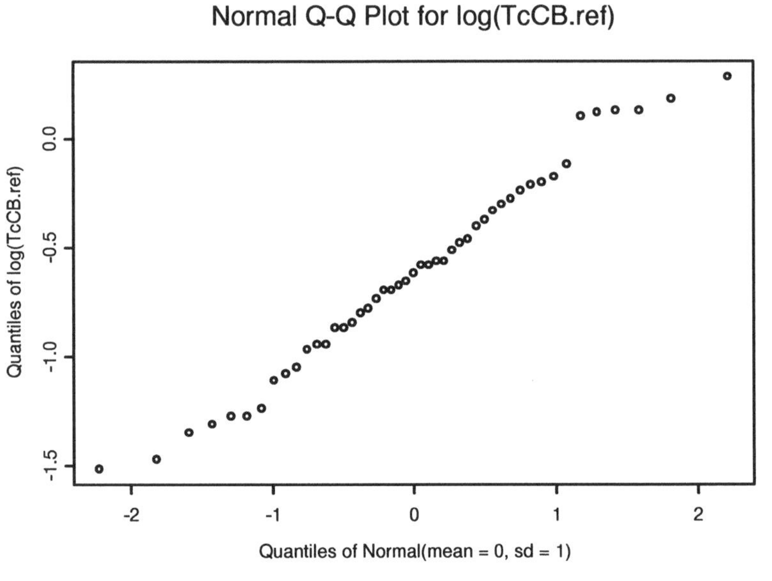

Figure 4.19: Normal probability plot for the log-transformed reference area TcCB data.

Example 4.13: Producing a Q-Q Plot for a Specified Distribution

In EnvironmentalStats for S-Plus, the function `qqplot` takes an argument called `distribution` that let's you specify the theoretical distribution to which you'd like to compare the data. Another way to produce a normal Q-Q plot based on the log-transformed data is to use the original data in the first argument to `qqplot` and specify a lognormal distribution:

```
> qqplot(TcCB.ref, distribution = "lnorm")
```

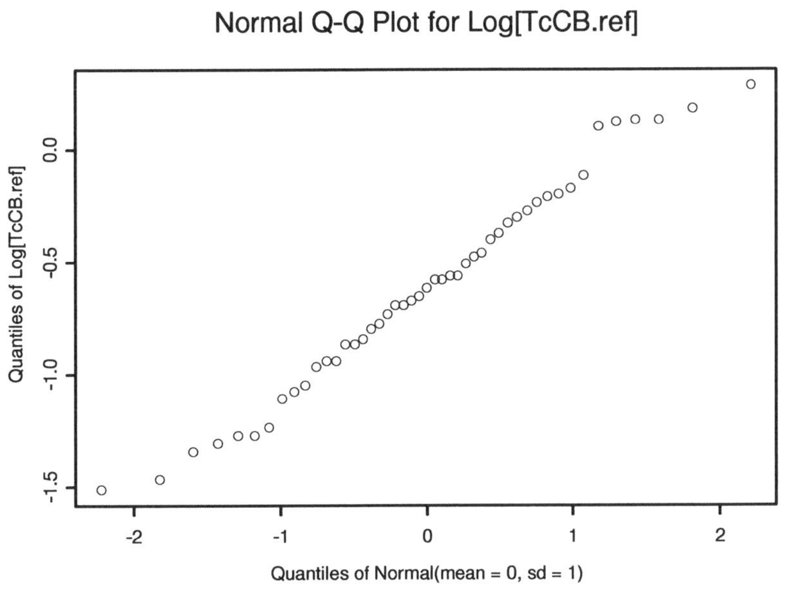

Figure 4.20: Normal probability plot for the log-transformed reference area TcCB data.

Figure 4.20 shows that in this case the data are automatically log-transformed before being compared to quantiles from a standard normal distribution.

Example 4.14: Adding a Regression Line to the Q-Q Plot

It is usually easier to assess deviations from a straight line by adding a straight line to the plot. We'll do this for the two probability plots we just created. The results are shown in **Figure 4.21** and **Figure 4.22**.

By default, the line produced with `qqplot` when `add.line=T` is a simple linear regression line. For the case of a normal Q-Q plot where quantiles from the standard normal distribution are used, the slope of this line is an estimate of the standard deviation and the intercept is an estimate of the mean.

```
> qqplot(TcCB.ref, add.line = T)
```

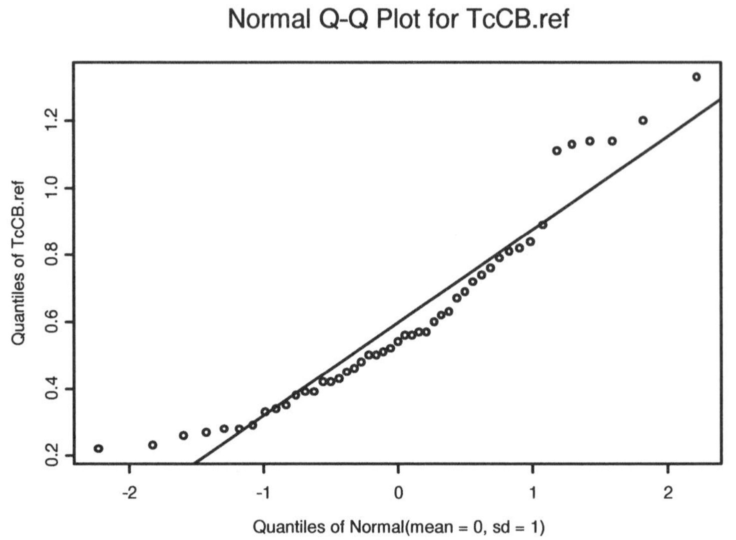

Figure 4.21: Figure 4.18 with a regression line added.

```
> qqplot(TcCB.ref, distribution = "lnorm",
+    add.line = T)
```

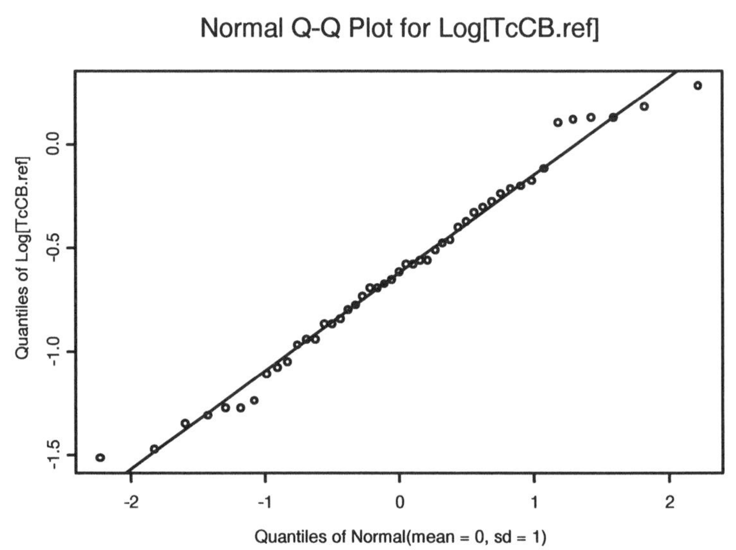

Figure 4.22: Figure 4.20 with a regression line added.

Example 4.15: Estimating the Distribution Parameters and Adding a 0-1 Line

You can also instruct qqplot to estimate the parameters of the assumed theoretical distribution and plot the data against the quantiles from this distribution. In this case, you want to specify a 0-1 line (that is, a line with slope 1 and intercept 0):

```
> qqplot(TcCB.ref, dist = "lnorm",
+    estimate.params = T, add.line = T,
+    qq.line.type = "0-1")
```

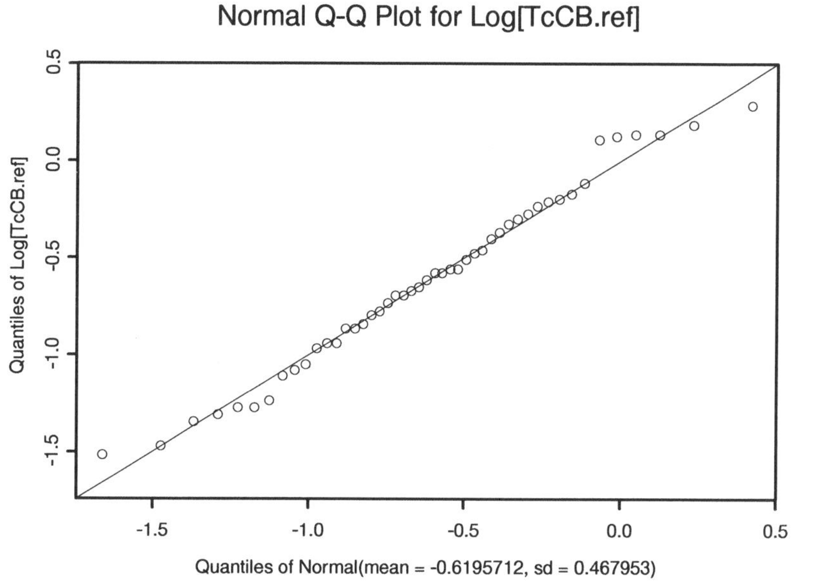

Figure 4.23: Normal Q-Q plot based on estimated parameters with 0-1 line.

Note that in **Figure 4.23** the values of the estimated parameters are indicated in the *x*-axis label.

Tukey Mean-Difference Q-Q Plots

An even better way to assess deviation from a straight line on a Q-Q plot is to use a Tukey mean-difference Q-Q plot, also called an m-d plot (Cleveland, 1993, pp.22-23). This is a plot of the difference between the quantiles on the *y*-axis vs. the average of the quantiles on the *x*-axis. If the two sets of quantiles come from the same parent distribution, then the points in an m-d plot should fall roughly along the horizontal line *y*=0. If one set of quantiles come from the same distribution with a shift in median, then the points in this plot should fall along a horizontal line above or below the line *y*=0. If the parent distributions of the quantiles differ in scale, then the points on this plot will fall at an angle.

A Tukey mean-difference Q-Q plot enhances our perception of how the points in the Q-Q plot deviate from a straight line, because it is easier to judge deviations from a horizontal line than from a line with a non-zero slope.

Example 4.16: Mean-Difference Plots for the Reference Area TcCB Data

Let's create an m-d plot for the original and log-transformed reference area TcCB data. In this case, we need to specify the argument `estimate.params=T` because we don't want to assume that the standard deviation of the TcCB data is necessarily 1 (as is the case for a standard normal distribution).

The two m-d plots are shown in **Figure 4.24** and **Figure 4.25**. The deviation from the horizontal line y=0 is quite evident for the untransformed TcCB data.

```
> qqplot(TcCB.ref, estimate.params = T,
+    plot.type = "Tukey", add.line = T)
```

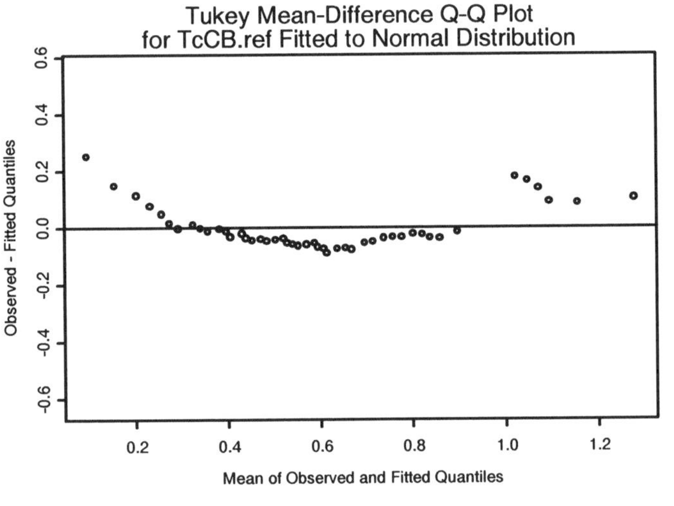

Figure 4.24: Tukey mean-difference Q-Q plot for reference area TcCB data.

```
> qqplot(TcCB.ref, dist = "lnorm",
+    estimate.params = T, plot.type = "Tukey",
+    add.line = T)
```

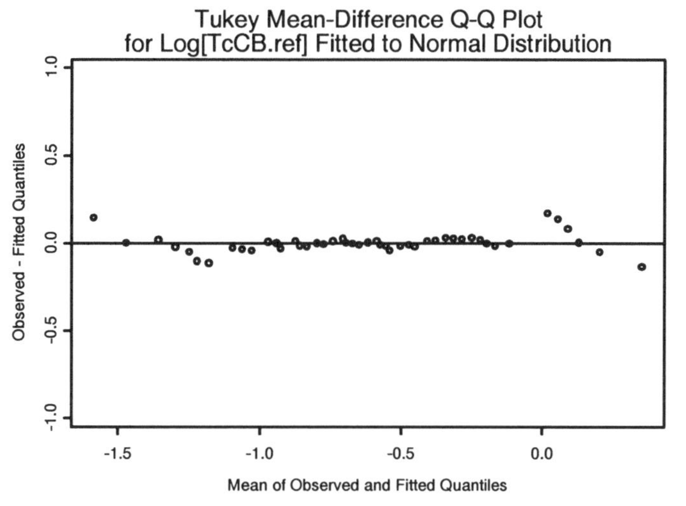

Figure 4.25: Tukey m-d plot for log-transformed reference area TcCB data.

Q-Q Plots for Other Distributions

So far in our discussion of Q-Q plots, we've only used normal Q-Q plots; that is, we've been interested in determining whether our data appear to come from a normal or lognormal distribution. You can also use the `qqplot` function to assess whether the data appear to come from a different theoretical distribution, such as a Poisson, gamma, extreme value, etc.

Example 4.17: A Poisson Q-Q Plot for Benzene Data

The guidance document **Statistical Analysis of Ground-Water Monitoring Data at RCRA Facilities: Addendum to Interim Final Guidance** (USEPA, 1992c, pp.34-40) contains a data set of benzene concentrations (ppb) from water samples collected over six months from six different background monitoring wells. These data are stored in the data frame `epa.92c.benzene1.df` (see the help file **Datasets: USEPA (1992c)**).

```
> epa.92c.benzene1.df
   Benzene.orig Benzene Censored Month Well
 1           <2       2        T     1    1
 2           <2       2        T     2    1
 3           <2       2        T     3    1
 4           <2       2        T     4    1
 .
33           <2       2        T     3    6
34           <2       2        T     4    6
35           10      10        F     5    6
36           <2       2        T     6    6
```

Out of the 36 observations, 33 are reported as "<2", and the other three observations are 10, 12, and 15. The example in the guidance document proposes to model these data as having come from a Poisson distribution, and sets each nondetect to 1 ppb (half the detection limit). Let's look at a Poisson Q-Q plot for these data to assess how well these data appear to come from a Poisson distribution.

```
> attach(epa.92c.benzene1.df)

> Benzene[Censored] <- 1

> qqplot(Benzene, dist = "pois", estimate.params = T,
+    duplicate.points.method = "number",
+    add.line = T, qq.line.type = "0-1")
```

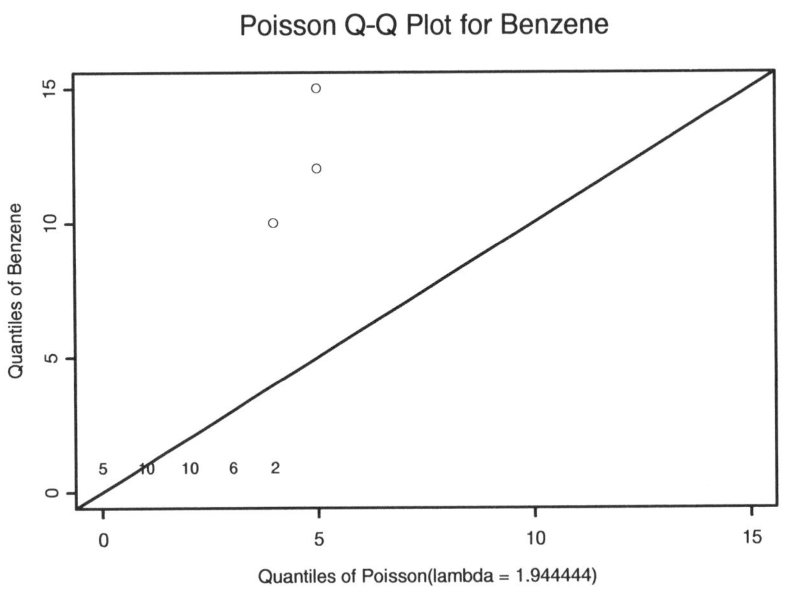

Figure 4.26: Poisson probability plot for benzene data.

The resulting plot is shown in **Figure 4.26**. The plot shows that the assumption of a Poisson distribution for these data is questionable: there are too many observations with the value 1 (recall the nondetects were set to 1), and the detected observations are too large.

In the commands above to create **Figure 4.26**, we set the argument `duplicate.points.method` to `"number"` to tell `qqplot` to use a number to indicate how many duplicate (x,y) pairs appear at a point (the default behavior of S-PLUS is to plot only one point even if there are several duplicate (x,y) pairs). Alternatively, we can tell `qqplot` to jitter duplicate points, as shown in **Figure 4.27**.

```
> qqplot(Benzene, dist = "pois", estimate.params = T,
+    duplicate.points.method = "jitter",
+    add.line = T, qq.line.type = "0-1")
```

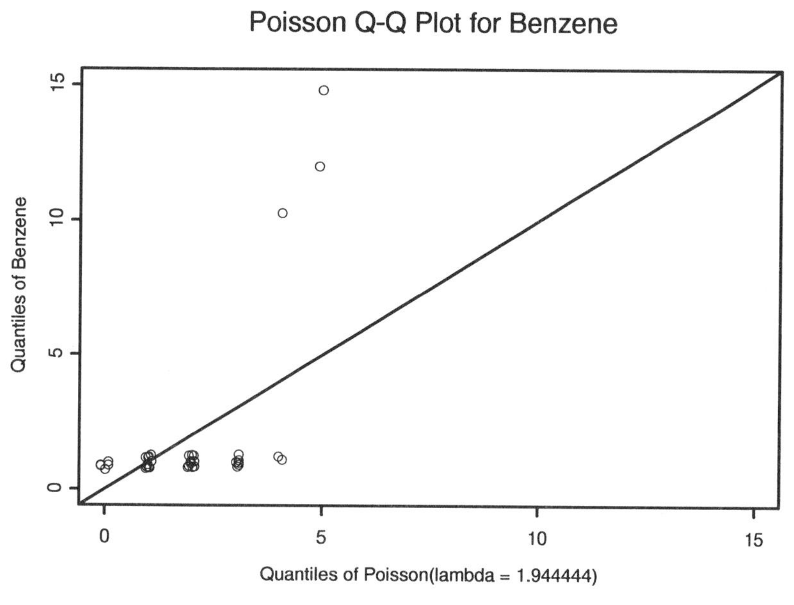

Figure 4.27: Poisson probability plot for benzene data with jittered points.

Building an Internal Gestalt for Q-Q Plots

Probability plots or Q-Q plots are a graphical, subjective way of assessing the goodness-of-fit of a data set to a specified theoretical distribution. In order to be able to assess the goodness-of-fit, you need to have an internal baseline image (a gestalt) of what a "typical" Q-Q plot looks like when in fact the data come from the specified distribution. The ENVIRONMENTALSTATS for S-PLUS function `qqplot.gestalt` lets you produce numerous Q-Q plots to build up such an internal gestalt (see the help file for `qqplot.gestalt` for more information).

Example 4.18: Building a Gestalt for Normal Q-Q Plots

Let's produce one page of typical normal (Gaussian) Q-Q plots:

```
> set.seed(426)

> qqplot.gestalt(num.pages = 1, add.line = T)
```

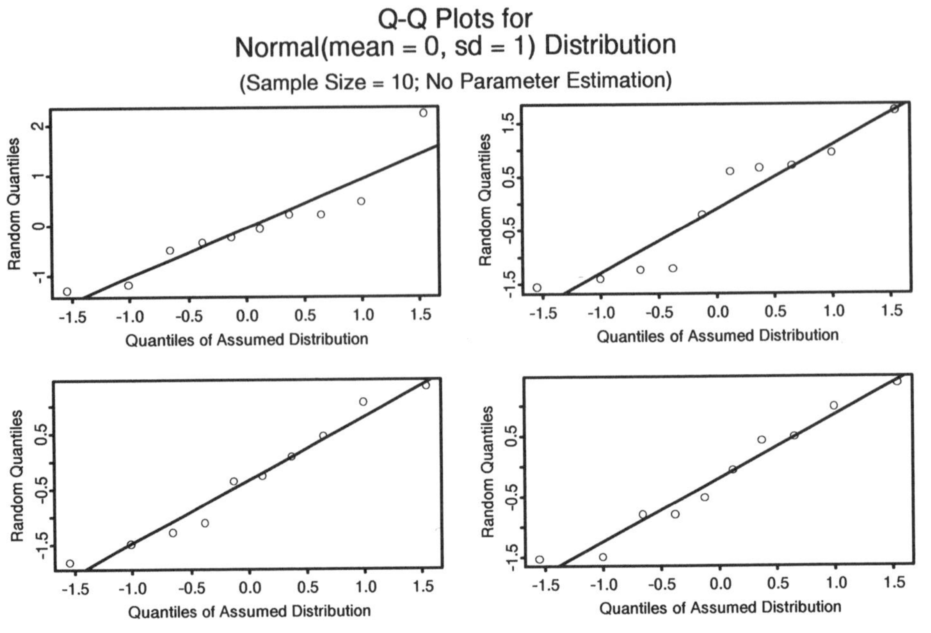

Figure 4.28: Four "typical" normal Q-Q plots.

Figure 4.28 shows the result of calling `qqplot.gestalt`. (The call to `set.seed` sets the random number seed so you can reproduce this figure.) Note that with only ten observations, there can be a bit of spread about the fitted regression line.

You can also produce a page of Tukey mean-difference Q-Q plots, as shown in **Figure 4.29**, with the following commands:

```
> set.seed(426)

> qqplot.gestalt(num.pages = 1, add.line = T,
+   plot.type = "Tukey", estimate.params = T)
```

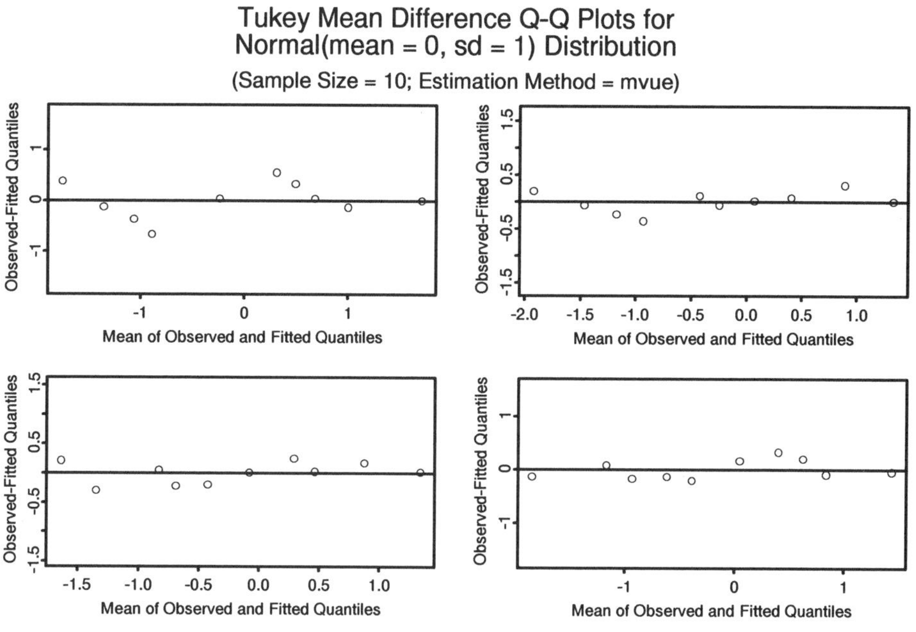

Figure 4.29: Four "typical" Tukey mean-difference normal Q-Q plots.

Example 4.19: Building a Gestalt for Poisson Q-Q Plots

You can also use `qqplot.gestalt` to produce Q-Q plots for other distributions besides the normal distribution. Here is a set of commands to produce the page of Poisson Q-Q plots shown in **Figure 4.30**. In this case, you must specify the mean of the Poisson distribution via the `param.list` argument.

```
> set.seed(357)

> qqplot.gestalt(dist = "pois",
+    param.list = list(lambda = 2),
+    num.pages = 1, add.line = T, sample.size = 15,
+    duplicate.points.method = "jitter")
```

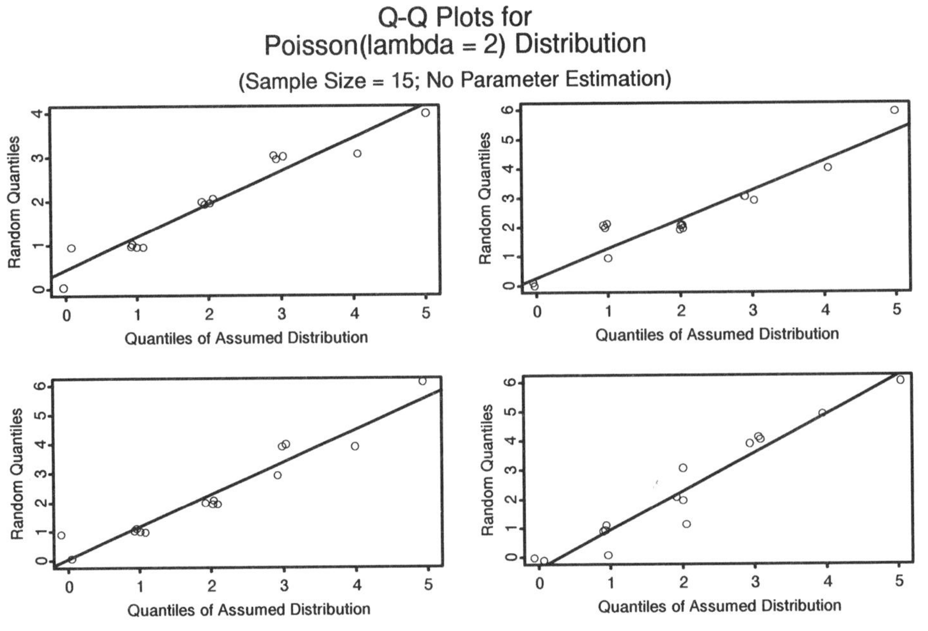

Figure 4.30: Four "typical" Poisson probability plots.

Using Q-Q Plots to Compare Two Data Sets

Besides using Q-Q plots or probability plots to assess whether a set of data appear to come from a particular probability distribution, you can use a Q-Q plot to assess whether two sets of data appear to have the same parent distribution (i.e., the same shape but not necessarily the same location or scale). If the distributions have the same shape (but not necessarily the same location or scale parameters), then the plot will fall roughly on a straight line. If the distributions are exactly the same, then the plot will fall roughly on the straight line $y=x$.

Example 4.20: Comparing the Cleanup and Reference Area TcCB Data

Figure 4.31 shows a Q-Q plot constructed from the TcCB data for the reference and cleanup areas.

```
> qqplot(log(TcCB.ref), log(TcCB.cleanup),
+   add.line = T, equal.axes = T,
+   qq.line.type = "0-1",
+   main = paste("Q-Q Plot Comparing Cleanup",
+   "and Reference Areas"))
```

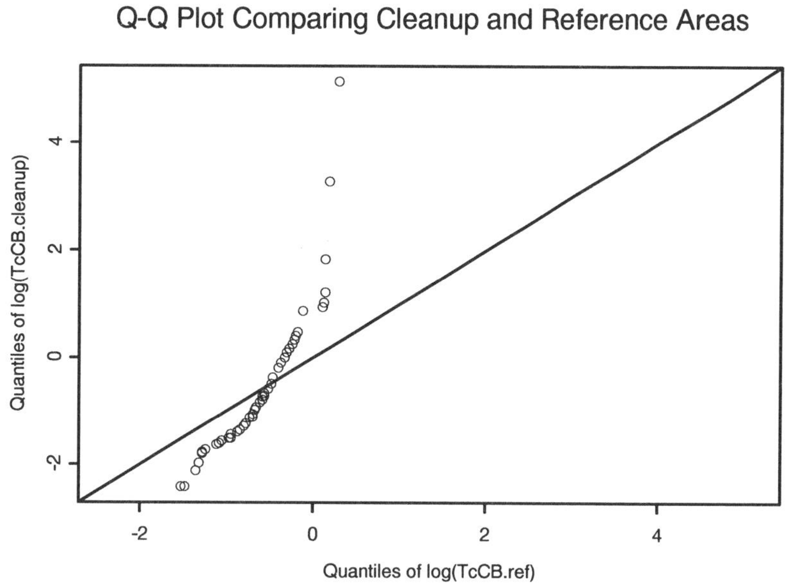

Figure 4.31: Q-Q plot comparing cleanup and reference areas for the TcCB data.

Compare **Figure 4.31** to **Figure 4.7**, **Figure 4.9**, **Figure 4.13**, and **Figure 4.17**, which compare these data sets using strip plots, histograms, boxplots, and empirical cdf's, respectively. Like these three other plots, the Q-Q plot indicates that the distribution of TcCB in the cleanup area has a larger spread and is more skewed to the right compared to the reference area, as well as having a few observations that are quite large.

Autocorrelation Plots

Classical statistical methods for constructing confidence intervals and hypothesis tests are based on two critical assumptions: that the observations are independent of one another and that they come from the same population (probability distribution). Parametric methods add the third assumption that the observations come from a particular probability distribution, usually the normal distribution (see **Chapter 6: Estimating Population Parameters and Quantiles** and **Chapter 8: Hypothesis Tests**). Of these three assumptions, the most important one with respect to classical hypothesis tests and confidence intervals is the first, the assumption of independent observations (e.g., Millard et al., 1985).

When observations are collected sequentially over time or close together in space, the assumption of independent observations may not be valid. The S-PLUS module S+SPATIALSTATS from MathSoft, Inc. contains functions for modeling and analyzing observations that may be spatially correlated. S-PLUS contains several functions for creating, displaying, and analyzing time series data (see the S-PLUS help file **Time Series** and the S-PLUS documentation on time series analysis). Commonly used time series functions include `ts.plot` to plot a time series, `lag.plot` to plot the original observations against lagged observations, and `acf` to plot the estimated autocorrelation at various lags.

> **Note:** See the ENVIRONMENTALSTATS for S-PLUS help file **Glossary: Independence and Serial Correlation** for more information about these concepts.

Example 4.21: Looking at Autocorrelation in Daily Air Quality Measurements

S-PLUS contains two built-in data frames called `air` and `environmental` that contain observations on ozone (ppb for `environmental`, ppb$^{1/3}$ for `air`), solar radiation (langleys), temperature (degrees Fahrenheit), and wind speed (mph) for a period of 111 consecutive days (about four months). Ignoring the information contained in the other three variables, let's look at the behavior of ozone over time using the cube-root values contained in the `air` data frame:

```
> attach(air)

> ts.plot(ozone, xlab = "Day",
+    ylab = "Ozone (ppb^1/3)", main = "Ozone Data")
```

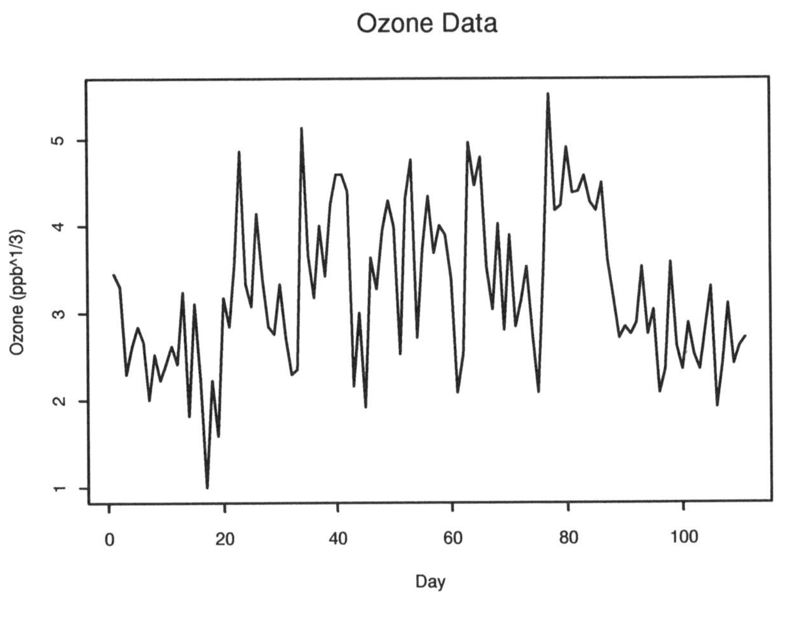

Figure 4.32: Time series plot of ozone data.

```
> acf(ozone)
```

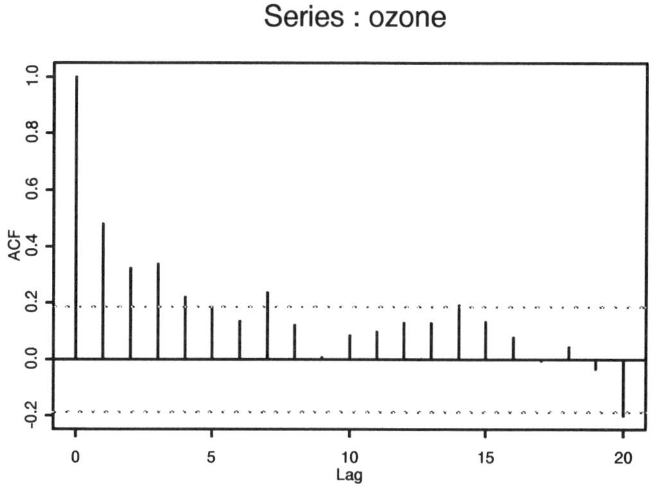

Figure 4.33: Autocorrelation plot for ozone data.

The results of calling `ts.plot` and `acf` are shown in **Figure 4.32** and **Figure 4.33**. In **Figure 4.33**, the y-axis label "ACF" stands for "Autocorrelation Function". This plot shows the estimated correlation between the ith observation and the $(i+m)^{th}$ observation (where m denotes the lag number) on the y-axis vs. the lag number on the x-axis. (**Note:** the lag-0 autocorrelation is always 1.)

The dashed lines in **Figure 4.33** represent an approximate 95% confidence interval for an individual estimated autocorrelation (of lag $\geq$ 1), assuming the true autocorrelation is in fact 0. Thus, for independent observations, a particular autocorrelation (e.g., the lag-1 autocorrelation) should fall within the approximate confidence limits about 95% of the time. In this case, the autocorrelation plot strongly indicates the presence of serial correlation in the ozone data.

Note: Because of multiple comparisons, we would ***not*** expect all 19 estimated autocorrelations of lag $\geq$ 1 to fall within the dashed lines about 95% of the time.

As already noted, we are ignoring the information in the other three variables (radiation, wind speed, and temperature). Including this information in a model for the ozone data may reduce or eliminate the serial correlation. Also note that even though these data span about four months, we have not attempted to account for any kind of seasonality.

Graphs for Bivariate and Multivariate Data

S-PLUS includes several functions for plotting bivariate and multivariate data, including functions to create scatterplots, matrix scatterplots, three-dimensional plots, brush and spin plots, and multi-panel displays (conditioning plots).

Table 4.2 shows the names of the original S-PLUS functions for producing certain bivariate and multivariate data plots and the analogous Trellis functions. Some of the original S-PLUS functions may eventually be replaced by the Trellis functions, but in this section we will show you how to use both types of functions. See the S-PLUS documentation on Trellis graphics for more information on using the Trellis graphics functions.

Plot Type	**S-PLUS Function**	**Trellis Function**
Scatterplot	`plot`, `scatter.smooth`	`xyplot`
Scatterplot matrix	`pairs`	`splom`
Interactive scatterplot matrix	`brush`	
3-D point cloud	`brush`, `spin`	`cloud`
Contour plot	`contour`	`contourplot`
Image or Level plot	`image`	`levelplot`
Wireframe plot	`persp`	`wireframe`
Conditioning plot (Multi-Panel Display)	`coplot`	Any Trellis function

Table 4.2: A list of S-PLUS and Trellis bivariate and multivariate plotting functions.

Scatterplots

The S-PLUS function `plot` and the Trellis function `xyplot` produce simple two-dimensional scatterplots. Scatterplots are used to look at the relationship between two variables. Often, we are interested in determining whether there is some sort of deterministic relationship between the two variables (e.g., linear or curvilinear) that may be hidden by noise in the data (Helsel and Hirsch, 1992, pp.45-50).

Example 4.22: Scatterplot of Ozone vs. Temperature

Recall the built-in data frame called `air` that contains observations on ozone, solar radiation, temperature, and wind speed for a period of 111 consecutive days. **Figure 4.34** shows the scatterplot of ozone vs. temperature. Note that for the `plot` function the x-variable is the first argument and the y-variable is the second argument. **Figure 4.35** shows the same plot based on using the Trellis function `xyplot`. These scatterplots show that ozone is positively correlated with temperature: as temperature increases, so does the ozone concentration.

```
> attach(air)

> plot(temperature, ozone, ylab = "Ozone (ppb^1/3)",
+    xlab = "Temperature (degrees F)",
+    main = "Ozone vs. Temperature for the air data")
```

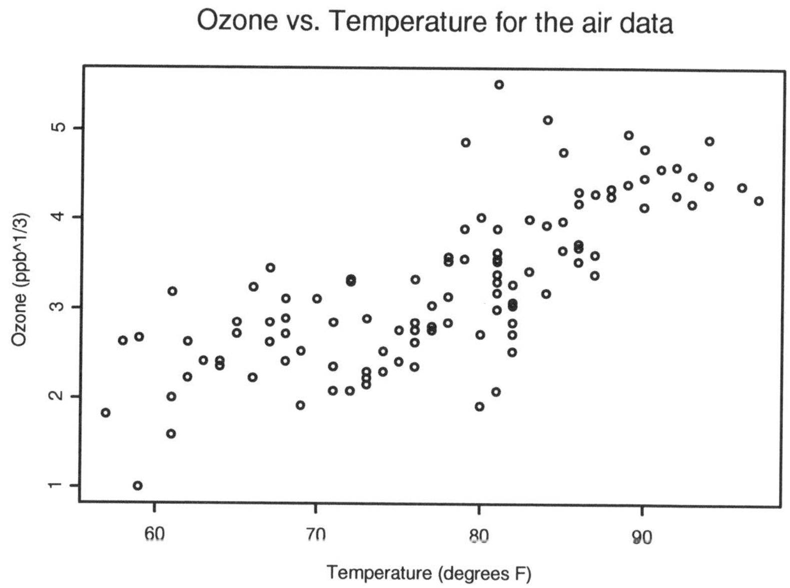

Figure 4.34: Scatterplot of ozone vs. temperature for the `air` data.

```
> xyplot(ozone ~ temperature, data = air,
+    ylab = "Ozone (ppb^1/3)",
+    xlab = "Temperature (degrees F)",
+    main = list(paste("Ozone vs. Temperature",
+    "for the air data"), cex = 1.5))
```

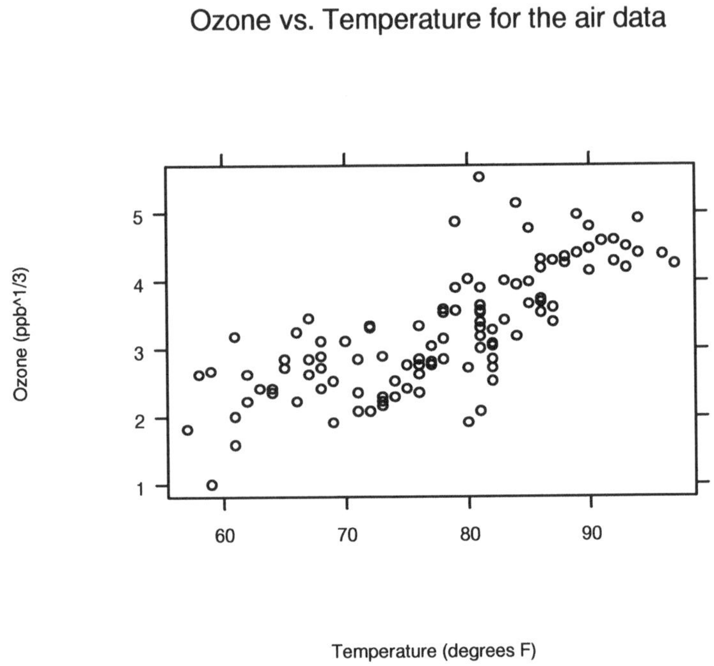

Figure 4.35: Scatterplot of ozone vs. temperature created by `xyplot`.

Adding a Loess Smooth to a Scatterplot

To asses whether the relationship between ozone and temperature is truly linear, we could fit a straight line using simple linear regression and assess the goodness of the fit by doing the usual things like looking at the residuals and testing whether the slope is significantly different from 0. (See the S-PLUS documentation for more information on fitting linear models in S-PLUS.)

A very handy method of looking for a "signal" in the "noise" without having to impose a specific model on the data is to use the locally weighted smoother called loess (Cleveland, 1993, Chapter 3; Helsel and Hirsch, 1992, pp.45-46). A loess curve is constructed as follows.

1. For each *x*-value, a window (local neighborhood or bin) is constructed about that value.

2. A straight line (or optionally a quadratic line) is fit only to the observations within the window using weighted regression so that observations close to the *x*-value receive more weight than observations farther away from the *x*-value.

3. Once the line is fit, the y-value is "predicted" based on the current x-value. Usually, a robust regression procedure is used in which during an iterative fitting process observations with large residuals ("outliers") are given less weight than the other observations, and the fit is computed again.

The S-PLUS function `scatter.smooth` creates a scatterplot and adds a loess curve to the plot. You can also use the Trellis function `xyplot` along with the `panel` argument to create a scatterplot and add a loess smooth, although the commands are a bit trickier because they are generalized for multi-panel displays (see the sub-section **Multi-Panel Conditioning Plots (Coplots)** later in this section).

Example 4.23: Adding a Loess Smooth with `scatter.smooth`

Figure 4.36 shows that in fact the relationship between ozone and temperature appears not to be linear, but curvilinear. Perhaps it could even be modeled as two separate straight lines, with one line describing the relationship when the temperature is less than 75° and a second line describing the relationship when the temperature is over 75°. Of course, as we stated before, we are ignoring the information available in the other variables (wind speed and radiation).

```
> scatter.smooth(temperature, ozone,
+    xlab = "Temperature (degrees F)",
+    ylab = "Ozone (ppb^1/3)",
+    main = "Ozone vs. Temperature with Loess Smooth")
```

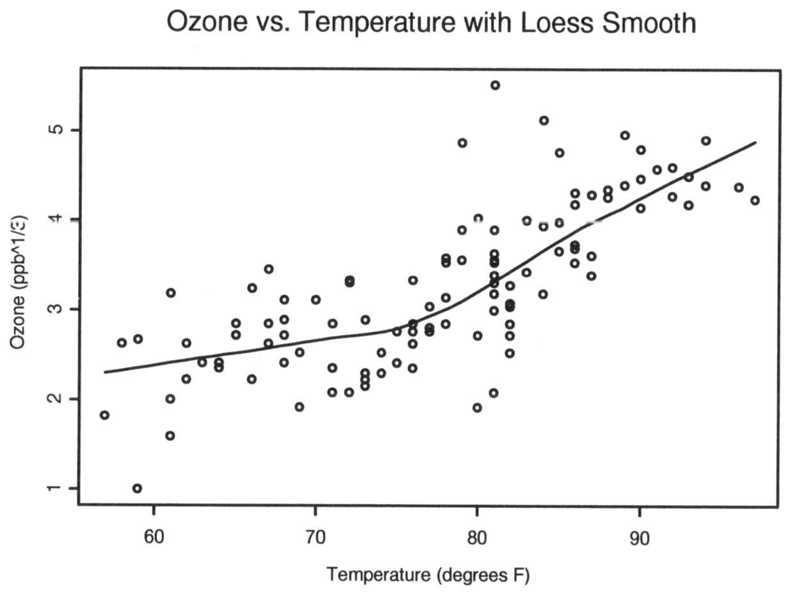

Figure 4.36: Ozone vs. temperature with loess smooth created by `scatter.smooth`.

Example 4.24: Adding a Loess Smooth with `xyplot` and the `panel` Argument

Figure 4.37 shows the same thing as **Figure 4.36**, but was created with the trellis function `xyplot`. See the Trellis documentation for more information on how to use the `panel` argument to trellis functions.

```
> xyplot(ozone ~ temperature, data = air,
+    xlab = "Temperature (degrees F)",
+    ylab = "Ozone (ppb^1/3)",
+    panel = function(x,y) {
+      panel.xyplot(x,y)
+      panel.loess(x,y) },
+    main = list(paste("Ozone vs. Temperature",
+    "with Loess Smooth", cex = 1.5))
```

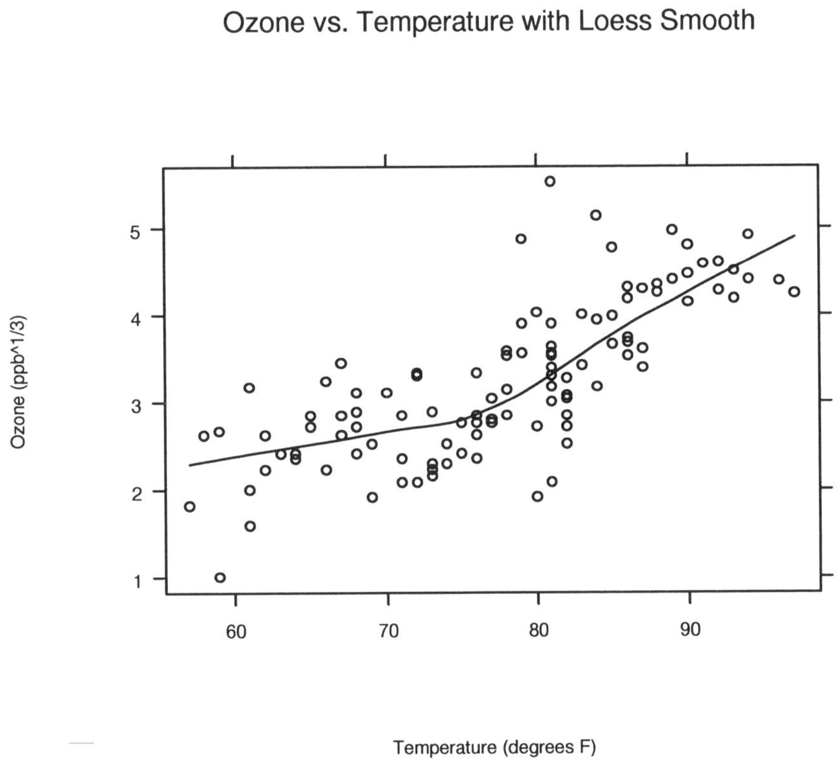

Figure 4.37: Ozone vs. temperature with loess smooth created by `xyplot`.

Scatterplot Matrix

When you have multivariate data, you can use a scatterplot matrix to look at scatterplots of each variable plotted against the others (Cleveland, 1993, pp.273-275). The S-PLUS function `pairs` and the Trellis function `splom` produce a scatterplot matrix.

Example 4.25: Scatterplot Matrix for the Ozone Data

Figure 4.38 and **Figure 4.39** show the scatterplot matrix produced when we consider all of the variables available in the `air` data frame. These figures show that ozone is positively correlated with radiation and temperature, and negatively correlated with wind speed. Also, temperature and wind speed are negatively correlated. There is no apparent relationship between radiation and temperature.

```
> pairs(air)
```

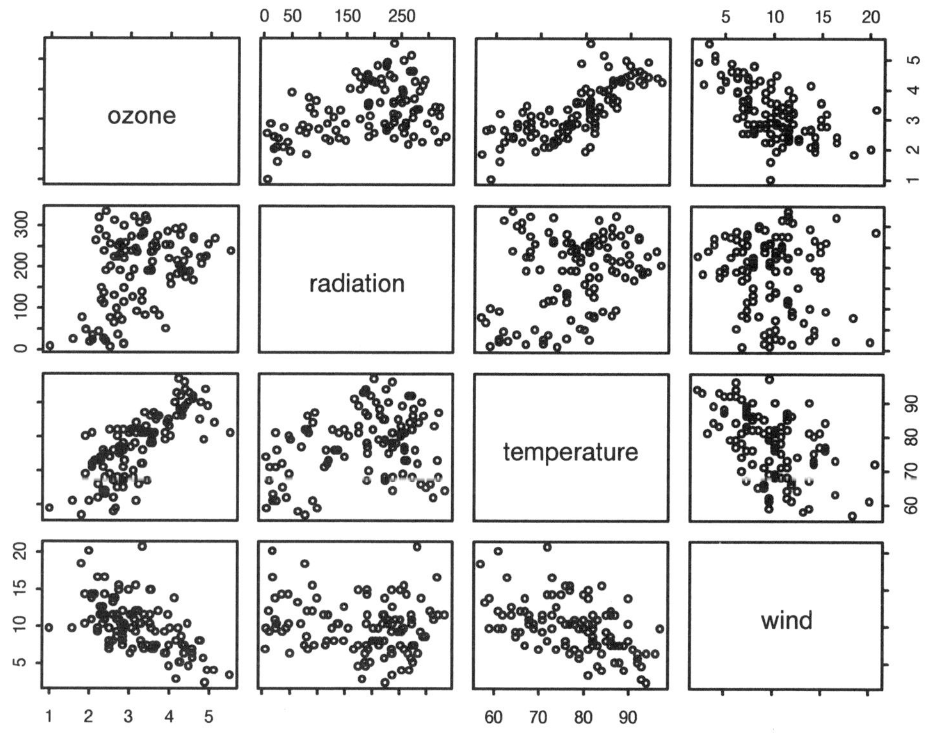

Figure 4.38: Scatterplot matrix for the `air` data created with `pairs`.

```
> splom(~air)
```

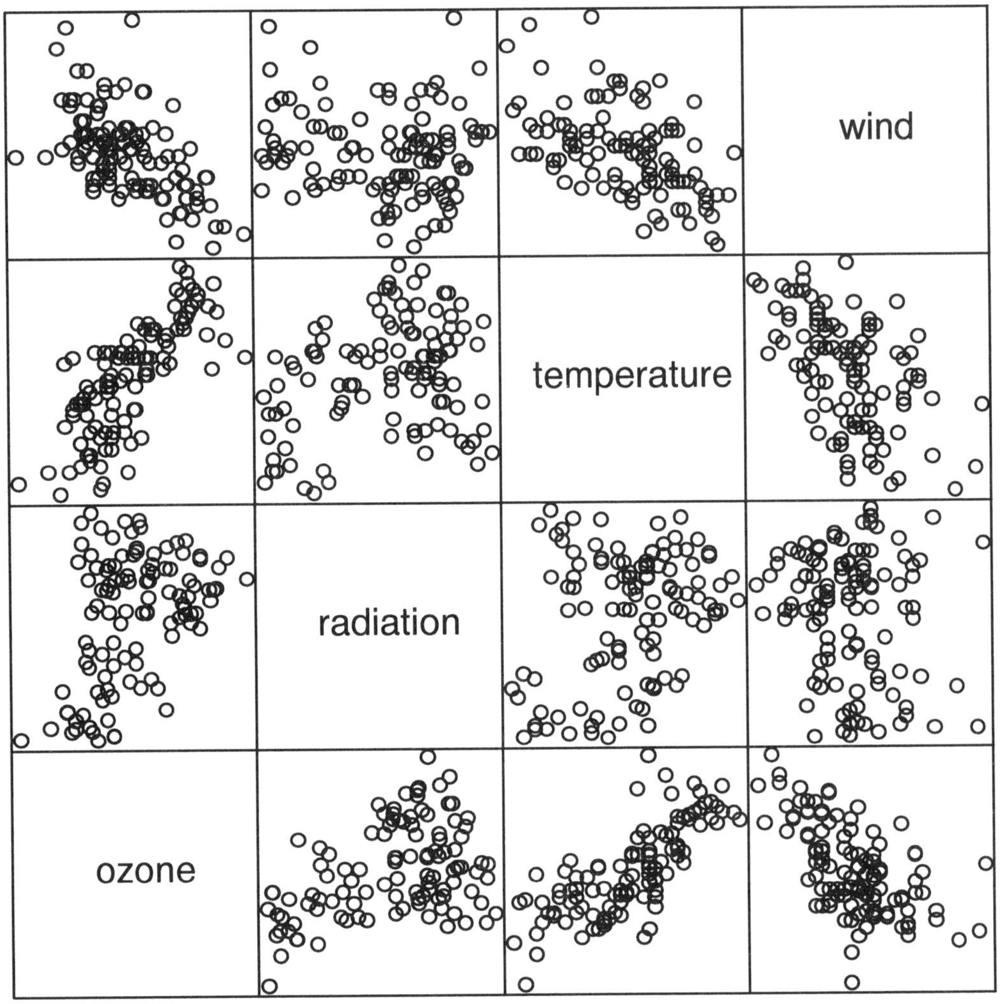

Figure 4.39: Scatterplot matrix for the `air` data created with `splom`.

Brushing A Scatterplot Matrix

The S-PLUS function `brush` lets you look at a scatterplot matrix interactively. You can highlight points in one of the scatterplots and these same cases will show up highlighted in the other scatterplots. To invoke the brush function for the `air` data, you type

```
> brush(air)
```

See the S-PLUS help file for `brush` and the S-PLUS documentation for more information on the `brush` function.

Three-Dimensional Plots

Contour plots, image plots (also called level plots), wireframe plots, and 3-D cloud plots are all used to display the relationship between three variables. Contour plots, image plots, and wireframe plots are best suited for data in which two of the variables are observed on a grid in the *x*-*y* plane and the third variable is then plotted on the *z*-axis. For cloud plots, however, evenly-spaced observations are not necessary.

Example 4.26: Cloud Plots

The Trellis function `cloud` creates static 3-D cloud plots. The S-PLUS functions `brush` and `spin` produce 3-D cloud plots that you can rotate in real time. **Figure 4.40** shows a cloud plot produced by the `cloud` function using the variables ozone, temperature and wind in the `air` data frame.

```
> cloud(ozone ~ wind * temperature, data = air,
+    xlab = "Wind Speed (mph)",
+    ylab = "Temperature (degrees F)",
+    zlab = "Ozone (ppb^1/3)",
+    main = list("Cloud Plot for Ozone Data",
+    cex = 1.5))
```

You can use the argument `screen` to change the angle from which you view the cloud plot. See the Trellis documentation and the S-Plus help file for `cloud` for more information.

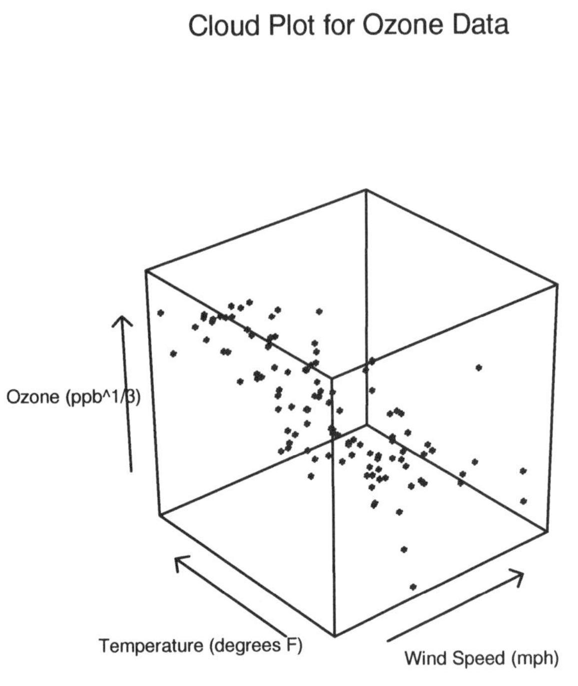

Figure 4.40: Cloud plot of ozone vs. wind speed and temperature for the `air` data.

Example 4.27: Contour Plots

The S-PLUS function `contour` and the Trellis function `contourplot` create contour plots. Both of these functions require triples (x, y, z) where the (x, y) pairs occur on an evenly-spaced grid. The wind and temperature data in the `air` data frame are observed variables and do not lie on an evenly-spaced grid in the `x-y` plane. To create a contour plot showing ozone as a function of wind and temperature, we have to somehow create interpolated values of ozone for gaps in the grid.

One way to create observations on an evenly-spaced grid is to use the `loess` smoothing function to fit a surface, and then get predicted values of the z-variable on a grid. Fitting a loess surface in three-dimensions is similar to fitting a loess curve in two dimensions (see **Example 4.23** above). In the case of three-dimensions, for each (x, y) pair, a neighborhood about that pair is defined and only points within that neighborhood are used to fit a plane or quadratic surface to the third variable.

Here is an example of how to create observations on an evenly-spaced grid using the ozone, temperature, and wind variables in the `air` data frame. **Figure 4.41** shows the contour plot generated by these steps.

1. Fit a smooth surface to the existing data using `loess`.

```
> ozone.fit <- loess(ozone ~ wind * temperature,
+   data = air, span = 0.5)
```

2. Specify the tick marks on the grid for the *x*- and *y*-variables.

```
> wind.ticks <- seq(min(wind), max(wind),
+   length = 50)

> wind.ticks
 [1]  2.300000  2.675510  3.051020  3.426531
    .
[47] 19.573469 19.948980 20.324490 20.700000

> temp.ticks <- seq(min(temperature),
+   max(temperature), length = 50)

> temp.ticks
 [1] 57.00000 57.81633 58.63265 59.44898
    .
[47] 94.55102 95.36735 96.18367 97.00000
```

3. Create a two-column matrix that contains all possible combinations of (*x*, *y*) pairs on the grid.

```
> grid <- expand.grid(list(wind = wind.ticks,
+   temperature = temp.ticks))

> dim(grid)
[1] 2500    2

> grid
          wind temperature
   1 2.300000          57
   2 2.675510          57
   .
2499 20.32449          97
2500 20.70000          97
```

4. Generate predicted values of ozone on the grid based on the loess fit constructed in Step 1.

```
> predicted.ozone <- predict(ozone.fit,
+   newdata = grid)
```

5. Create the contour plot.

```
> contour(wind.ticks, temp.ticks,
+   predicted.ozone, xlab = "Wind Speed (mph)",
+   ylab = "Temperature (degrees F)",
+   main = "Contour Plot of Ozone (ppb^1/3)")
```

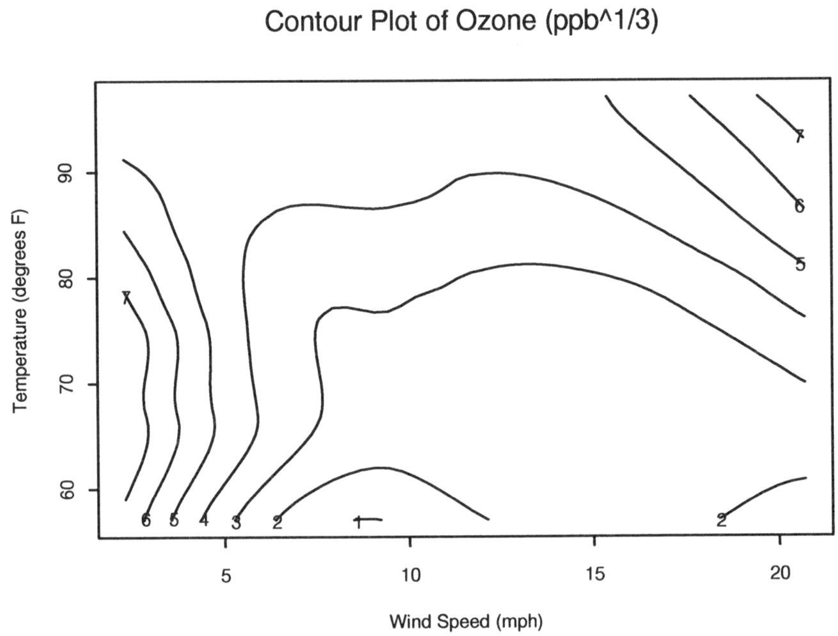

Figure 4.41: Contour plot of ozone as a function of wind speed and temperature.

You can also use the Trellis function `contourplot` to create a contour plot. The plot is shown in **Figure 4.42**.

```
> contourplot(predicted.ozone ~ wind * temperature,
+    data = grid, xlab = "Wind Speed (mph)",
+    ylab = "Temperature (degrees F)",
+    main = list("Contour Plot of Ozone (ppb^1/3)",
+    cex = 1.5))
```

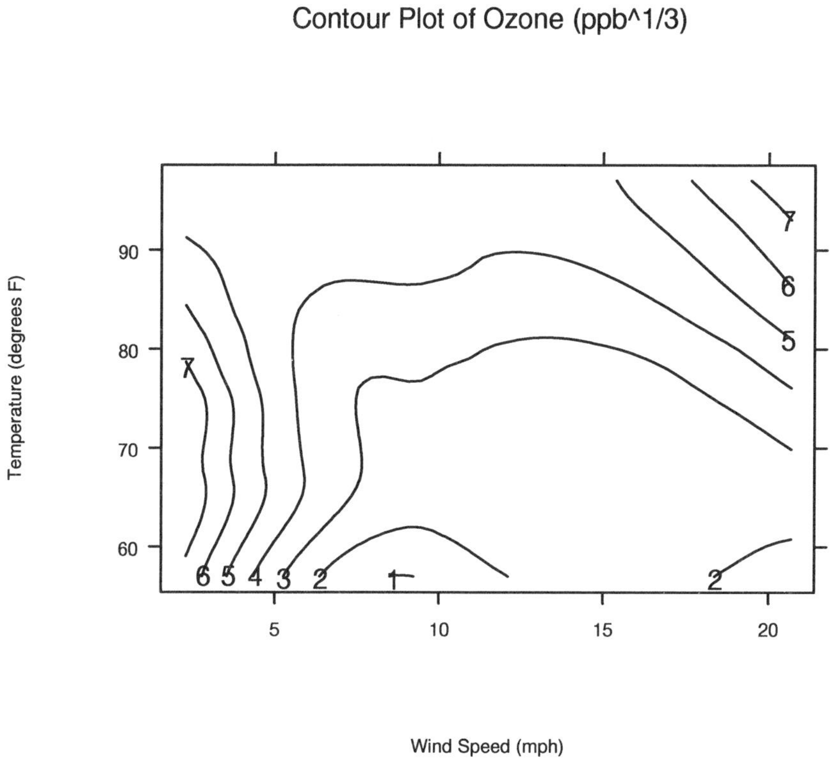

Figure 4.42: Contour plot of ozone data created with the Trellis function `contourplot`.

Example 4.28: Image or Level Plots

Image or level plots are similar to contour plots, but use color or gray-scaling to represent changes in the value of the *z*-variable over the range of the *x*- and *y*-variables. The S-PLUS function `image` creates an image plot, and the Trellis function `levelplot` creates a level plot.

Figure 4.43 and **Figure 4.44** show the image and level plots for the ozone, wind, and temperature data produced with `image` and `levelplot`, respectively, based on the color scheme "Trellis Black on White" (see the S-PLUS documentation for information on how to change the color scheme of your graphics device). Note that smaller values of ozone correspond to darker shades and larger values of ozone correspond to lighter shades. (See the S-PLUS documentation for information on how to change the color scheme of a plot.). Compare these figures to **Figure 4.41** and **Figure 4.42**.

```
> image(wind.ticks, temp.ticks, predicted.ozone,
+    xlab = "Wind Speed (mph)",
+    ylab = "Temperature (degrees F)",
+    main = "Image Plot of Ozone (ppb^1/3)")

> image.legend(predicted.ozone, -1, 107,
+    size = c(2.5, .5))
```

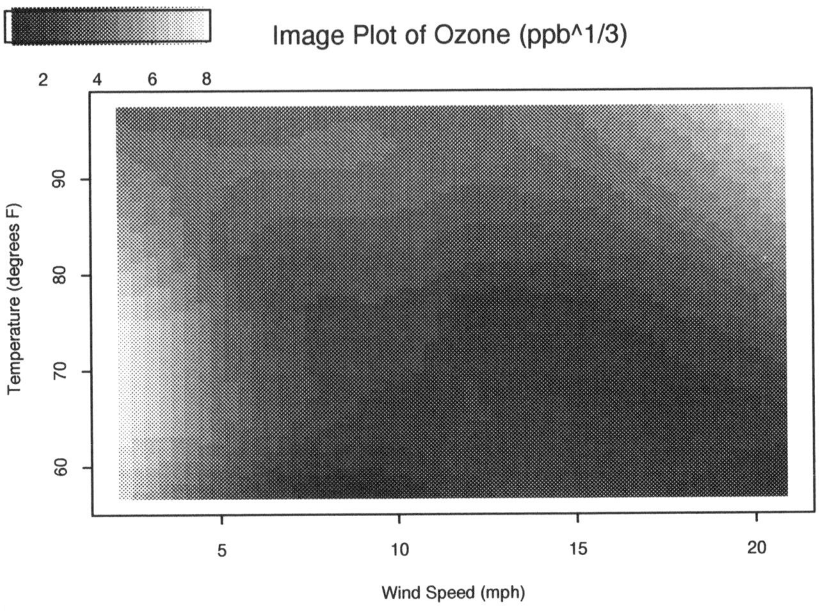

Figure 4.43: Image plot of ozone data.

```
> levelplot(predicted.ozone ~ wind * temperature,
+    at = 0:8,data = grid, xlab = "Wind Speed (mph)",
+    ylab = "Temperature (degrees F)",
+    main = list("Level Plot of Ozone (ppb^1/3)",
+    cex = 1.5))
```

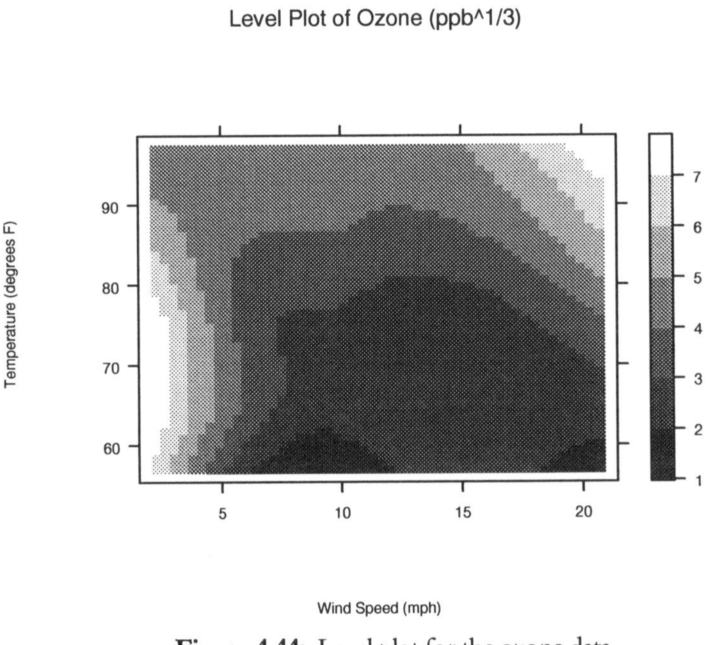

Figure 4.44: Level plot for the ozone data.

Example 4.29: Wireframe or Perspective Plots

Wireframe or perspective plots are used to display a three-dimensional surface. The S-PLUS function `persp` and the Trellis function `wireframe` both produce wireframe plots. **Figure 4.45** and **Figure 4.46** show the wireframe plots produced by these two functions using the same smooth surface we computed to create contour and level plots. **Figure 4.45** is produced by these commands:

```
> persp(wind.ticks, temp.ticks, predicted.ozone,
+    xlab = "Wind Speed (mph)",
+    ylab = "Temperature (degrees F)",
+    zlab = "Ozone (ppb^1/3)")

> title(main = "Perspective Plot for Ozone Data")
```

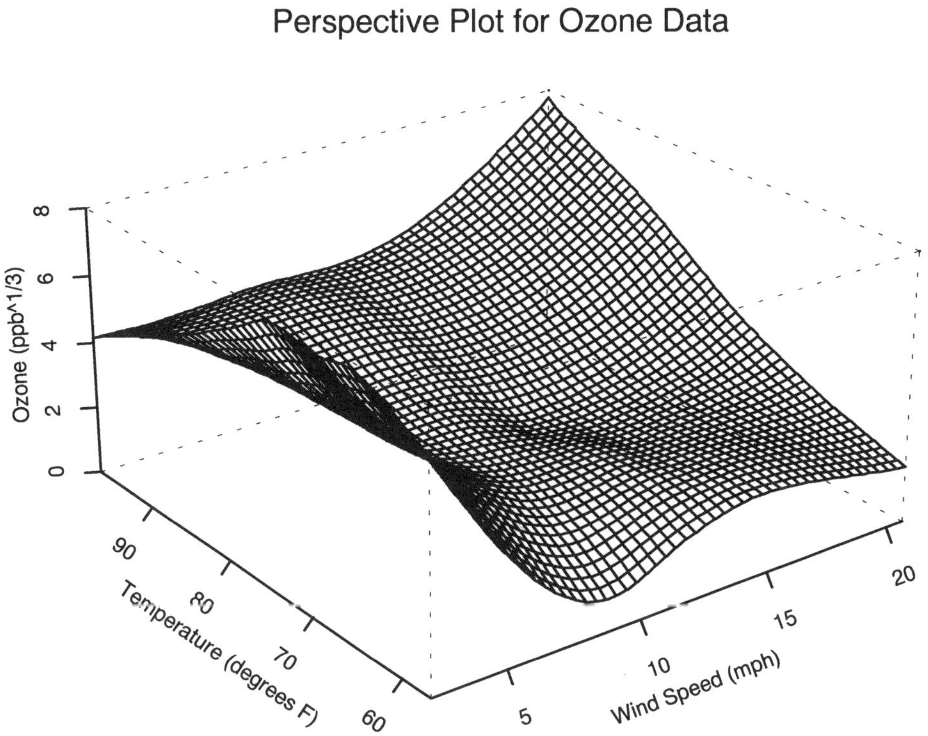

Figure 4.45: Wireframe (perspective) plot for the ozone data.

You can use the `eye` argument to `persp` to change the angle from which you view the plot.

Figure 4.46 is produced by these commands:

```
> wireframe(predicted.ozone ~ wind * temperature,
+    data = grid, xlab = "Wind Speed (mph)",
+    ylab = "Temperature (degrees F)",
+    zlab = "Ozone (ppb^1/3)",
+    main = list("Wireframe Plot of Ozone Data",
+    cex = 1.5))
```

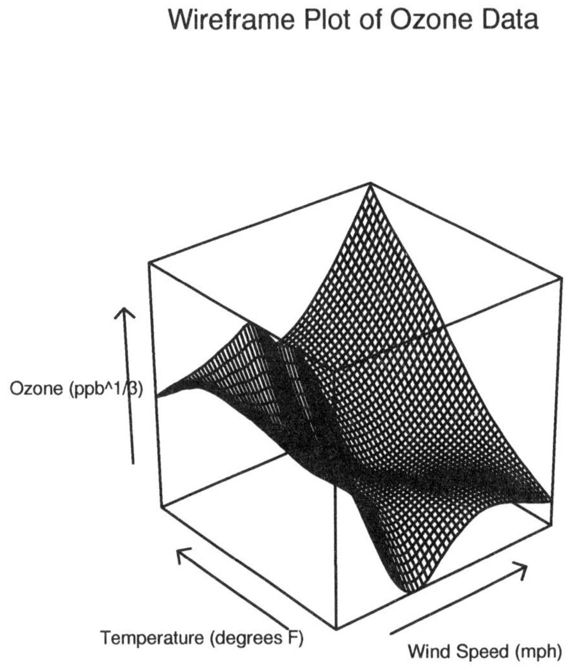

Figure 4.46: Wireframe plot of ozone data produced with the Trellis function `wireframe`.

You can use the `screen` argument to `wireframe` to change the angle from which you view the plot.

Multi-Panel Conditioning Plots (Coplots)

Multi-panel conditioning plots, also called coplots, let you view the relationship between two or more variables conditioned on the value of one or more other variables (Cleveland, 1993, Chapters 4 and 5). The S-Plus function `coplot` and all of the Trellis functions produce coplots.

Example 4.30: Multi-Panel Scatterplots with a Loess Smooth for Ozone as a Function of Temperature Given Wind Speed

Figure 4.47 shows the coplot produced by the S-Plus function `coplot` using the variables ozone, wind and temperature in the `air` data frame. In order to produce the coplot, we have to use the function `co.intervals` to create a two-column matrix (here we call it `wind.ints`) that holds a range of values (intervals) for the wind speed variable, since we are conditioning on that variable and it is continuous, not discrete. In our call to `co.intervals`, we specify that we want to break up wind speed into four equally-sized intervals (in terms of number of observations) that overlap each other about 25% (i.e., each interval shares about 25% of its points with the interval below it).

```
> attach(air)

> wind.ints <- co.intervals(wind, number = 4,
+ overlap=0.25)

> wind.ints
      [,1] [,2]
[1,]   2.3  8.0
[2,]   7.4 10.3
[3,]   9.7 12.0
[4,]  11.5 20.7

> coplot(ozone ~ temperature | wind, data = air,
+   given = wind.ints,
+   xlab = c("Temperature (degrees F)",
+   "Wind Speed (mph)"), ylab = "Ozone (ppb^1/3)",
+   panel = function(x,y) {
+     panel.smooth(x, y) })
```

The top panel in **Figure 4.47** is a legend. It shows that the lower left-hand scatterplot is a plot of ozone vs. temperature when the wind speed ranges from 2.3-8.0 mph. The lower right-hand scatterplot is a plot of ozone vs. temperature when the wind speed ranges from 7.4-10.3 mph. The upper left-hand scatterplot corresponds to wind speeds between 9.7-12.0 mph, and the upper right-hand scatterplot corresponds to wind speeds between 11.5-20.7 mph.

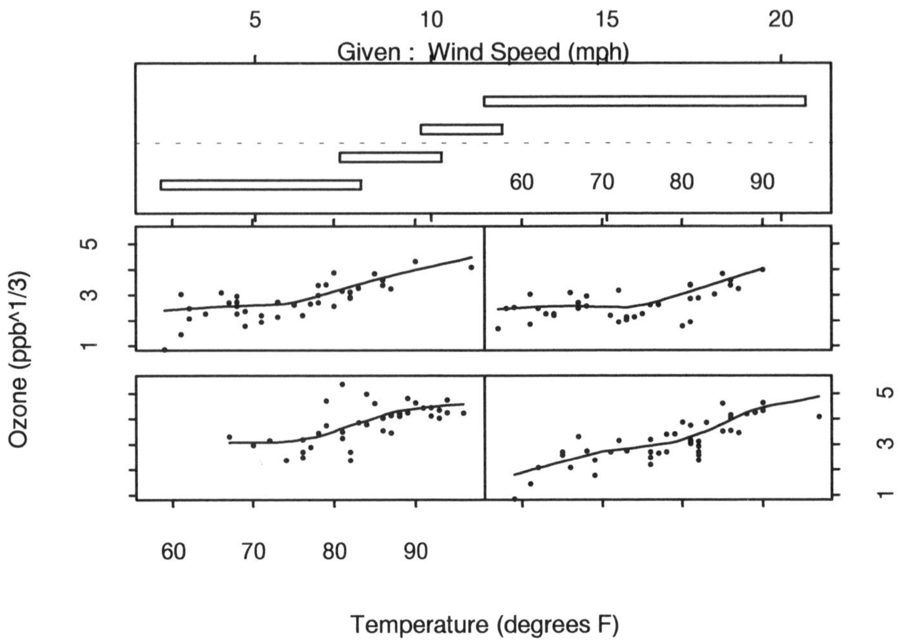

Figure 4.47: Coplot of ozone data produced with the S-PLUS function `coplot`.

Based on this conditioning plot, you can see that the relationship between ozone and temperature is pretty much the same for different values of wind speed, but the ozone concentrations tend to be smaller for larger values of wind speed.

Figure 4.48 shows the same coplot produced with the Trellis function `xyplot`. For Trellis functions, you need to use the function `equal.count` instead of `co.intervals` to create intervals for continuous data.

```
> Wind <- equal.count(wind, number = 4,
+ overlap = 0.25)

> xyplot(ozone ~ temperature | Wind, data = air,
+    xlab = "Temperature (degrees F)",
+    ylab = "Ozone (ppb^1/3)",
+    panel = function(x,y) {
+    panel.xyplot(x,y)
+    panel.loess(x,y) })
```

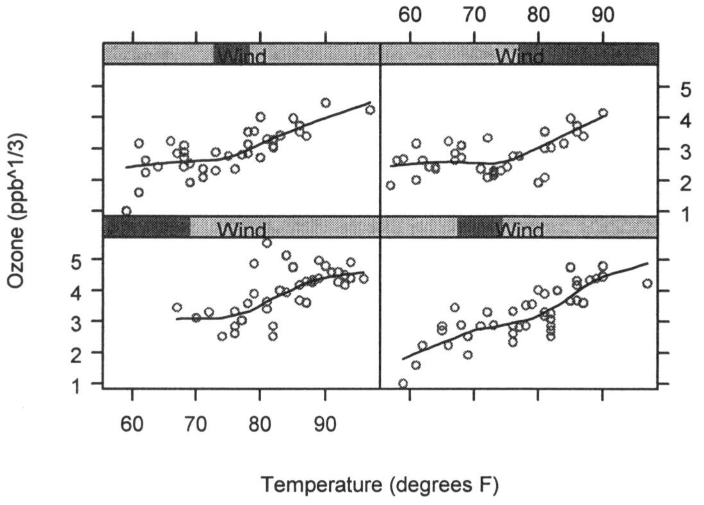

Figure 4.48: Coplot of ozone data produced with the Trellis function `xyplot`.

Example 4.31: Multi-Panel Scatterplots with a Loess Smooth for Ozone as a Function of Temperature Given Wind Speed and Radiation

We can extend the coplot of **Figure** 4.48 to include all three predictor variable for ozone (temperature, wind speed, and radiation) with the following commands:

```
> Radiation <- equal.count(radiation, number = 4,
+   overlap = 0.25)

> xyplot(ozone ~ temperature | Wind * Radiation,
+   data = air, xlab = "Temperature (degrees F)",
+   ylab = "Ozone (ppb^1/3)",
+   panel = function(x,y) {
+     panel.xyplot(x,y)
+     panel.loess(x,y) })
```

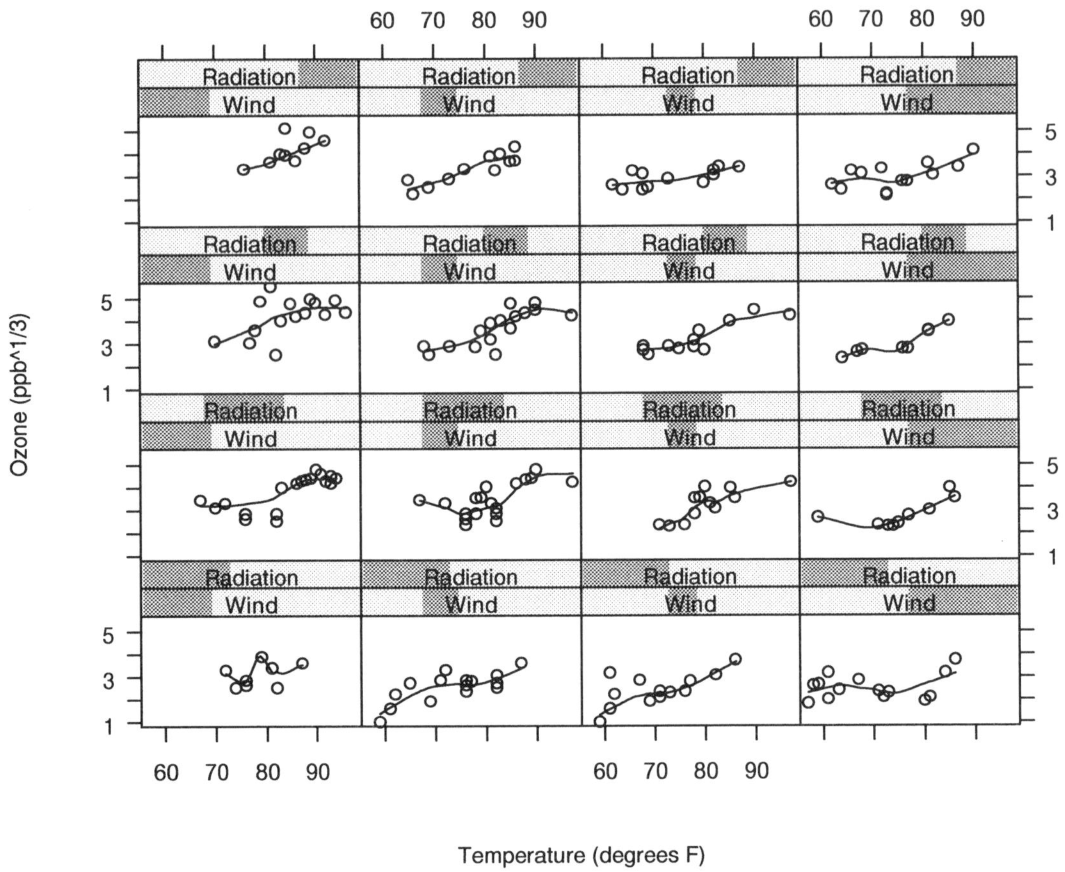

Figure 4.49: Coplot of ozone data with all three predictor variables included.

You can use any of the Trellis functions to produce coplots. For example, you can show histograms of one variable conditioned on a value or range of values of a second variable. See the Trellis documentation for more information.

Summary

- S-PLUS and ENVIRONMENTALSTATS for S-PLUS contain several functions for computing summary statistics (see the S-PLUS help files **Mathematical Operations** and **Robust/Resistant Techniques** and the ENVIRONMENTALSTATS for S-PLUS help file **Summary Statistics**).

- ENVIRONMENTALSTATS for S-PLUS includes functions for computing probability-weighted moments and *L*-moments (see the ENVIRONMENTALSTATS for S-PLUS help files **Abstract: Greenwood et al. (1979)** and **Abstract: Hosking (1990)**, and the help files for `pw.moment` and `l.moment`).

- Graphical displays are usually far superior to summary statistics for conveying information in a data set.

- S-PLUS and ENVIRONMENTALSTATS for S-PLUS contain several functions for creating graphical displays (see the S-PLUS help files **Computations Related to Plotting**, **Dynamic Graphics**, **Graphical Devices**, **High-Level Plots**, **Interacting with Plots**, **Multivariate Techniques**, and **Trellis Displays Library**, and the ENVIRONMENTALSTATS for S-PLUS help file **Plotting Probability Distributions**).

- Before you can create a plot in S-PLUS you must start a graphics device.

- Commonly used plots for univariate data include bar charts, dot charts, pie charts, histograms, boxplots, quantile plots (also called empirical cumulative distribution function plots), quantile-quantile (Q-Q) plots (also called probability plots), and autocorrelation plots.

- Commonly used plots for bivariate and multivariate data include scatterplots, matrix scatterplots, three-dimensional plots, brush and spin plots, and multi-panel displays (conditioning plots).

CHAPTER 5 :

Probability Distributions

As we stated in **Chapter 3**, a population is defined as the entire collection of measurements about which we want to make a statement, such as all possible measurements of dissolved oxygen in a specific section of a stream within a certain time period. Probability distributions are idealized mathematical models that are used to model the variability inherent in a population (e.g., all measures of dissolved oxygen will not exactly match each other). Certain probability distributions come up again and again in environmental statistics. This chapter discusses the concepts of a random variable and a probability distribution, talks about important probability distributions in environmental statistics, and shows you how to use various functions in ENVIRONMENTALSTATS for S-PLUS to plot and generate quantities associated with probability distributions.

Overview

Figure 5.1 displays the probability density functions for the probability distributions that are available in S-PLUS and ENVIRONMENTALSTATS for S-PLUS.

Table 5.1 lists the probability distributions available in S-PLUS and ENVIRONMENTALSTATS for S-PLUS (see the ENVIRONMENTALSTATS for S-PLUS help file **Probability Distributions and Random Numbers**). For each of these distributions, there are functions for computing the probability density function (pdf), the cumulative distribution function (cdf), quantiles, and random numbers. The form of the names of these functions are `d`*abb*, `p`*abb*, `q`*abb*, and `r`*abb*, where *abb* denotes the abbreviation of the distribution name (see column 2 of the table). For example, the functions `dnorm`, `pnorm`, `qnorm`, and `rnorm` compute the pdf, cdf, quantiles, and random numbers for the normal distribution.

The help file for `.Distribution.frame` contains a more extensive table that includes the distribution name, abbreviation, type (continuous, discrete, finite discrete, mixed), range (i.e., support), parameters, default values for the parameters, parameter ranges, and estimation methods available for the parameters.

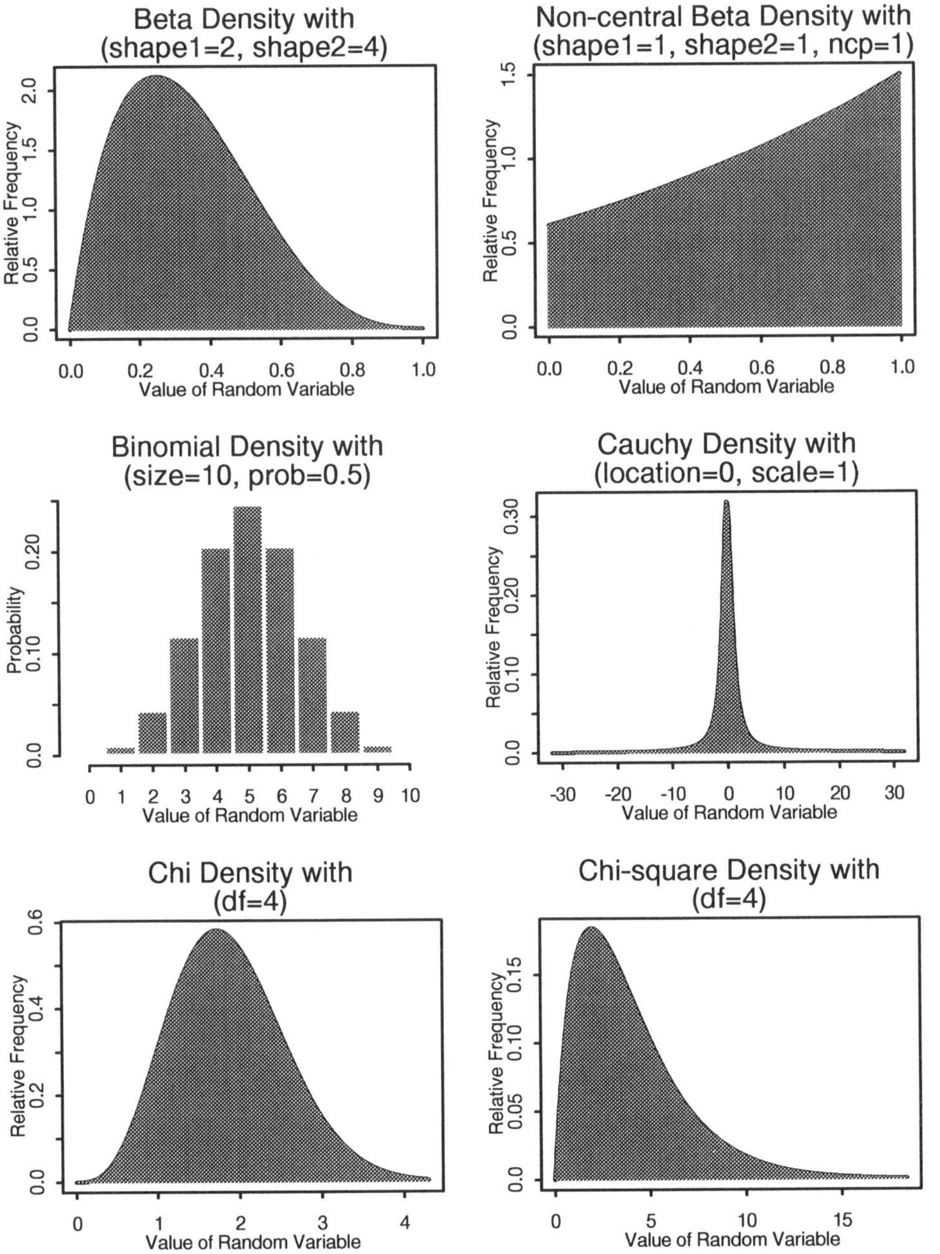

Figure 5.1: Probability distributions available in S-PLUS and ENVIRONMENTALSTATS for S-PLUS.

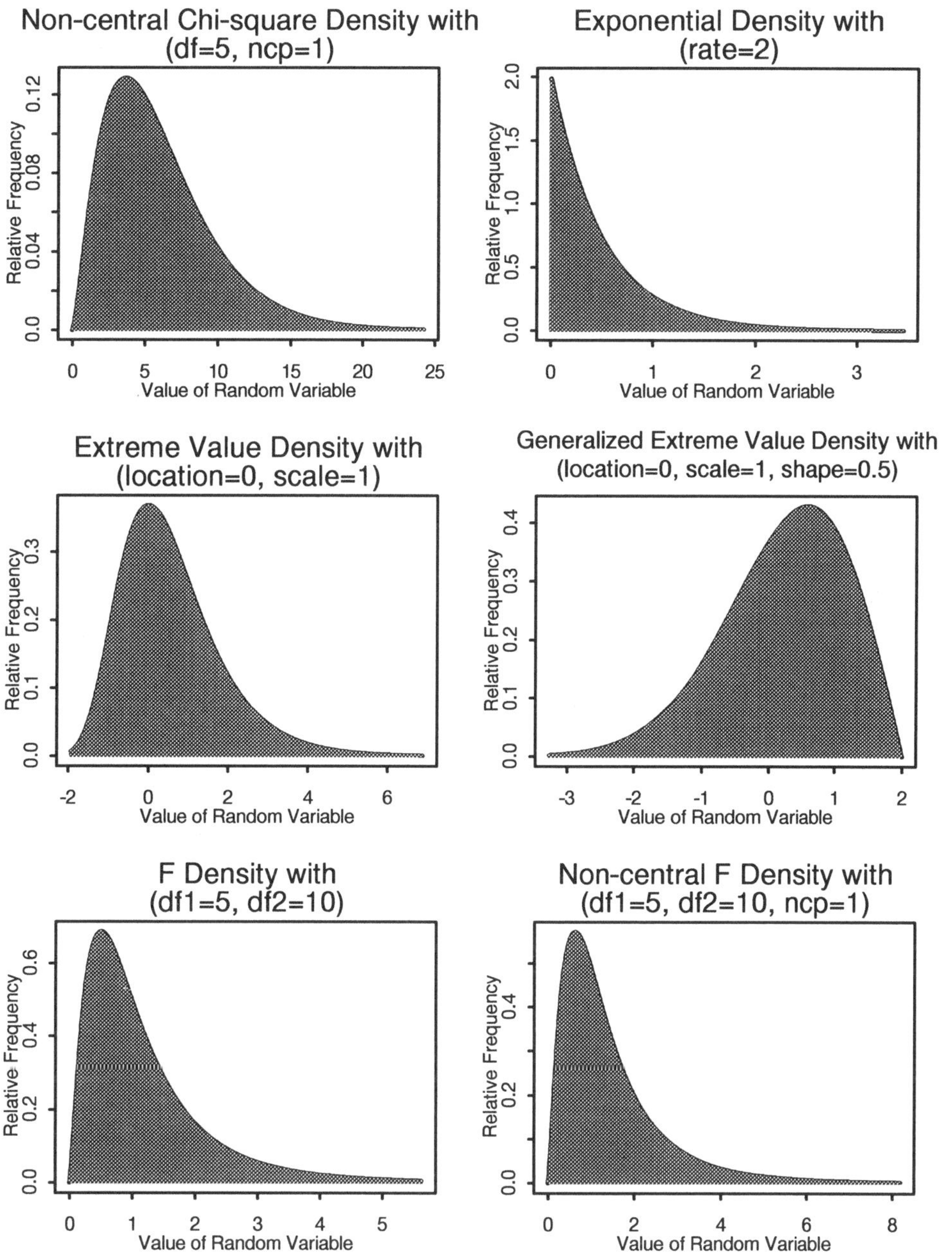

Figure 5.1 (continued): Probability distributions available in S-PLUS and ENVIRONMENTALSTATS for S-PLUS.

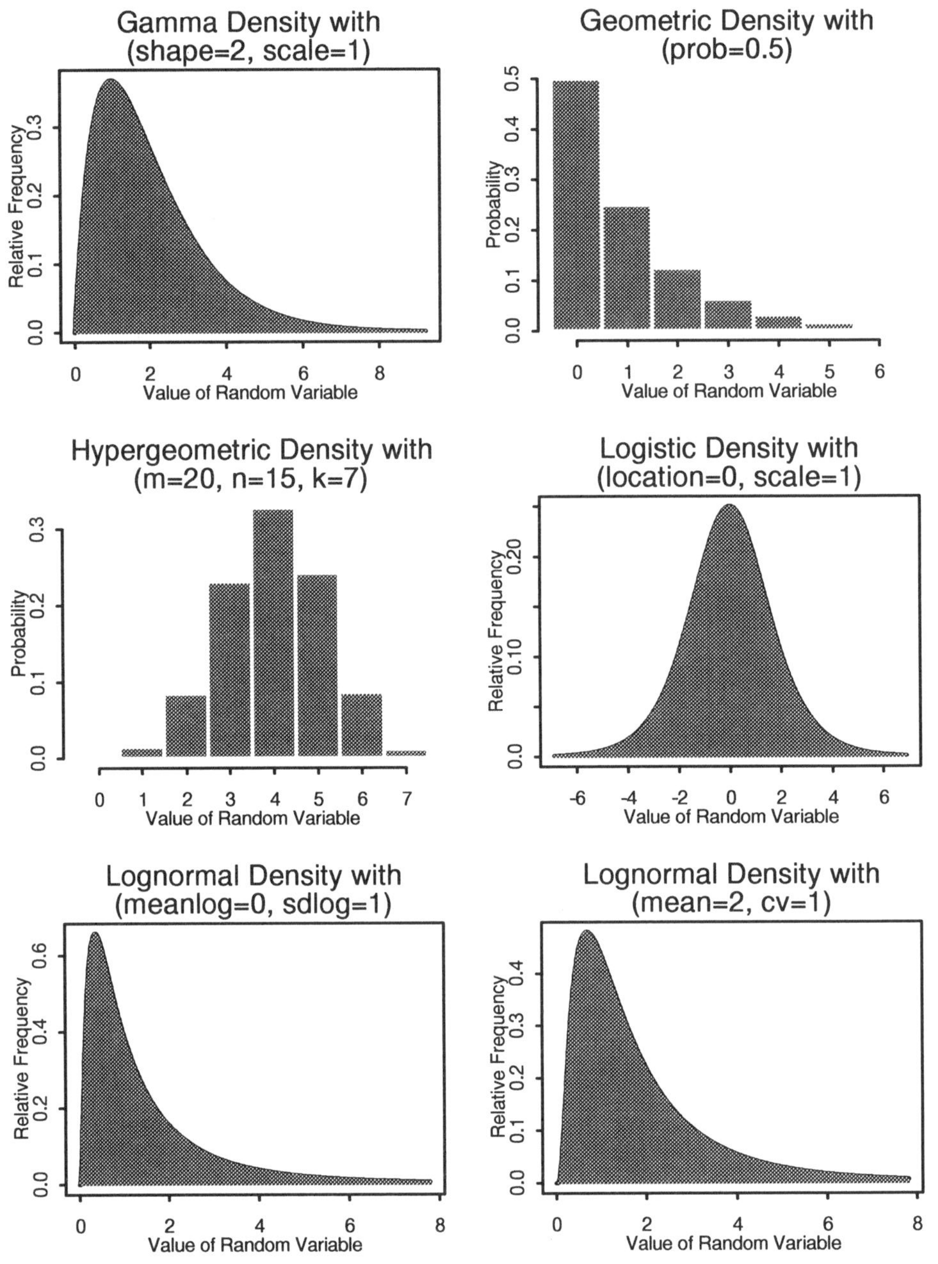

Figure 5.1 (continued): Probability distributions available in S-PLUS and ENVIRONMENTALSTATS for S-PLUS.

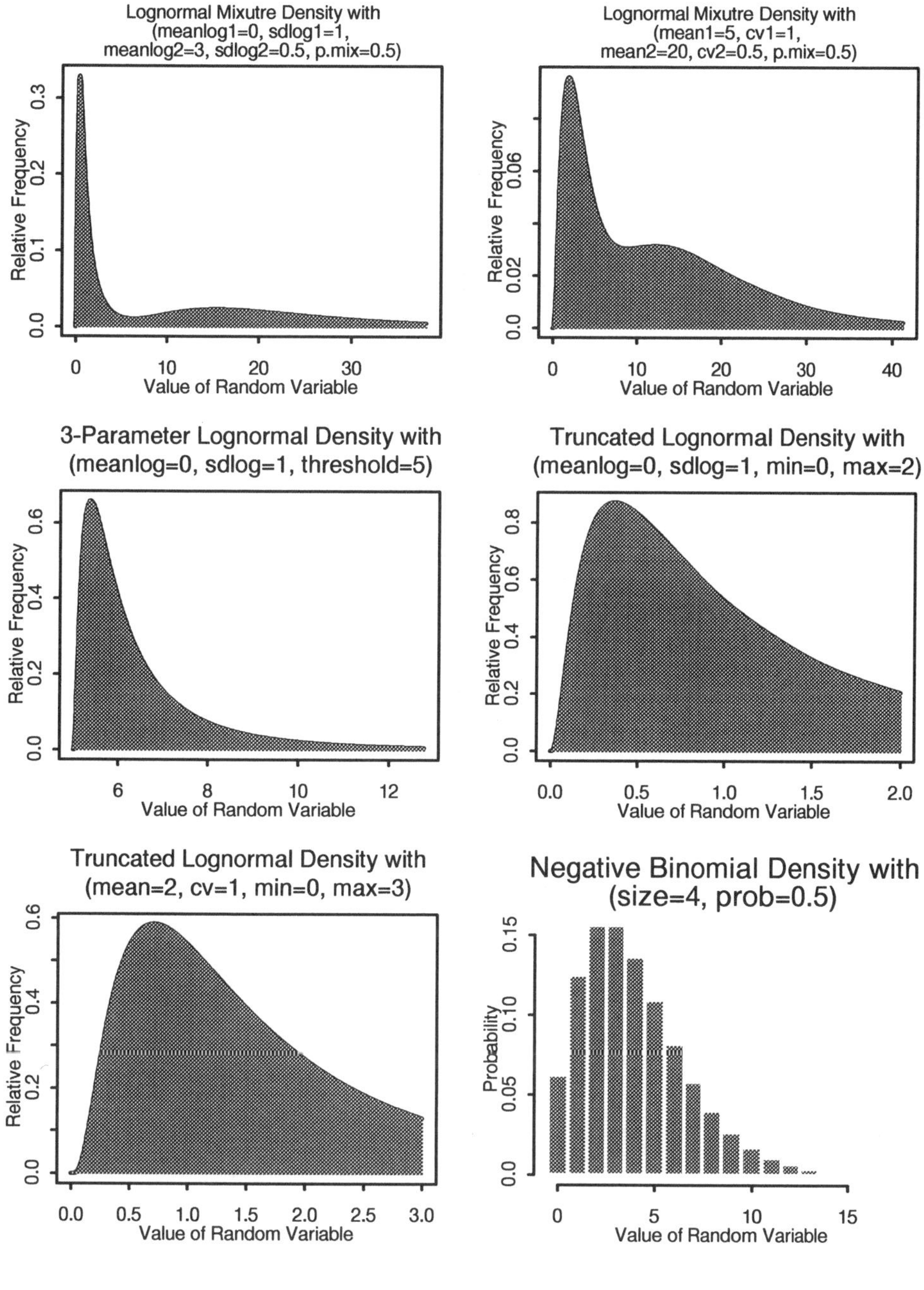

Figure 5.1 (continued): Probability distributions available in S-PLUS and ENVIRONMENTALSTATS for S-PLUS.

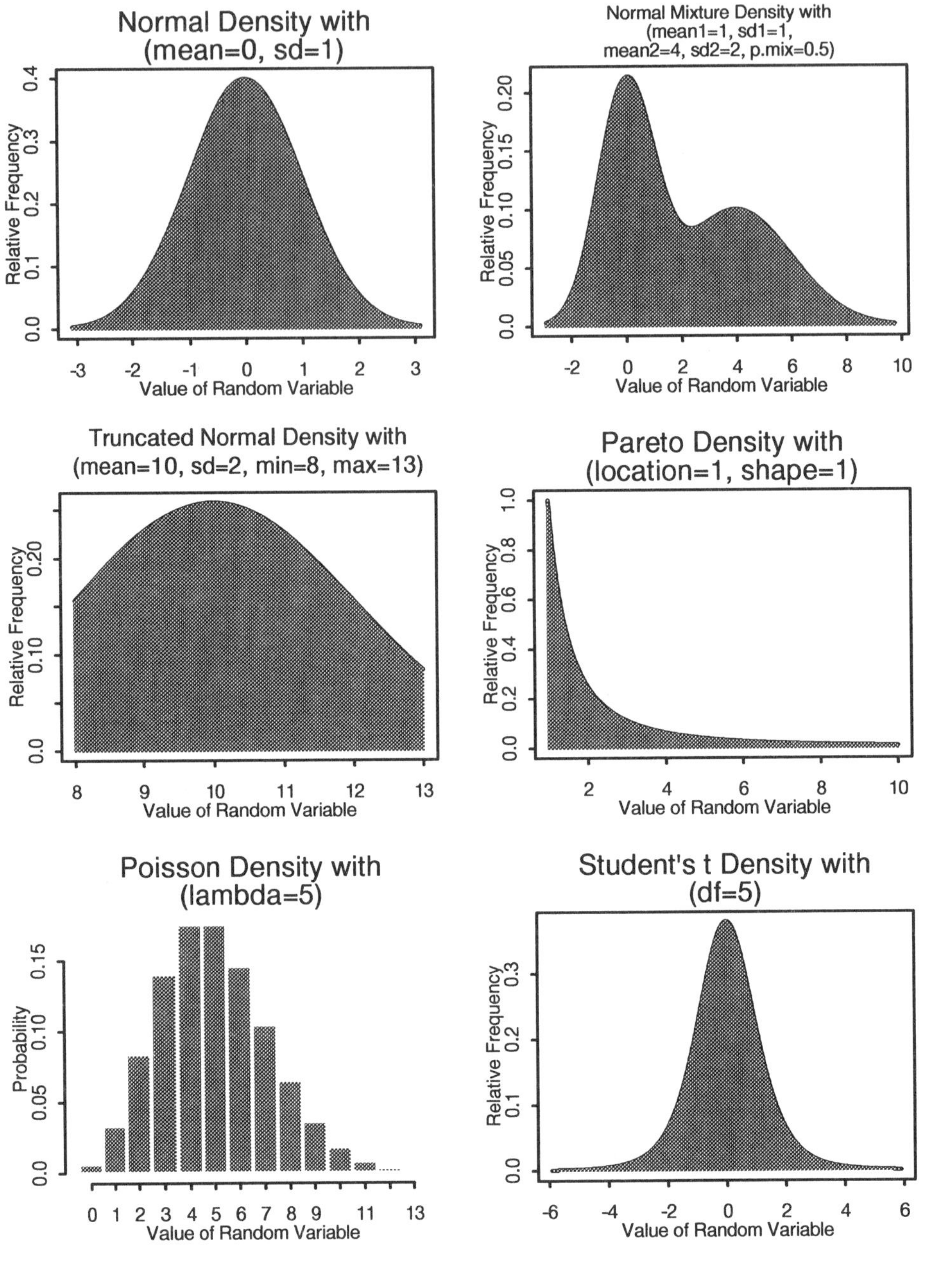

Figure 5.1 (continued): Probability distributions available in S-PLUS and ENVIRONMENTALSTATS for S-PLUS.

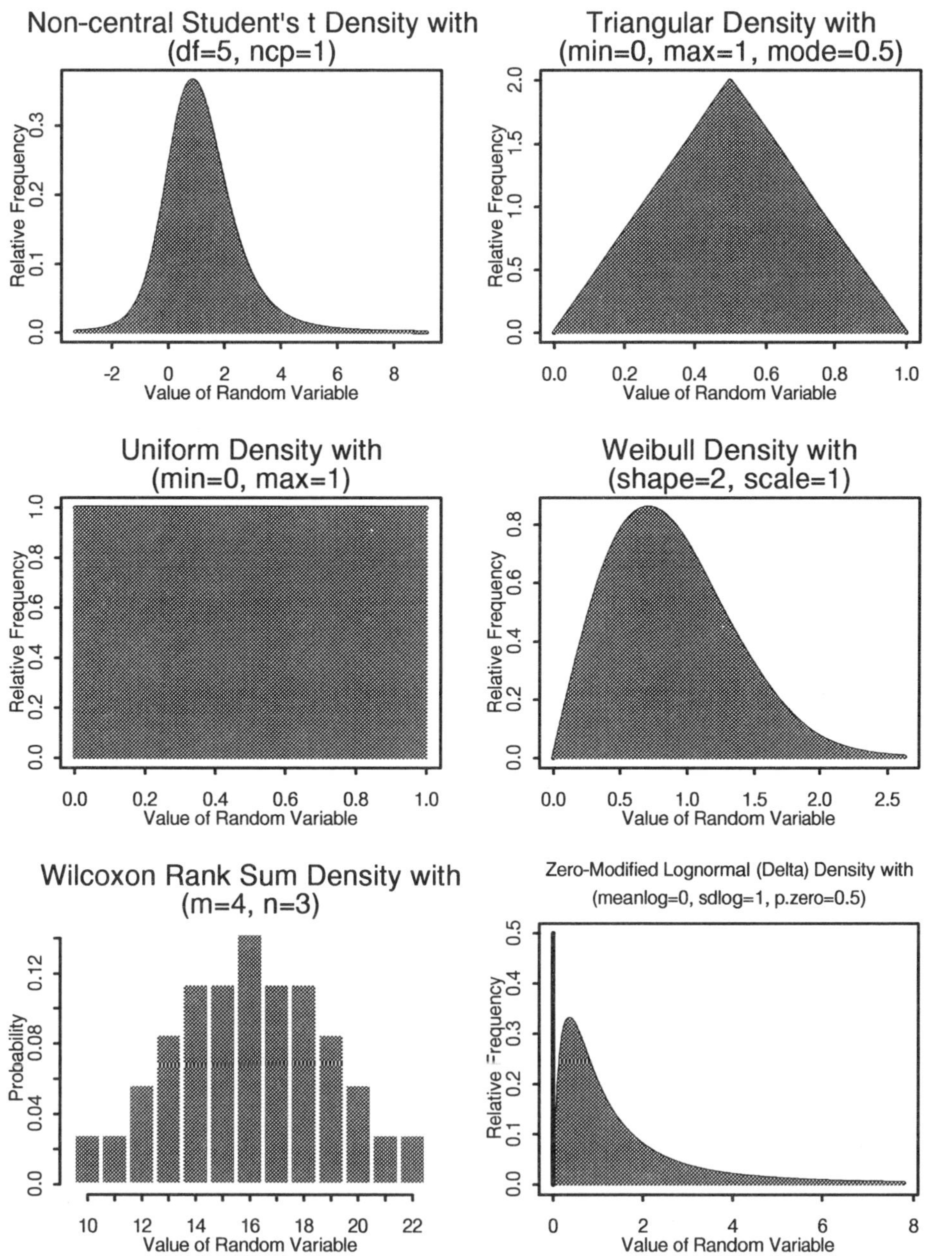

Figure 5.1 (continued): Probability distributions available in S-PLUS and ENVIRONMENTALSTATS for S-PLUS.

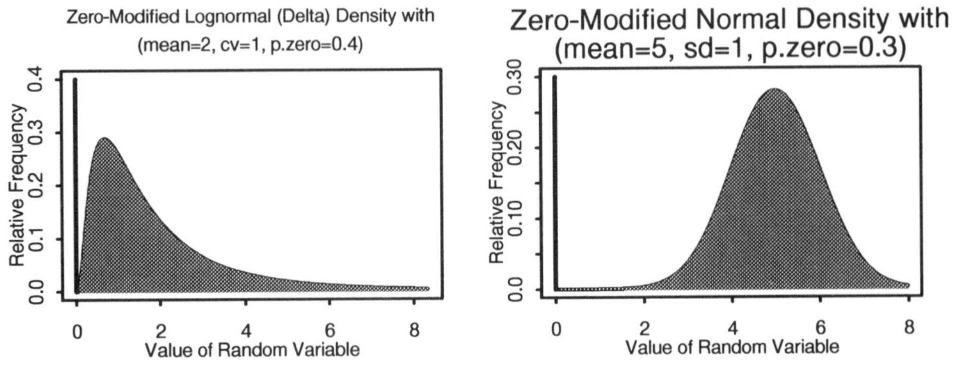

Figure 5.1 (continued): Probability distributions available in S-PLUS and ENVIRONMENTALSTATS for S-PLUS.

Distribution Name	Abbreviation	Parameter(s)
Beta*	`beta`	`shape1, shape2, ncp`
Binomial	`binom`	`size, prob`
Cauchy	`cauchy`	`location, scale`
Chi**	`chi`	`df`
Chi-square*	`chisq`	`df, ncp`
Empirical	`emp`	
Exponential	`exp`	`rate`
Extreme Value**	`evd`	`location, scale`
F*	`f`	`df1, df2, ncp`

* Modified in ENVIRONMENTALSTATS for S-PLUS.
** Only available in ENVIRONMENTALSTATS for S-PLUS.
*** Only random numbers are available for the stable distribution.

Table 5.1: Probability distributions available in S-PLUS and ENVIRONMENTALSTATS for S-PLUS.

Distribution Name	Abbreviation	Parameter(s)
Gamma*	`gamma`	`shape, scale`
Generalized Extreme Value**	`gevd`	`location, scale, shape`
Geometric	`geom`	`prob`
Hypergeometric	`hyper`	`m, n, k`
Logistic	`logis`	`location, scale`
Lognormal*	`lnorm`	`meanlog, sdlog`
Lognormal (alternative parameterization)**	`lnorm.alt`	`mean, cv`
Lognormal Mixture**	`lnorm.mix`	`meanlog1, sdlog1, meanlog2, sdlog2, p.mix`
Lognormal Mixture (alternative parameterization)**	`lnorm.mix.alt`	`mean1, cv1, mean2, cv2, p.mix`
Three-Parameter Lognormal**	`lnorm3`	`meanlog, sdlog, threshold`
Truncated Lognormal**	`lnorm.trunc`	`meanlog, sdlog, min, max`
Truncated Lognormal (alternative parameterization)**	`lnorm.trunc.alt`	`mean, cv, min, max`

* Modified in ENVIRONMENTALSTATS for S-PLUS.
** Only available in ENVIRONMENTALSTATS for S-PLUS.
*** Only random numbers are available for the stable distribution.

Table 5.1 (continued): Probability distributions available in S-PLUS and ENVIRONMENTALSTATS for S-PLUS.

Distribution Name	Abbreviation	Parameter(s)
Negative Binomial	nbinom	size, prob
Normal	norm	mean, sd
Normal Mixture**	norm.mix	mean1, sd1, mean2, sd2 p.mix
Truncated Normal	norm.trunc	mean, sd, min, max
Pareto**	pareto	location, shape
Poisson	pois	lambda
Stable***	stab	index, skewness
Student's t*	t	df, ncp
Triangular**	tri	min, max, mode
Uniform	unif	min, max
Weibull	weibull	shape, scale
Wilcoxon Rank Sum	wilcox	m, n
Zero-Modified Lognormal (Delta)**	zmlnorm	meanlog, sdlog, p.zero
Zero-Modified Lognormal (Delta) (alternative parameterization)**	zmlnorm.alt	mean, cv, p.zero
Zero-Modified Normal**	zmnorm	mean, sd, p.zero

* Modified in ENVIRONMENTALSTATS for S-PLUS.
** Only available in ENVIRONMENTALSTATS for S-PLUS.
*** Only random numbers are available for the stable distribution.

Table 5.1 (continued): Probability distributions available in S-PLUS and ENVIRONMENTALSTATS for S-PLUS.

Table 5.2 lists the functions available in ENVIRONMENTALSTATS for S-PLUS to create plots associated with theoretical and empirical probability distributions (see the help file **Plotting Probability Distributions**). The functions `cdf.compare`, `ecdfplot`, `qqplot`, and `qqplot.gestalt` were already discussed in **Chapter 4**. We'll discuss the functions `cdfplot` and `pdfplot` in this chapter.

Function	**Description**
`cdf.compare*`	Plot two cumulative distribution functions.
`cdfplot`	Plot a cumulative distribution function.
`ecdfplot`	Plot an empirical cumulative distribution function.
`pdfplot`	Plot a probability density function.
`qqplot*`	Produce a quantile-quantile (Q-Q) plot, (also called a probability plot).
`qqplot.gestalt`	Develop a gestalt of Q-Q plots for a specific distribution.

* Available in S-PLUS but modified in ENVIRONMENTALSTATS for S-PLUS.

Table 5.2: Functions for plotting probability distributions.

The rest of this chapter explains the concepts of random variable and probability distribution, discusses various quantities associated with probability distributions, explains how to use functions in ENVIRONMENTALSTATS for S-PLUS to compute these quantities, and talks about specific probability distributions that are frequently used in environmental statistics.

What is a Random Variable?

When you perform an environmental study, you take a sample from a population and record the measurements associated with the sample. For example, the population may be all possible measurements of dissolved oxygen in a certain section of a stream within a given time period, and you may take 50 physical samples of water and measure the dissolved oxygen in each sample. The actual numbers, the actual concentrations of dissolved oxygen, are realizations of a concept called a "random variable".

A ***random variable*** is "the value of the next observation in an experiment" (Watts, 1991, as quoted by Berthoux and Brown, 1994, p.7). It is the measurement associated with the next physical sample from the population. Outcomes of a random variable usually differ in value from observation to observation.

Example 5.1: Flipping a Fair Coin 10 Times

One of the simplest examples of a random variable involves flipping a coin and observing whether the coin lands "heads" or "tails". In this case, the population is the set of outcomes from flipping the coin an infinite number of times, and the random variable is the outcome from flipping the coin.

In S-PLUS, it is easy to simulate an experiment in which we flip a "fair" coin say 10 times (i.e., we take 10 samples from the population). Here, the term "fair" means that a head is just as likely to occur as a tail. The histogram (really a bar chart) in **Figure 5.2** shows the outcome from such an experiment, where 0 represents a tail and 1 represents a head. We can see that there were six tails and four heads in our experiment. Here are the commands to reproduce this figure. (We'll talk more about the `rbinom` and `set.seed` functions later in this chapter.)

```
> set.seed(482)

> flips.10 <- rbinom(10, size = 1, prob = 0.5)

> flips.10
[1] 0 0 0 1 0 1 0 1 0 1

> hist(flips.10, breaks = c(-0.5, 0.5, 1.5), axes = F,
+   xlab = "Outcome of Flip", ylab = "Frequency")

> axis(2); axis(1, at = c(0,1))
```

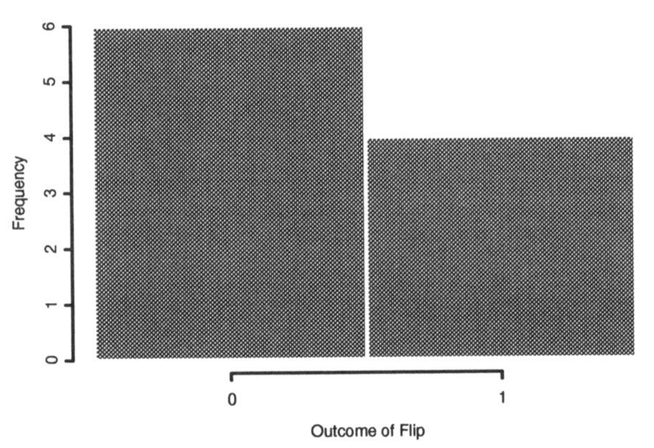

Figure 5.2: Result of flipping a fair penny 10 times.

Discrete vs. Continuous Random Variable

The outcome in the coin-tossing experiment is an example of a ***discrete*** random variable; it can take on only two possible values (either 0 or 1). In general, a discrete random variable can take on only a finite or countably infinite number of values.

On the other hand, a ***continuous*** random variable can take on an infinite number of values. For example, the concentration of 1,2,3,4-tetrachlorobenze (TcCB) in a physical sample of soil may theoretically take on an infinite number of values. In reality, however, the values of a "continuous" random variable are limited by the precision of the instrument used to measure the random variable and the level of mathematical rounding or truncation used to report the results.

Example 5.2: TcCB Measurements are a Continuous Random Variable

The data in the data frame `epa.94b.tccb.df` are associated with an experiment in which soil samples were taken from two different areas and the amount of TcCB (in ppb) was measured in each physical soil sample. In this case, the random variable is the concentration of TcCB that *will be* observed in a soil sample. The actual measures of TcCB in the reference are summarized in the histogram in **Figure 5.3**.

```
> attach(epa.94b.tccb.df)

> hist(TcCB[Area == "Reference"], xlab = "TcCB (ppb)",
+    xlim = c(0, 2), ylab = "Frequency",
+    main = "Histogram of Reference Area TcCB Data")
```

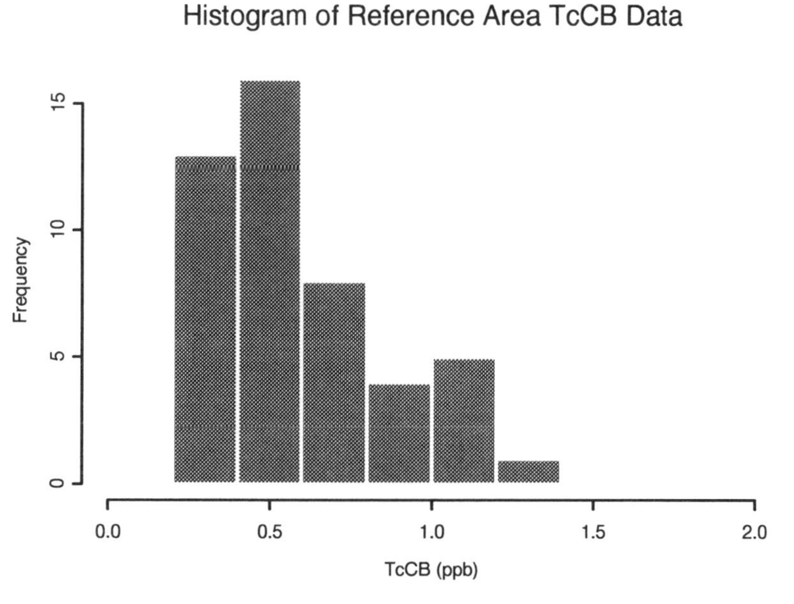

Figure 5.3: Histogram of the reference area TcCB data.

What is a Probability Distribution?

In our coin-tossing experiment, 40% of our tosses (4/10) resulted in a head. We can represent this with a ***relative frequency*** histogram by setting `prob=T` in the call to `hist`:

```
> hist(flips.10, prob = T, breaks = c(-0.5, 0.5, 1.5),
+    axes = F, xlab = "Outcome of Flip",
+    ylab = "Relative Frequency")

> axis(2); axis(1, at = c(0,1))
```

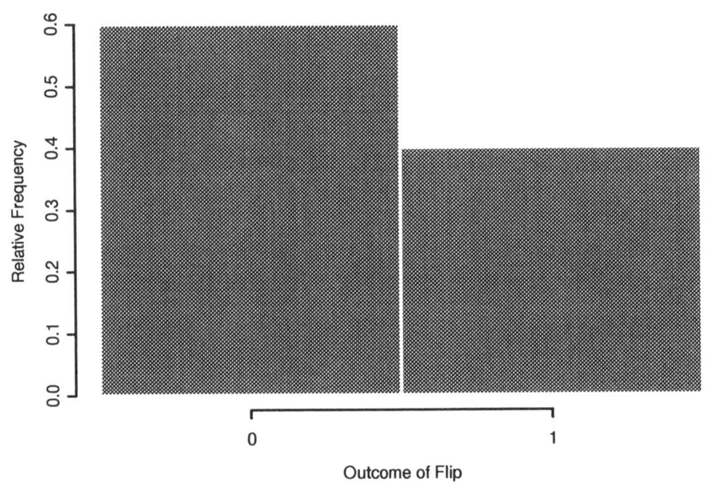

Figure 5.4: Relative frequency histogram for the coin toss experiment.

Note that **Figure 5.4** is exactly the same as **Figure 5.2** except we have divided the numbers on the *y*-axis by the total number of times we flipped the coin.

Now let's repeat this experiment, except let's flip the coin 100 times. The result of this experiment is shown in **Figure 5.5**. In this case a little over 50% of our tosses resulted in a head.

```
> set.seed(256)

> flips.100 <- rbinom(100, size = 1, prob = 0.5)

> hist(flips.100, prob = T, breaks = c(-0.5, 0.5, 1.5),
+    axes = F, xlab = "Outcome of Flip",
+    ylab = "Relative Frequency")
```

```
> axis(2); axis(1, at = c(0,1))
```

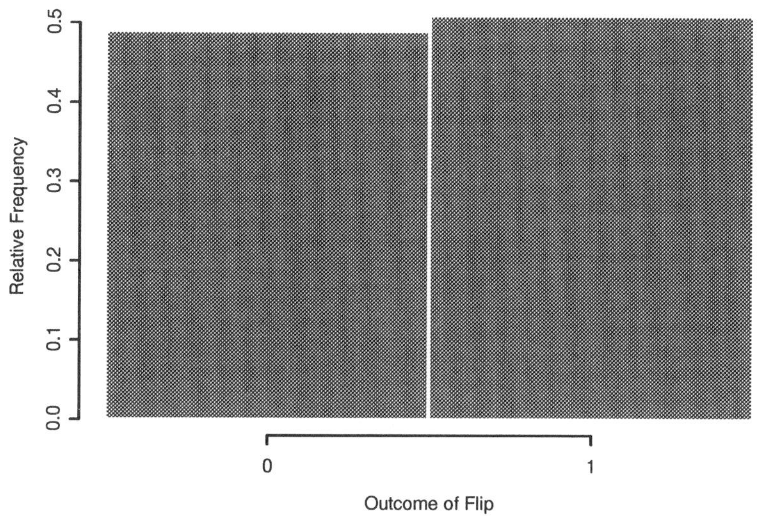

Figure 5.5: Relative frequency histogram for 100 coin flips.

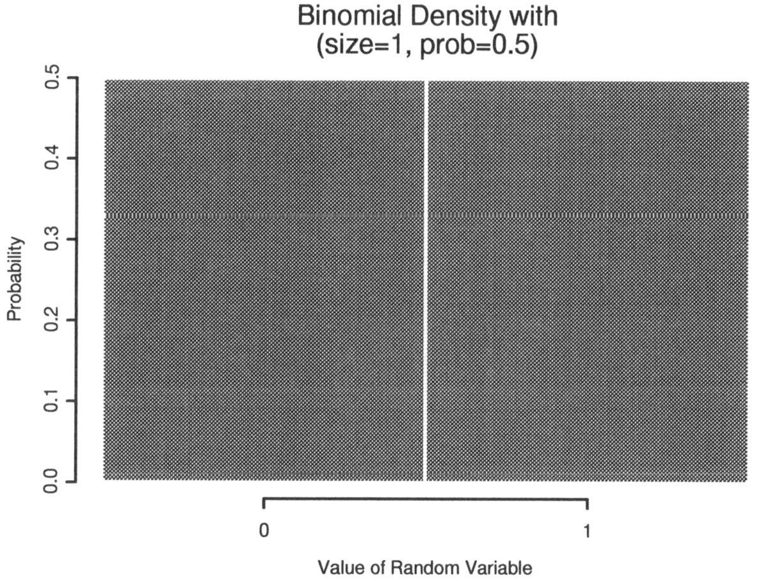

Figure 5.6: Probability distribution for the coin-tossing experiment.

Because we are tossing a fair coin, we would expect that as we toss the coin more and more times, the percentage of heads should "get close" to 50%. This is called the "Law of Large Numbers". A ***probability distribution*** can be thought of as what a relative frequency histogram of outcomes would look like if you could take a very large or infinite number of samples.

Figure 5.6 shows the probability distribution for the coin-tossing experiment. Here is the command to produce the plot of this probability distribution:

```
> pdfplot("binom", list(size = 1, prob = 0.5))
```

The function `pdfplot` plots probability density functions for all of the probability distributions available in ENVIRONMENTALSTATS for S-PLUS (see the help file for `pdfplot` more information).

Probability Density Function (pdf) and Probability Mass Function (pmf)

A ***probability density function (pdf)*** is a mathematical formula that describes the relative frequency of a random variable. Sometimes the picture of this formula (e.g., **Figure 5.6**) is called the pdf.

If a random variable is discrete, its probability density function is sometimes called a ***probability mass function***, since it shows the "mass" of probability at each possible value of the random variable. The probability density (mass) function for the coin-tossing experiment is given by:

$$f(x) = \Pr[X = x] = \begin{cases} 0.5 & \text{if } x = 0 \\ 0.5 & \text{if } x = 1 \end{cases} \quad \textbf{(5.1)}$$

In this equation, the letter f stands for the probability density (mass) function, the letters Pr stand for the phrase "the probability that", the upper case X stands for the random variable (the outcome of the coin toss), and the lower case x stands for the actual value that the random variable takes on after one toss of the coin.

Figure 5.7 shows the relative frequency histogram for outcomes of a continuous random variable: the reference area TcCB data. The command to create this figure is as follows:

```
> hist(TcCB[Area == "Reference"], xlab = "TcCB (ppb)",
+    ylab = "Relative Frequency", prob = T,
+    xlim = c(0, 2), ylim = c(0, 2),
+    main = paste("Relative Frequency Histogram of",
+    "Reference Area TcCB Data"))
```

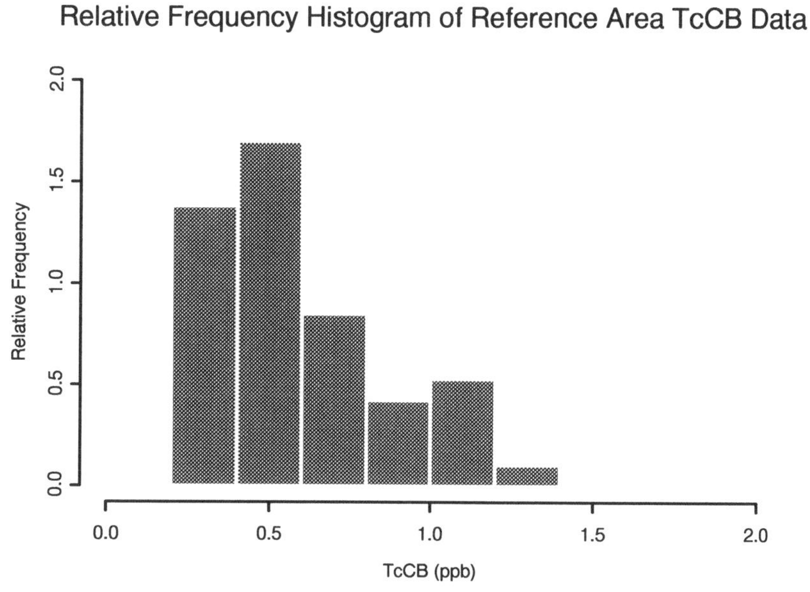

Figure 5.7: Relative frequency histogram for the reference area TcCB data.

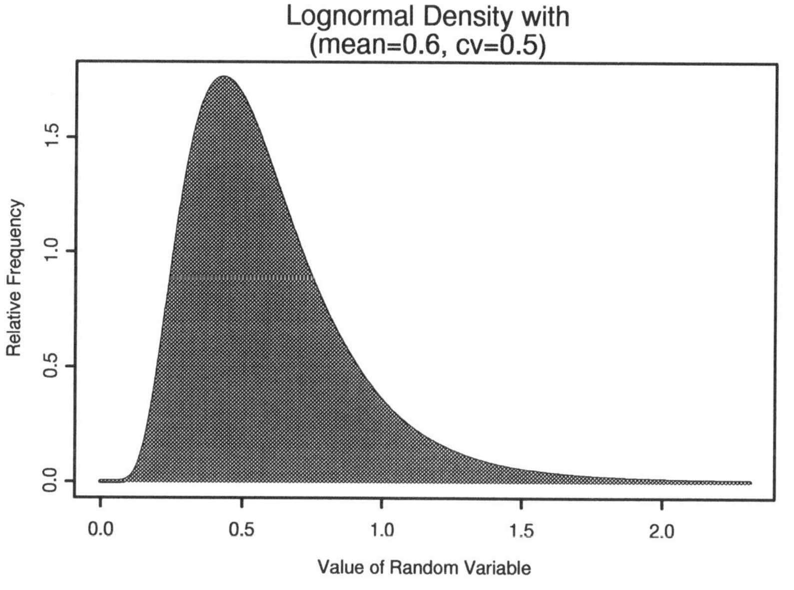

Figure 5.8: A lognormal probability distribution.

For each class (bar) of this histogram, the proportion of observations falling in that class is equal to the area of the bar; that is, it is the width of the bar times the height of the bar. If we could take many, many more samples and create relative frequency histograms with narrower and narrower classes, we might end up with a picture that looks like **Figure 5.8**.

Figure 5.8 shows the probability density function of a lognormal random variable with a mean of 0.6 and a coefficient of variation of 0.5. It was produced with the command:

```
> pdfplot("lnorm.alt", list(mean = 0.6, cv = 0.5))
```

Note the close resemblance to the histogram in **Figure 5.7** (we'll talk more about this in **Chapter 6: Estimating Population Parameters and Quantiles**). The mathematical equation for this pdf is given by:

$$f(x) = \frac{1}{x\sigma\sqrt{2\pi}} \exp\left\{-\frac{1}{2\sigma^2}\left[\log(x) - \mu\right]^2\right\}, \quad x > 0 \tag{5.2}$$

where

$$\begin{aligned} \mu &= \log\left(\frac{\theta}{\sqrt{\tau^2 + 1}}\right) \\ \sigma &= \left[\log\left(\tau^2 + 1\right)\right]^{1/2} \\ \theta &= 0.6 \\ \tau &= 0.5 \end{aligned} \tag{5.3}$$

(see the help file for **Lognormal**).

For a continuous random variable, the probability that the random variable falls into some interval, say between 0.5 and 1, is simply the area under the pdf between these two interval endpoints. Mathematically, this is written as:

$$\Pr(0.5 \le X \le 1) = \int_{0.5}^{1} f(x)\,dx \tag{5.4}$$

For the pdf shown in **Figure 5.8**, we can compute the probability that the random variable falls into the interval [0.5, 1] with the following command:

```
> plnorm.alt(1, mean = 0.6, cv = 0.5) -
+    plnorm.alt(0.5, mean = 0.6, cv = 0.5)
[1] 0.4657052
```

We see there is about at 47% chance that an observation will fall in this interval if the true underlying probability distribution is described by **Figure 5.8**. The function `plnorm.alt` computes values of the cumulative distribution function (cdf) for the lognormal distribution (see the section **Cumulative Distribution Function (cdf)** below).

Computing Values of the Probability Density Function

For all of the distributions listed in **Table 5.1** (except the stable distribution), there are functions of the form `d`*`abb`*, where *`abb`* denotes the abbreviation of the distribution name (column 2 in **Table 5.1**). These functions compute the pdf of the distribution. The first argument to these functions is the value(s) of the random variable at which you want to compute the pdf. Subsequent arguments specify the values of the parameters associated with the probability distribution (see column 3 in **Table 5.1**).

Example 5.3: Computing the pdf of a Binomial Distribution

Consider the binomial distribution shown in **Figure 5.6**. You can compute the value of the pdf (pmf in this case) by using the function `dbinom`:

```
> dbinom(0:1, size = 1, prob = 0.5)
[1] 0.5 0.5
```

We see that the probability that the random variable equals 0 is 0.5, and the probability that it equals 1 is also 0.5. We'll talk more about the binomial distribution later in this chapter.

Example 5.4: Computing the pdf of a Lognormal Distribution

As another example, you can compute values of the pdf shown in **Figure 5.8** with the function `dlnorm.alt`:

```
> dlnorm.alt(c(0.5, 1, 1.5), mean = 0.6, cv = 0.5)
[1] 1.67023226 0.35452674 0.05277346
```

We see that the value of this pdf at 0.5 is about 1.7, the value at 1 is about 0.35, and the value at 1.5 is about 0.05 (see **Figure 5.8**).

Plotting the Probability Density Function

The function `pdfplot` plots the pdf of any of the probability distributions listed in **Table 5.1** (except the stable distribution). The first argument to `pdfplot` is a character string that specifies the distribution abbreviation (see column 2 of **Table 5.1**). The second argument is a list that specifies the values of the parameters associated with the distribution (see column 3 of **Table 5.1**). As already noted above, **Figure 5.6** was produced with the command

```
> pdfplot("binom", list(size = 1, prob = 0.5))
```

and **Figure 5.8** was produced with the command

```
> pdfplot("lnorm.alt", list(mean = 0.6, cv = 0.5))
```

Cumulative Distribution Function (cdf)

The ***cumulative distribution function (cdf)*** of a random variable X, sometimes called simply the distribution function, is the function F such that

$$F(x) = \Pr(X \le x) \tag{5.5}$$

for all values of x. That is, $F(x)$ is the probability that the random variable X is less than or equal to some number x.

The cdf can also be defined or computed in terms of the probability density function (pdf) f as:

$$F(x) = \Pr(X \le x) = \int_{-\infty}^{x} f(x)\, dx \tag{5.6}$$

for a continuous random variable and

$$F(x) = \Pr(X \le x) = \sum_{x_i \le x} f(x_i) \tag{5.7}$$

for a discrete random variable.

Computing Values of the Cumulative Distribution Function

For all of the distributions listed in **Table 5.1** (except the stable distribution), there are functions of the form p*abb*, where *abb* denotes the abbreviation of the distribution name (column 2 in **Table 5.1**). These functions compute the cdf of the distribution. The first argument to these functions is the value(s) of the random variable at which you want to compute the cdf. Subsequent arguments specify the values of the parameters associated with the probability distribution (see column 3 in **Table 5.1**).

Example 5.5: Computing the cdf of a Binomial Distribution

Consider the binomial distribution shown in **Figure 5.6** and defined in **Equation (5.1)**. You can compute the value of the cdf by using the function `pbinom`:

```
> pbinom(c(-1, 0, 0.4, 1), size = 1, prob = 0.5)
[1] 0.0 0.5 0.5 1.0
```

We see that the probability that the random variable is less than or equal to -1 is 0, the probability that it is less than or equal to 0 is 0.5, the probability that it is less than or equal to 0.4 is 0.5, and the probability that it is less than or equal to 1 is 1. We'll talk more about the binomial distribution later in this chapter.

Example 5.6: Computing the cdf of a Lognormal Distribution

As another example, you can compute values of the cdf of the lognormal distribution shown in **Figure 5.8** with the function `plnorm.alt`:

```
> plnorm.alt(c(0.5, 1, 1.5), mean = 0.6, cv = 0.5)
[1] 0.4404719 0.9061771 0.9852194
```

We see that the value of this pdf at 0.5 is about 0.4, the value at 1 is about 0.9, and the value at 1.5 is about 0.99 (see **Figure 5.8**).

Plotting the Cumulative Distribution Function

The function `cdfplot` plots the cdf of any of the probability distributions listed in **Table 5.1** (except the stable distribution). The first argument to `cdfplot` is a character string that specifies the distribution abbreviation (see column 2 of **Table 5.1**). The second argument is a list that specifies the values of the parameters associated with the distribution (see column 3 of **Table 5.1**).

Example 5.7: Plotting the cdf of a Binomial Distribution

Figure 5.9 shows the cdf for the coin-tossing experiment (see **Figure 5.6**). Note that this is a step function: the probability that the random variable is less than 0 is 0%, the probability that it is greater than or equal to 0 and less than 1 is 50%, and the probability that it is less than or equal to 1 is 100%.

Example 5.8: Plotting the cdf of a Lognormal Distribution

Figure 5.10 shows the cdf for the lognormal pdf shown in **Figure 5.8** and defined in **Equations (5.2)** and **(5.3)**.

```
> cdfplot("binom", list(size = 1, prob = 0.5))
```

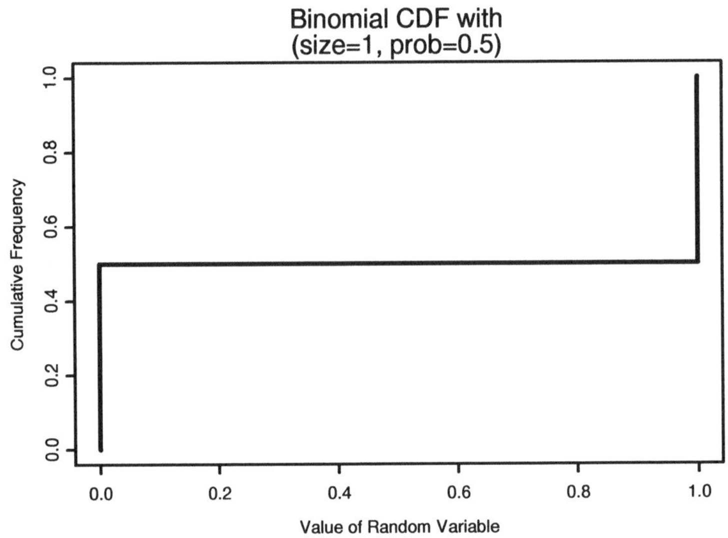

Figure 5.9: Cumulative distribution function for the coin-tossing experiment.

```
> cdfplot("lnorm.alt", list(mean = 0.6, cv = 0.5))
```

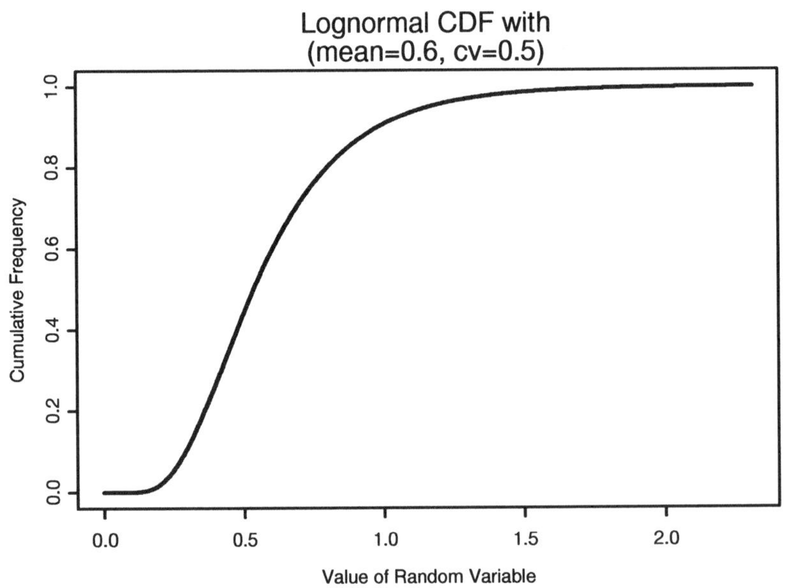

Figure 5.10: Cumulative distribution function of a lognormal distribution.

Quantiles and Percentiles

Let X denote a random variable with some specified distribution. The **p^{th} *quantile*** of the distribution of X, denoted x_p, is a (the) number that satisfies:

$$\Pr[X < x_p] \le p \le \Pr[X \le x_p] \tag{5.8}$$

Note that p must be a number between 0 and 1, inclusive.

If there is more than one number that satisfies the above condition, the pth quantile of X is often taken to be the average of the smallest and largest numbers that satisfy the condition. The S-PLUS functions for computing quantiles, however, return the smallest number that satisfies the above condition (see **Computing Quantiles of Probability Distributions** below).

If X is a continuous random variable, the pth quantile of X is simply defined as the value such that:

$$\Pr[X \le x_p] = F(x_p) = p \tag{5.9}$$

The **$100p^{th}$ *percentile*** is another name for the pth quantile. That is, the $100p^{th}$ percentile is the (a) number such that $100p\%$ of the distribution lies below this number.

Computing Quantiles of Probability Distributions

A plot of the cumulative distribution function makes it easy to visually pick out important quantiles, such as the median (50th percentile) or the 95th percentile. Looking at **Figure 5.9**, it is easy to see that any number less than 0 is a 0th percentile, any number greater than or equal to 0 and less than 1 is a 50'th percentile, and any number greater than or equal to 1 is a 100th percentile. (By convention, most textbooks and practitioners would say that 0.5 is the 50th percentile.) Similarly, looking at **Figure 5.10**, the median (50th percentile) is about 0.5 and the 95th percentile is about 1.125.

You can compute exact quantiles for all of the distributions listed in **Table 5.1** (except the stable distribution) using functions of the form `q`*`abb`*, where *`abb`* denotes the abbreviation of the distribution name (column 2 in **Table 5.1**). The first argument to these functions is the value(s) of the probabilities for which you want to compute the associated quantile(s). Subsequent arguments specify the values of the parameters associated with the probability distribution (see column 3 in **Table 5.1**).

Example 5.9: Computing Quantiles for a Binomial Distribution

Consider the binomial distribution with parameters `size=1` and `prob=0.5`, whose cdf is shown in **Figure 5.9** and whose pdf is defined in **Equation (5.1)**. You can compute quantiles from this distribution by using the function `qbinom`:

```
> qbinom(c(0, 0.25, 0.5, 0.75, 1), size = 1, prob = 0.5)
[1] 0 0 0 1 1
```

We see that the 0th, 25th, and 50th percentiles are all the same number (0), and the 75th and 100th percentiles are the number 1.

Example 5.10: Computing Quantiles for a Lognormal Distribution

As another example, you can compute quantiles of the lognormal distribution whose cdf is shown in **Figure 5.10** with the function `qlnorm.alt`:

```
> qlnorm.alt(c(0.5, 0.95), mean = 0.6, cv = 0.5)
[1] 0.5366563 1.1671907
```

We see that the median (50th percentile) is about 0.54 and the 95th percentile is about 1.17.

Characteristics of Probability Distributions

Every probability distribution has certain characteristics. Statisticians have defined various quantities to describe the characteristics of a probability distribution, such as the mean, median, mode, variance, standard deviation, coefficient of variation, skew, and kurtosis. The following sub-sections present detailed definitions of these quantities.

Median and Mode

The ***median*** of a probability distribution is simply the 50th percentile of that distribution. That is, it is the (a) number such that half of the area under the pdf is to the left of that number and half of the area is to the right. The median is a measure of the central tendency of the probability distribution.

Note that for discrete distributions percentiles are not necessarily unique so neither is the median necessarily unique. As we saw earlier, *a* median of the binomial distribution shown in **Figure 5.6** and **Figure 5.9** is 0, and *the* median of the lognormal distribution shown in **Figure 5.8** and **Figure 5.10** is 0.54.

The ***mode*** of a probability distribution is simply the quantile associated with the largest value or values of the pdf. The mode of the lognormal distribution shown in **Figure 5.8** is about 0.5. The binomial distribution shown in **Figure 5.6** does not have a unique mode.

Mean or Expected Value

With respect to a probability distribution, the terms ***mean***, ***average***, and ***expected value*** all denote the same quantity. Like the median, the mean is a measure of central tendency. If you think of a picture of a probability density function and imagine it is made out of a thin sheet of metal, the mean is the point along the bottom of the distribution (the x-axis) at which the distribution balances.

The mean of a probability distribution is often denoted by the Greek letter μ (mu). Mathematically, the mean is defined as:

$$\mu = E(X) = \int_{-\infty}^{\infty} x\, f(x)\, dx \tag{5.10}$$

for a continuous random variable and as

$$\mu = E(X) = \sum_{x} x\, f(x) \tag{5.11}$$

for a discrete random variable.

Note that the mean may be equal to a value that the random variable never assumes. For example, for the binomial distribution shown in **Figure 5.6**, the mean of the distribution is 0.5, even though the random variable can only take on the values 0 (tails) or 1 (heads).

The mean is often used as a distribution parameter to define the probability distribution (see column 3 in **Table 5.1**). See the help file **Glossary: Mean, Median, and Mode** for more information on these quantities.

Variance and Standard Deviation

The ***variance*** of a probability distribution is the average squared distance between an observed value of X and the mean μ. It is a measure of the spread of the distribution about the mean. Because the variance is an average *squared* distance, it is in squared units (e.g., ppb^2). The ***standard deviation*** is the square root of the variance and its units are the same as the original units of the random variable. The standard deviation is usually denoted by the Greek letter σ (sigma). Note that both the variance and standard deviation are always positive quantities.

Mathematically, the variance is defined as

$$\sigma^2 = Var(X) = E\left[(X - \mu)^2\right] = \int_{-\infty}^{\infty} (x - \mu)^2 f(x)\, dx \qquad (5.12)$$

for a continuous random variable and as

$$\sigma^2 = Var(X) = E\left[(X - \mu)^2\right] = \sum_x (x - \mu)^2 f(x) \qquad (5.13)$$

for a discrete random variable.

The standard deviation is often used as a distribution parameter to define the probability distribution (see column 3 in **Table 5.1**). See the help file **Glossary: Variance, Standard Deviation, and Interquartile Range** for more information on these quantities.

Coefficient of Variation

The coefficient of variation (sometimes denoted CV) of a distribution is defined as the ratio of the standard deviation to the mean. That is:

$$CV = \sigma/\mu \qquad (5.14)$$

The coefficient of variation is a measure of the spread of the distribution relative to the size of the mean. Note that since μ and σ have identical units (e.g., ppb, degrees, mg/L, etc.), the coefficient of variation is a unit-less measure. It is usually used to characterize positive, right-skewed distributions such as the lognormal distribution.

Skew and Kurtosis

In the past, the skew and kurtosis of a distribution were often used to compare the distribution to a normal (Gaussian) distribution. These quantities and their estimates are not so widely reported anymore, although they are still used to fit distributions in the system of Pearson curves (Johnson et al., 1994, pp.15-25).

Skew

The rth central moment of a random variable X is defined as:

$$\mu_r = E\left[(X - \mu)^r\right] \qquad (5.15)$$

The first central moment of any distribution is 0. The second central moment is the same thing as the variance.

The ***coefficient of skewness*** of a distribution is defined to be the third central moment scaled by the cube of the standard deviation:

$$Skew = \mu_3/\sigma^3 \tag{5.16}$$

The skew measures the distribution of values about the mean. If small values of the distribution are bunched close to the mean and large values of the distribution extend far above the mean (as in **Figure 5.8**), the skew will be positive and the distribution is said to be ***positively skewed*** or ***right-skewed***. Conversely, if small values of the distribution extend well below the mean and large values of the distribution tend to bunch close to the mean, the skew will be negative and the distribution is said to be ***negatively skewed*** or ***left-skewed***.

Kurtosis

The ***coefficient of kurtosis*** of a distribution is the fourth central moment scaled by the square of the variance:

$$Kurtosis = \mu_4/\sigma^4 \tag{5.17}$$

The kurtosis is a measure of the proportion of values in the tails (extreme ends) of the distribution. The kurtosis is independent of the mean and variance of the distribution.

For a normal distribution, the coefficient of kurtosis is 3. Distributions with kurtosis less than 3 are called ***platykurtic*** — they have shorter tails than a normal distribution. Distributions with kurtosis greater than 3 are called ***leptokurtic*** — they have heavier tails than a normal distribution. Sometimes, the term "kurtosis" is used to mean the "coefficient of excess kurtosis", which is the coefficient of kurtosis minus 3.

Generating Random Numbers From Probability Distributions

With the advance of modern computers, experiments and simulations that just a decade ago would have required an enormous amount of time to complete using large-scale computers can now be easily carried out on personal computers. Simulation is fast becoming an important tool in environmental statistics and all fields of statistics in general (see **Chapter 10: Monte Carlo Simulation and Risk Assessment**).

For all of the distributions listed in **Table 5.1** (except the stable distribution), there are functions of the form `rabb`, where *abb* denotes the abbreviation of the distribution name (column 2 in **Table 5.1**). These functions produce random numbers from the distribution. The first argument to these functions is the number of the random numbers you want to produce. Subsequent arguments specify the values of the parameters associated with the probability distribution (see column 3 in **Table 5.1**).

Example 5.11: Generating Random Numbers from a Lognormal Distribution

You can produce random numbers from the lognormal distribution shown in **Figure 5.8** by using the function `rlnorm.alt`:

```
> set.seed(243)

> rlnorm.alt(n = 4, mean = 0.6, cv = 0.5)
[1] 0.5856242 0.7944820 0.8068604 0.9684402
```

The function `set.seed` lets you set the random seed that the random number generator starts with. This way, you can reproduce all of the examples in this manual and the help files by setting the random seed to the same starting value as shown in each example.

Important Distributions in Environmental Statistics

Random variables are random because they are affected by physical, chemical, and/or biological processes that we cannot measure or control with absolute precision. Certain probability distributions are used extensively in environmental statistics because they model the outcomes of physical, chemical, and/or biological processes fairly well. The rest of this section discusses some of these distributions.

Normal Distribution

The normal or Gaussian distribution is by far the most frequently used distribution in probability and statistics. It sometimes is a good model for environmental data, but in a lot of cases the lognormal distribution or some other positively-skewed distribution provides a better model (see the next sub-section).

The normal distribution often provides a good model for data because it can be shown that this is the distribution that results when you add up or average lots of different random variables, no matter what the underlying distribution of any of these random variables. The technical name for this phenomenon is the Central Limit Theorem (see the help file **Glossary: Normal Distribution** for an example of the Central Limit Theorem in action).

The probability density function of the normal distribution is given by:

$$f(x) = \frac{1}{\sigma\sqrt{2\pi}} \exp\left[\frac{(x - \mu)^2}{\sigma^2}\right], \quad -\infty < x < \infty \tag{5.18}$$

where μ denotes the mean of the distribution and σ denotes the standard deviation. This distribution is often denoted by $N(\mu, \sigma)$.

A standard normal distribution is a normal distribution with a mean of 0 and a standard deviation of 1, denoted N(0,1). **Figure 5.11** shows the pdf of the standard normal distribution, and **Figure 5.12** shows the corresponding cdf.

```
> pdfplot()
```

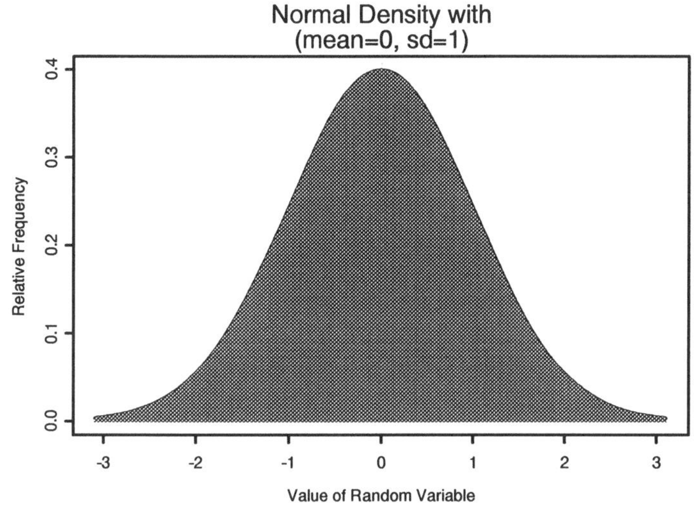

Figure 5.11: The probability density function of a standard normal distribution.

```
> cdfplot()
```

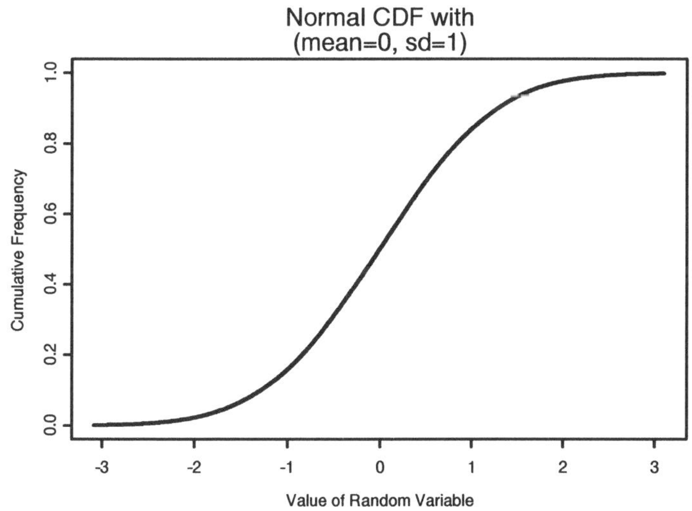

Figure 5.12: The cumulative distribution function of a standard normal distribution.

The Mean and Standard Deviation of a Normal Distribution

A normal probability distribution (probability density function) is a bell-shaped curve, regardless of the value of μ and σ. The value of μ determines where the center of the distribution lies along the *x*-axis, and the value of σ determines the spread of the distribution. **Figure 5.13** compares a N(0,1) pdf with a N(3,2) pdf.

```
> pdfplot(xlim = c(-4, 10), ylim = c(0, 0.42),
+    pdf.col = 2, curve.fill = F,
+    main = "Two Normal Distributions")

> pdfplot(param.list = list(mean = 3, sd = 2),
+    add = T, pdf.lty = 8)

> legend(2.5, 0.4, c("N(0,1)", "N(3,2)"),
+    col = 2:1, lty = c(1, 8), lwd = 2)
```

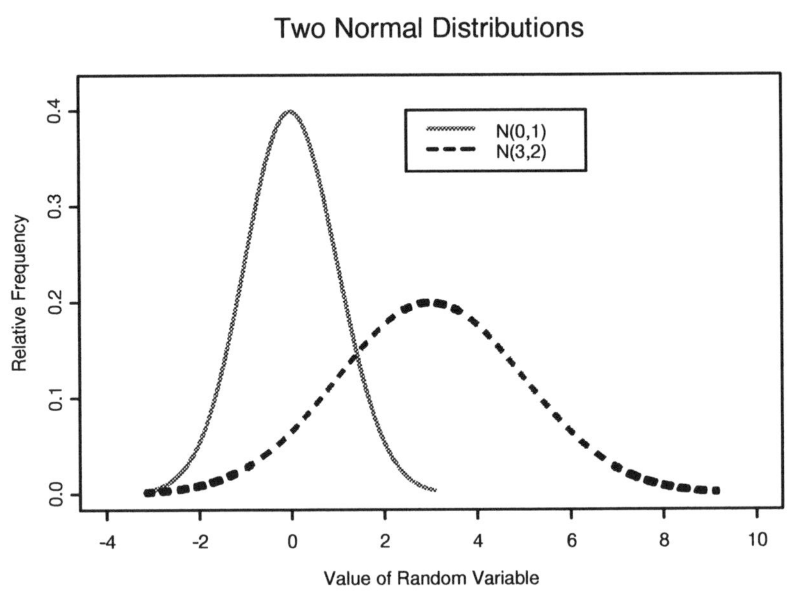

Figure 5.13: Two normal probability density functions.

Transforming Back to a Standard Normal Distribution: The Z-transformation

Any normal random variable can be transformed back to a standard normal random variable by subtracting its mean and dividing by its standard deviation. That is, if X is a random variable with a N(μ,σ) distribution, then

$$Z = \frac{X - \mu}{\sigma} \tag{5.19}$$

is a random variable with a N(0,1) distribution. This relationship is very important in the theory of confidence intervals and hypothesis tests.

Generating Quantities from a Normal Distribution

You can use the functions `dnorm`, `pnorm`, `qnorm`, and `rnorm` to generate pdf values, cdf values, quantiles, and random numbers from the normal distribution (see the S-PLUS help file for **Normal** for more information). For example, let's compute the 2.5th, 5th, 10th, 25th, 50th, 75th, 90th, 95th and 97.5th percentiles of the standard normal distribution

```
> round(qnorm(c(.025, .05, .1, .25, .5)), 2)
[1] -1.96 -1.64 -1.28 -0.67  0.00

> round(qnorm(c(.975, .95, .9, .75, .5)), 2)
[1] 1.96 1.64 1.28 0.67 0.00
```

Note that the 97.5th percentile is 1.96 and the (1 - 97.5) = 2.5th percentile is -1.96. In general, if x_p denotes the pth quantile, then

$$x_p = -x_{1-p} \tag{5.20}$$

This property is called symmetry; the normal distribution is said to be ***symmetric about its mean***.

Chapter 13 of Johnson et al. (1994) gives extensive details about the history and application of the normal distribution.

Lognormal Distribution

The two-parameter lognormal distribution is the distribution of a random variable whose logarithm is normally distributed. That is, if Y=`log(`X`)` (where `log` denotes the natural logarithm) and Y has a normal distribution with mean μ and standard deviation σ, then X is said to have a lognormal distribution. The distribution of X is often denoted as $\Lambda(\mu,\sigma)$ where μ and σ denote the mean of the *transformed* random variable Y. (**Note:** Λ denotes the upper-case Greek letter lambda.)

Figure 5.14 shows the pdf of a lognormal distribution with parameters μ=0 and σ=1; that is, X=`exp(`Z`)`, where Z is a standard normal random variable. **Figure 5.15** shows the corresponding cdf.

```
> pdfplot("lnorm", list(meanlog = 0, sdlog = 1),
+    right.tail.cutoff = .01)
```

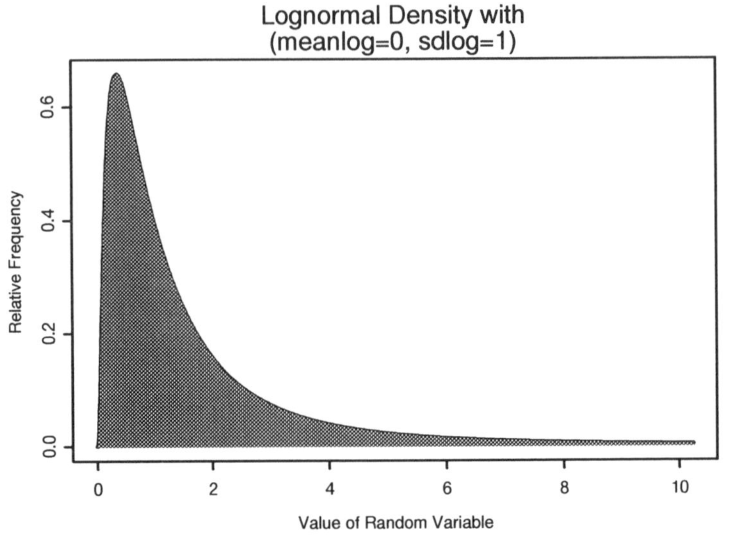

Figure 5.14: The probability density function of a lognormal distribution.

```
> cdfplot("lnorm", list(meanlog = 0, sdlog = 1),
+    right.tail.cutoff = .01)
```

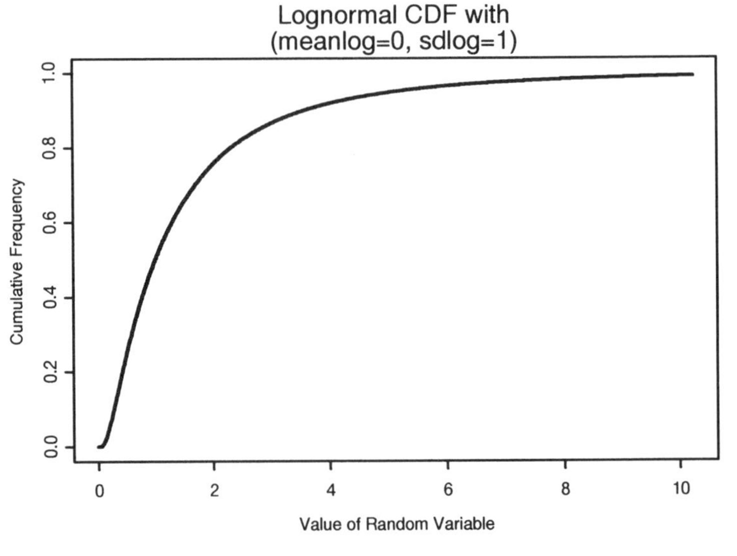

Figure 5.15: The cumulative distribution function of a lognormal distribution.

The two major characteristics of the lognormal distribution (probability density function) are that it is bounded below at 0 and it is skewed to the right. These characteristics are typical of environmental data. Several authors have illustrated various mechanisms to explain the occurrence of the lognormal distribution in nature (e.g., Ott, 1990, Chapters 8-9). The basic idea is that since the Central Limit Theorem states that the result of summing up random variables will produce a normal distribution, multiplying random variables will produce a lognormal distribution (recall that $\log(ab) = \log(a) + \log(b)$).

The probability density function of a lognormal distribution is given by:

$$f(x) = \frac{1}{x\sigma\sqrt{2\pi}} \exp\left\{-\frac{1}{2\sigma^2}\left[\log(x) - \mu\right]^2\right\}, \quad x > 0 \tag{5.21}$$

Alternative Parameterization: Mean and CV

Sometimes, a lognormal distribution is characterized by its mean and coefficient of variation rather than the mean and standard deviation of the log-transformed random variable. Let $Y=\log(X)$ be distributed as a $N(\mu,\sigma)$ random variable, so that X is distributed as a $\Lambda(\mu,\sigma)$ random variable. Let θ (theta) denote the mean of X and let τ (tau) denote the coefficient of variation of X. The relationship between μ, σ, θ, and τ is given by:

$$\begin{aligned} \theta &= E(X) = \exp\left(\mu + \frac{\sigma^2}{2}\right) \\ \tau &= CV(X) = \left[\exp\left(\sigma^2\right) - 1\right]^{1/2} \\ \mu &= E(Y) = \log\left(\frac{\theta}{\sqrt{\tau^2 + 1}}\right) \\ \sigma &= \sqrt{Var(Y)} = \left[\log\left(\tau^2 + 1\right)\right]^{1/2} \end{aligned} \tag{5.22}$$

There are two very important points to note about these relationships. The first is that even though $X=\exp(Y)$, it is *not* true that $\theta=\exp(\mu)$. In fact, the quantity $\exp(\mu)$ is the *median* of the lognormal distribution, so the median is always smaller than the mean. The second point is that τ, the coefficient of variation of X, only depends on σ, the standard deviation of Y. Large values of τ correspond to very skewed distributions. As τ gets small, the distribution becomes less skewed and starts to resemble a normal distribution.

Figure 5.16 shows two different lognormal distributions characterized by the mean (θ) and cv (τ).

```
> pdfplot('lnorm.alt', list(mean = 2, cv = 0.5),
+    curve.fill = F, pdf.col = 2,
+    main = "Two Lognormal Distributions")

> pdfplot('lnorm.alt', list(mean = 4, cv = 2),
+    pdf.lty = 8, add = T)

> text(2.5, 0.4, "mean = 2\ncv = 0.5", adj = 0)

> text(6, 0.1, "mean = 4\ncv = 2", adj = 0)
```

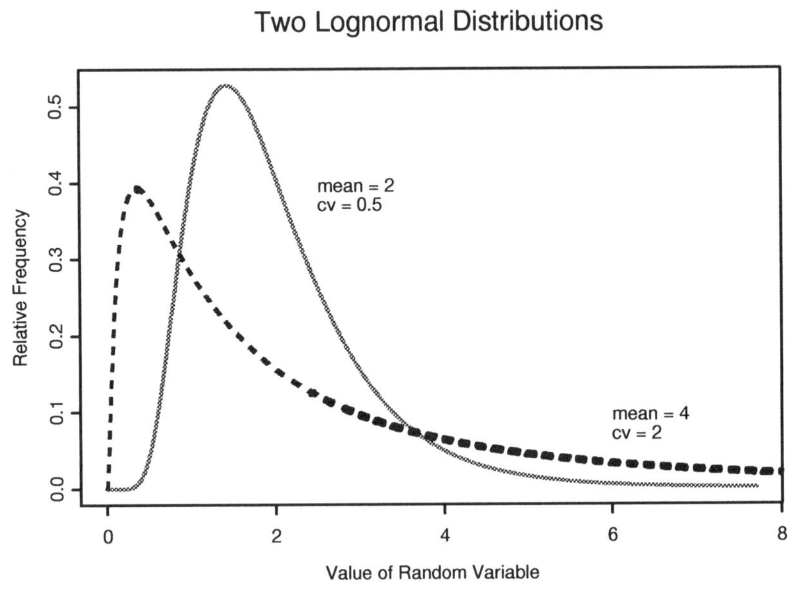

Figure 5.16: The probability density functions of two lognormal distributions.

Generating Quantities from a Lognormal Distribution

You can use the functions `dlnorm`, `plnorm`, `qlnrom`, and `rlnorm` to generate pdf values, cdf values, quantiles, and random numbers from the lognormal distribution, where the distribution is characterized by the mean and standard deviation of the log-transformed random variable. Similarly, you can use the functions `dlnorm.alt`, `plnorm.alt`, `qlnrom.alt`, and `rlnorm.alt` to generate pdf values, cdf values, quantiles, and random numbers from the lognormal distribution, where the distribution is characterized by the alternative parameterization, the mean and coefficient of variation of the original random variable. See the help file for **Lognormal** for more information.

For example, consider a lognormal distribution with mean θ=4 and coefficient of variation τ=2. By **Equation (5.22)**, the mean and standard deviation of the log-transformed random variable are given as follows:

```
> theta <- 4

> tau <- 2

> mu <- log(theta / sqrt(tau^2 + 1))

> sigma <- sqrt( log(tau^2 + 1) )

> mu
[1] 0.5815754

> sigma
[1] 1.268636
```

Let's confirm that the quantity exp(μ) is in fact the median of this lognormal distribution:

```
> exp(mu)
[1] 1.788854

> qlnorm(0.5, meanlog = mu, sdlog = sigma)
[1] 1.788854

> qlnorm.alt(0.5, mean = 4, cv = 2)
[1] 1.788854
```

so the median is about 1.8, while the mean is 4 (see **Figure 5.16**).

Now let's determine what quantile the mean corresponds to:

```
> plnorm.alt(4, mean = 4, cv = 2)
[1] 0.7370634
```

so the mean is about the 74th percentile.

Three-Parameter Lognormal Distribution

The two-parameter lognormal distribution is bounded below at 0. The three-parameter lognormal distribution includes a threshold parameter γ (gamma) that determines the lower boundary of the random variable. That is, if Y=log(X-γ) has a normal distribution with mean μ and standard deviation σ, then X is said to have a three-parameter lognormal distribution. The threshold parameter γ affects only the location of the three-parameter lognormal distribution; it has no effect on the variance or the shape of the distribution. Note that when γ=0, the three-parameter lognormal distribution reduces to the two-parameter lognormal distribution. The three-parameter lognormal distribution is sometimes used in hydrology to model extreme events (Stedinger et al., 1993).

```
> pdfplot('lnorm3', list(meanlog=0, sdlog=1, threshold=5),
+    right.tail.cutoff = 0.01)
```

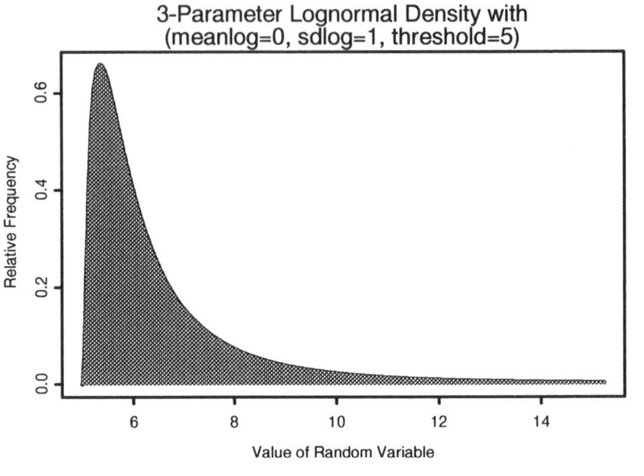

Figure 5.17: The probability density function of a three-parameter lognormal distribution.

```
> cdfplot('lnorm3', list(meanlog=0, sdlog=1, threshold=5),
+    right.tail.cutoff = 0.01)
```

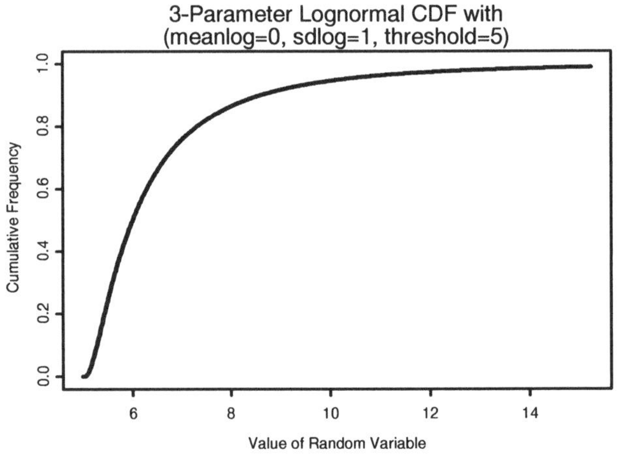

Figure 5.18: The cumulative distribution function of a three-parameter lognormal distribution.

Figure 5.17 shows the pdf of a three-parameter lognormal distribution with parameters $\mu=0$, $\sigma=1$, and $\gamma=5$. **Figure 5.18** shows the corresponding cdf. Compare these figures to **Figure 5.14** and **Figure 5.15**. See the ENVIRONMENTALSTATS for S-PLUS help file for **Lognormal3** for more information on this distribution.

Binomial Distribution

After the normal distribution, the binomial distribution is one of the most frequently used distributions in probability and statistics. It is used to model the number of occurrences of a specific event in n independent trials. The outcome for each trial is binary: yes/no, success/failure, 1/0, etc. The binomial random variable X represents the number of "successes" out of the n trials.

In environmental monitoring, sometimes the binomial distribution is used to model the proportion of observations of a pollutant that exceed some ambient or cleanup standard, or to compare the proportion of detected values at background and compliance units (USEPA, 1989a, Chapters 7-8; USEPA, 1989b, Chapter 8; USEPA, 1992b, p.5-29; Ott, 1995, Chapter 4).

The probability density (mass) function of a binomial random variable X is given by:

$$f(x) = \binom{n}{x} p^x (1-p)^{n-x}, \quad x = 0, 1, 2, \ldots, n \tag{5.23}$$

where n denotes the number of trials and p denotes the probability of "success" for each trial. It is common notation to say that X has a $B(n,p)$ distribution.

The first quantity on the right-hand side of **Equation (5.23)** is called the binomial coefficient. It represents the number of different ways you can arrange the x "successes" to occur in the n trials. The formula for the binomial coefficient is:

$$\binom{n}{x} = \frac{n!}{x!(n-x)!} \tag{5.24}$$

The quantity $n!$ is called "n factorial" and is the product of all of the integers between 1 and n. That is,

$$n! = n(n-1)(n-2)\cdots 2\,1 \tag{5.25}$$

The S-PLUS functions `choose` and `fact` compute the binomial coefficient and factorials, respectively. These functions have been modified in ENVIRONMENTALSTATS for S-PLUS to accept vectors as arguments and to accept missing values.

Figure 5.6 above shows the pdf of a B(1,0.5) random variable and **Figure 5.9** shows the associated cdf. **Figure 5.19** and **Figure 5.20** below show the pdf's of a B(10, 0.5) and B(10, 0.2) random variable, respectively.

```
> pdfplot("binom", list(size = 10, prob = 0.5))
```

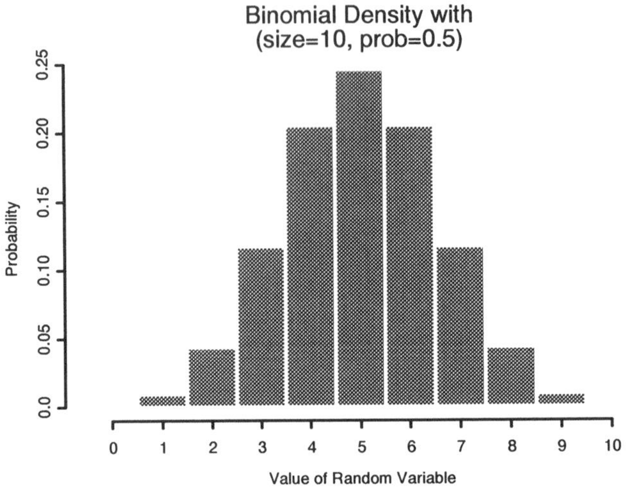

Figure 5.19: The probability density function of a B(10, 0.5) random variable.

```
> pdfplot("binom", list(size = 10, prob = 0.2))
```

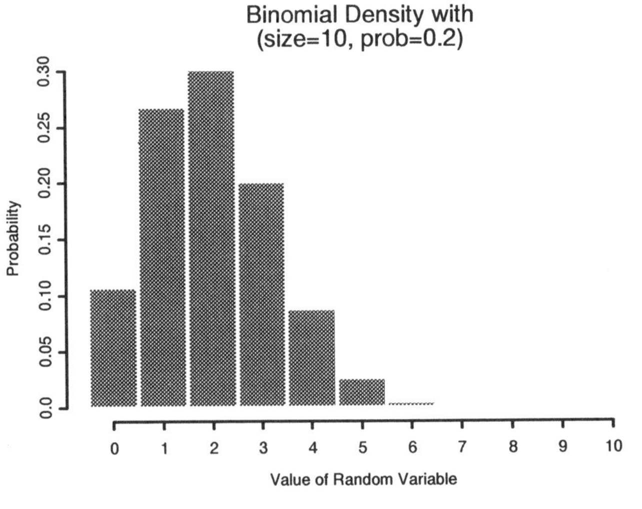

Figure 5.20: The probability density function of a B(10, 0.2) random variable.

The Mean and Variance of the Binomial Distribution

The mean and variance of a random variable X with a $B(n,p)$ distribution are given by:

$$E(X) = np \tag{5.26}$$

$$Var(X) = np(1 - p)$$

The average number of successes in n trials is simply the probability of a success for one trial multiplied by the number of trials. The variance depends on the probability of success. **Figure 5.21** shows the function $p(1-p)$ as a function of p. The variance of a binomial random variable is greatest when the probability of success is ½, and the variance decreases to 0 as the probability of success decreases to 0 or increases to 1.

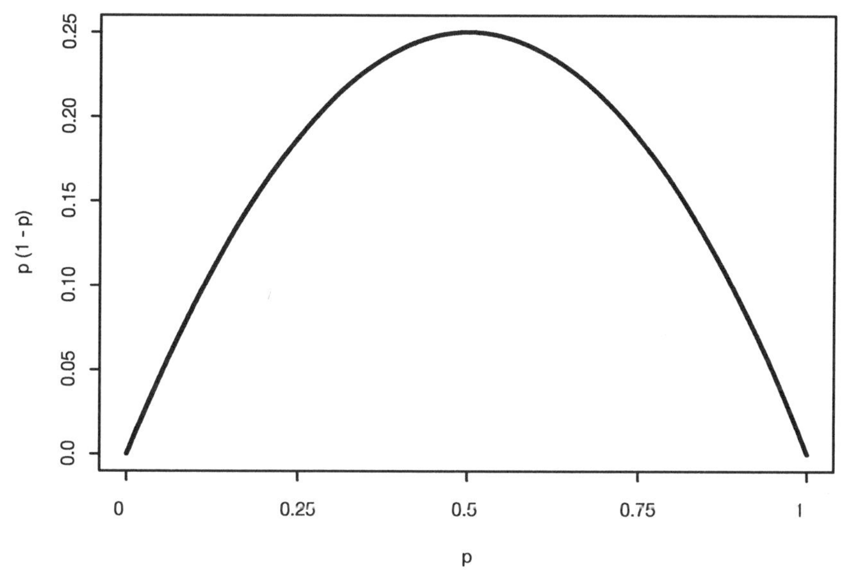

Figure 5.21: The variance of a $B(1, p)$ random variable as a function of p.

Generating Quantities from a Binomial Distribution

You can use the functions `dbinom`, `pbinom`, `qbinom`, and `rbinom` to generate pdf values, cdf values, quantiles, and random numbers from the binomial distribution (see the S-PLUS help file for **Binomial** for more information). For example, let's compute the probability of getting exactly 5 heads out of 10 flips when we use a "fair" coin:

```
> dbinom(5, size = 10, prob = 0.5)
[1] 0.2460938
```

So there is about a 25% chance of this happening. What's the probability of getting at least 5 heads?

```
> 1 - pbinom(4, size = 10, prob = 0.5)
[1] 0.6230469
```

There is about a 62% chance of getting at least 5 heads out of 10 flips with a fair coin.

See the ENVIRONMENTALSTATS for S-PLUS help file **Glossary: Binomial Distribution** for more information on this distribution.

Poisson Distribution

The Poisson distribution is named after its discoverer, who derived this distribution as the limiting distribution of a B(n,p) distribution, where n tends to infinity, p tends to 0, and np stays constant. This distribution is useful for modeling the number of occurrences of a "rare" event (e.g., radioactive particle emissions, number of cars arriving at an intersection, number of people getting in line, etc.) within a given time period, or the number of "rare" objects (e.g., plants, animals, microorganisms, etc.) within a specific geographic area.

In environmental monitoring, the Poisson distribution has been used to model the number of times a pollution standard is violated over a given time period (Ott, 1995, Chapter 5). Gibbons (1987b) used the Poisson distribution to model the number of detected compounds per scan of the 32 volatile organic priority pollutants (VOC), and also to model the distribution of chemical concentration (in ppb).

The probability density (mass) function of a Poisson random variable is given by:

$$f(x) = \frac{\lambda^x e^{-\lambda}}{x!}, \quad x = 0, 1, 2, \ldots \tag{5.27}$$

where λ (lambda) denotes the average number of events that occur within the given time period. This distribution is often denoted by Poisson(λ) or P(λ).

Figure 5.22 and **Figure 5.23** show the pdf's of a Poisson(1) and Poisson(3) random variable, respectively.

```
> pdfplot("pois", list(lambda = 1), ylim = c(0, 0.4))
```

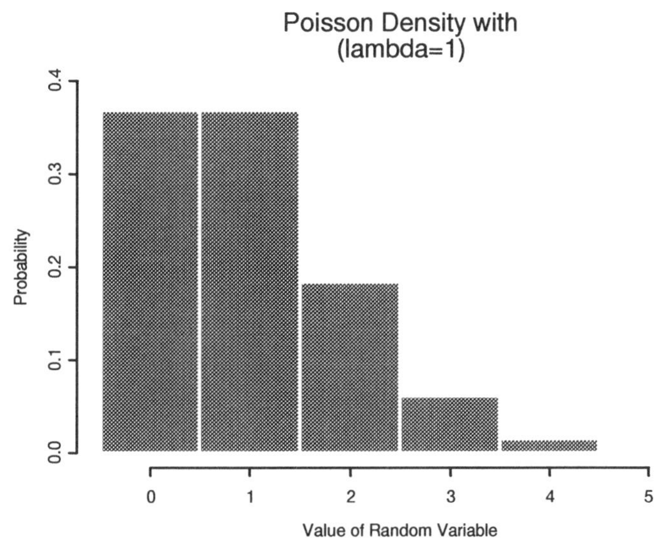

Figure 5.22: The probability density function of a Poisson(1) random variable.

```
> pdfplot("pois", list(lambda = 3), ylim = c(0, 0.25))
```

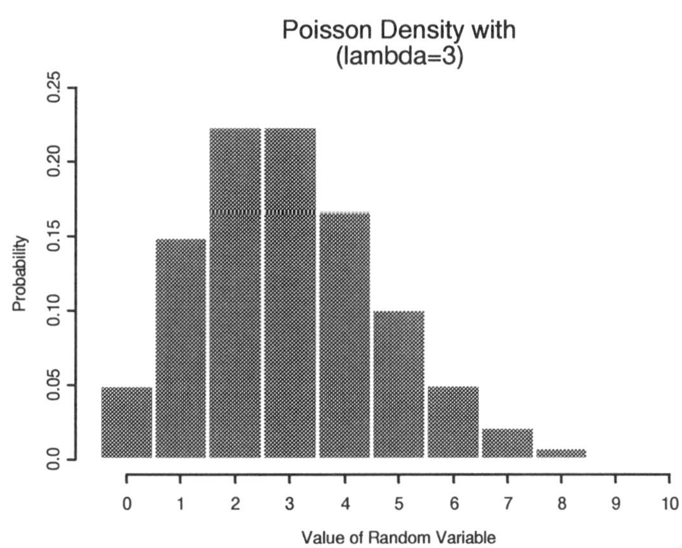

Figure 5.23: The probability density function of a Poisson(3) random variable.

The Mean and Variance of the Poisson Distribution

The mean and variance of a random variable X with a Poisson(λ) distribution are given by:

$$E(X) = \lambda \tag{5.28}$$

$$Var(X) = \lambda$$

Note that the variance depends upon (in fact is exactly equal to) the mean. Thus, the larger the mean, the more variable the distribution.

Generating Quantities from a Poisson Distribution

You can use the functions `dpois`, `ppois`, `qpois`, and `rpois` to generate pdf values, cdf values, quantiles, and random numbers from the Poisson distribution (see the S-PLUS help file for **Poisson** for more information). For example, let's compute the probabilities of seeing exactly 0, 1, 2, ..., 10 events for a Poisson(3) random variable:

```
> round(dpois(0:10, lambda = 3), 2)
 [1] 0.05 0.15 0.22 0.22 0.17 0.10 0.05 0.02 0.01
 [9] 0.00 0.00
```

These numbers correspond to the heights of the bars in **Figure 5.23**. As another example, let's compute the probability of seeing at least one event for a Poisson(1) and a Poisson(3) random variable:

```
> 1 - ppois(0, lambda = c(1,3))
[1] 0.6321206 0.9502129
```

So there is about a 63% chance of seeing at least one event when there is an average of one event per time period, and a 95% chance of seeing at least on event when there is an average of three events per time period.

Extreme Value Distribution

The extreme value distribution is used to model extreme values (maxima or minima) of some process or series of measurements, such as daily maximum temperatures, daily maximum concentrations of air pollutants (e.g., Kuchenhoff and Thamerus, 1996), and annual maximum of daily precipitation (e.g., Stedinger et al., 1993). The extreme value distribution gets its name from the fact that this distribution is the limiting distribution (as n approaches infinity) of the greatest (smallest) value among n independent random variables each having the same continuous distribution.

The probability density function of a random variable X with a two-parameter extreme value distribution is given by:

$$f(x) = \frac{1}{\theta} e^{-(x-\eta)/\theta} \exp\left[- e^{-(x-\eta)/\theta} \right], \tag{5.29}$$

$$-\infty < x < \infty$$
$$-\infty < \eta < \infty$$
$$\theta > 0$$

where η (eta) denotes the location parameter and θ (theta) denotes the scale parameter. This distribution is the *largest* extreme value distribution; it is used to model maxima. A different pdf is used for the smallest extreme value distribution.

There are three families of extreme value distributions. The one described here is the Type I, also called the Gumbel extreme value distribution or simply ***Gumbel distribution***. This distribution is sometimes denoted EV(η,θ).

Figure 5.24 shows the pdf of an EV(10,1) distribution and **Figure 5.25** shows the corresponding cdf.

```
> pdfplot("evd", list(location=10, scale=1),
+   left.tail.cutoff = .0001)
```

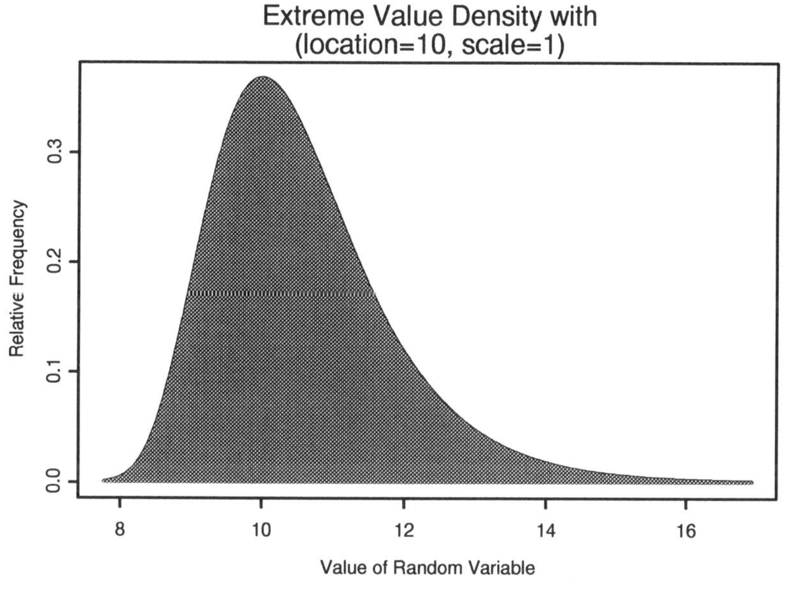

Figure 5.24: The probability density function of a two-parameter extreme value distribution.

```
> cdfplot("evd", list(location=10, scale=1),
+   left.tail.cutoff = .0001)
```

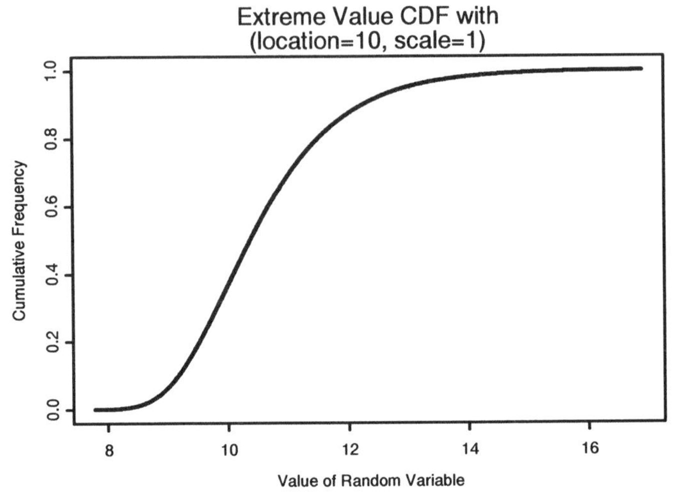

Figure 5.25: The cumulative distribution function of a two-parameter extreme value distribution.

The Mode, Mean, and Variance of an Extreme Value Distribution

The mode, mean, and variance of the two-parameter extreme value distribution are:

$$Mode(X) = \eta \tag{5.30}$$

$$E(X) = \eta + \varepsilon\,\theta$$

$$Var(X) = \frac{\theta^2\pi^2}{6}$$

where ε denotes a number called Euler's constant and is approximately equal to

```
> Eulers.constant
[1] 0.5772157
```

(see the help file for `Eulers.contant`).

Generating Quantities from an Extreme Value Distribution

You can use the ENVIRONMENTALSTATS for S-PLUS functions `devd`, `pevd`, `qevd`, and `revd` to generate pdf values, cdf values, quantiles, and random numbers from the two-parameter extreme value distribution. For example, let's find the 10th, 50th, and 90th percentiles of an EV(10,1) distribution:

```
> qevd(c(.1,.5,.9), location = 10, scale = 1)
[1]  9.165968 10.366513 12.250367
```

See the ENVIRONMENTALSTATS for S-PLUS help file for **EVD** for more information on the extreme value distribution.

Generalized Extreme Value Distribution

A generalized extreme value (GEV) distribution includes a shape parameter κ (kappa), as well as the location and scale parameters η and θ. The three families of extreme value distributions are all special kinds of GEV distributions. When the shape parameter $\kappa=0$, the GEVD reduces to the Type I extreme value (Gumbel) distribution. When $\kappa > 0$, the GEVD is the same as the Type II extreme value distribution, and when $\kappa < 0$ it is the same as the Type III extreme value distribution. See the ENVIRONMENTALSTATS for S-PLUS help file **GEVD** for more information on this distribution.

Mixture Distributions

A mixture distribution results when some portion of the population comes from one probability distribution and the rest of the population comes from another distribution. Mixture distributions are often used to model data sets that contain a few or several gross "outliers"; that is, most of the observations appear to come from one distribution, but some appear to come from a distribution with a shifted mean and possibly shifted standard deviation.

In environmental statistics, mixture distributions are useful for modeling data from a remediated site that contains residual contamination. Gilliom and Helsel (1986) used a lognormal mixture distribution (as well as several others) to assess the performance of various estimators of population parameters.

ENVIRONMENTALSTATS for S-PLUS contains functions for computing the pdf, cdf, quantiles, and random numbers from the mixture of two normal distributions or the mixture of two lognormal distributions (see the help files for **NormalMix** and **LognormalMix**). **Figure 5.26** shows the pdf of a normal mixture distribution where the first distribution is a N(5,1), the second distribution is a N(10,2), and the mixing proportion is 30% (i.e., 70% of the first distribution and 30% of the second distribution).

```
> pdfplot("norm.mix", list(mean1 = 5, sd1 = 1,
+    mean2 = 10, sd2 = 2, p.mix = 0.3))
```

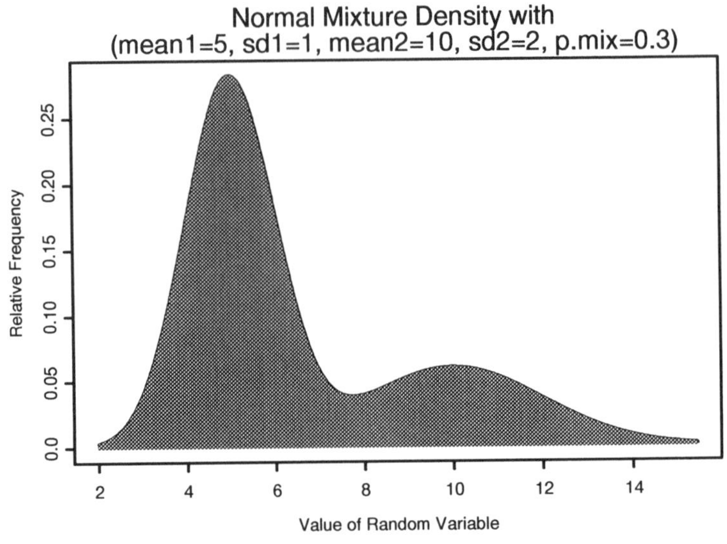

Figure 5.26: The probability density function of a normal mixture distribution.

Zero-Modified Distributions

A zero-modified distribution is a distribution that has been modified to put extra probability mass at 0. That is, it is a special kind of mixture distribution in which a portion of the population comes from the original distribution, but the rest of the population is all zeros (Johnson et al., 1992, p.312). In environmental statistics, the zero-modified lognormal distribution, also called the ***delta distribution***, is sometimes used as a model for data sets with nondetects, where the nondetects are assumed to actually be 0 values (e.g., Owen and DeRouen, 1980). USEPA (1992c, pp.28-34) contains an example where the data appear to come from a zero-modified normal distribution.

ENVIRONMENTALSTATS for S-PLUS contains functions for computing the pdf, cdf, quantiles, and random numbers from the zero-modified normal and zero-modified lognormal distributions (see the help files for **Zero-Modified Normal** and **Zero-Modified Lognormal**). **Figure 5.27** shows the pdf of a zero-modified lognormal distribution with a mean of 3 and a cv of 0.5 for the lognormal part of the distribution, and a probability mass of 20% at 0.

```
> pdfplot("zmlnorm.alt", list(mean = 3, cv = 0.5,
+   p.zero = 0.2))
```

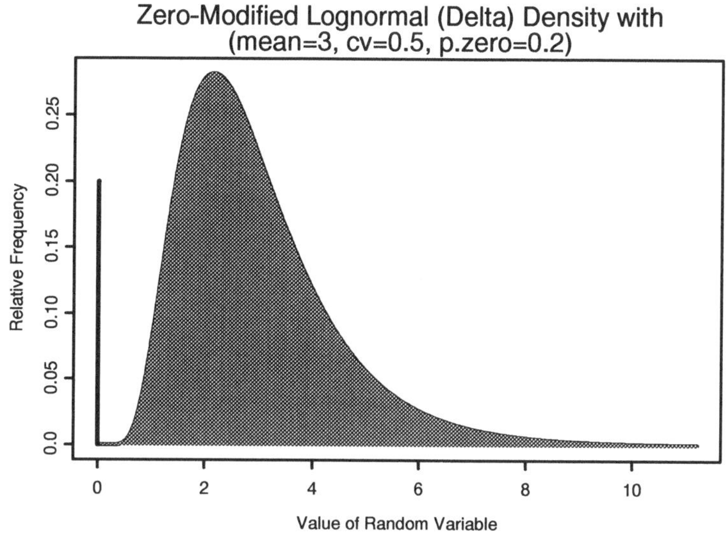

Figure 5.27: The probability density function of a delta distribution.

Summary

- **Figure 5.1** shows examples of probability density functions for all of the probability distributions available in S-PLUS and ENVIRONMENTALSTATS for S-PLUS.

- **Table 5.1** lists all of the probability distributions available in S-PLUS and ENVIRONMENTALSTATS for S-PLUS. For each of these distributions, there are functions for computing the probability density function (pdf), the cumulative distribution function (cdf), quantiles, and random numbers. The form of the names of these functions are d*abb*, p*abb*, q*abb*, and r*abb*, where *abb* denotes the abbreviation of the distribution name (see column 2 of the table).

- **Table 5.2** lists the functions available in ENVIRONMENTALSTATS for S-PLUS to create plots associated with theoretical and empirical probability distributions (see the help file **Plotting Probability Distributions**). The function `pdfplot` plots probability density functions, and the function `cdfplot` plots cumulative distribution functions.

- A ***random variable*** is the measurement associated with the next physical sample from the population. Outcomes of a random variable usually differ in value from observation to observation.

- A ***discrete*** random variable can take on only a finite or countably infinite number of values. A ***continuous*** random variable can take on an infinite number of values.

- A ***probability density function (pdf)*** is a mathematical formula that describes the relative frequency of a random variable. Sometimes the picture of this formula (e.g., **Figure 5.6**) is called the pdf. If a random variable is discrete, its probability density function is sometimes called a ***probability mass function***.

- The ***cumulative distribution function (cdf)*** of a random variable X is the probability that the random variable X is less than or equal to some number x.

- The pth ***quantile***, also called the $100p^{th}$ ***percentile***, is the (a) number such that $100p\%$ of the distribution lies below this number.

- Every probability distribution has certain characteristics. Statisticians have defined various quantities to describe the characteristics of a probability distribution, such as the mean, median, mode, variance, standard deviation, coefficient of variation, skew, and kurtosis.

- Important probability distributions that are used in environmental statistics include the normal, lognormal, binomial, Poisson, and extreme value distributions, as well as mixture and zero-modified distributions.

CHAPTER 6 :

Estimating Distribution Parameters and Quantiles

Chapter 3 discussed the ideas of a population and a sample. Chapter 5 described probability distributions, which are used to model populations. Based on using the tools discussed in Chapter 4 to look at your data, and based on knowledge of the mechanism producing the data, you can model the data from your sampling program as having come from a particular kind of probability distribution. Once you decide on what probability distribution to use (if any), you'll need to estimate the parameters associated with that distribution. This chapter discusses the tools available in ENVIRONMENTALSTATS for S-PLUS to let you estimate the parameters and quantiles of a probability distribution, and to optionally construct confidence intervals for the parameters or quantiles.

Overview

Table 6.1 lists the probability distributions in ENVIRONMENTALSTATS for S-PLUS that have associated functions for estimating the distribution parameters (see the help file **Estimating Distribution Parameters and Quantiles**). For each of these distributions, the form of the name of the estimating function is `e`*`abb`*, where *`abb`* denotes the abbreviation of the distribution name (see column 2 of **Table 5.1**). For example, the function `enorm` estimates the mean and standard deviation of a normal distribution based on a sample of observations.

The help file for `.Distribution.frame` contains a more extensive table that includes the distribution name, abbreviation, type (continuous, discrete, finite discrete, mixed), range (i.e., support), parameters, default values for the parameters, parameter ranges, and estimation methods available for the parameters.

Distribution Name	Estimation Function	Parameter(s)
Beta	ebeta	shape1, shape2, ncp
Binomial	ebinom	size, prob
Exponential	eexp	rate
Extreme Value	eevd	location, scale
Gamma	egamma	shape, scale
Generalized Extreme Value	egevd	location, scale, shape
Geometric	egeom	prob
Hypergeometric	ehyper	m, n, k
Logistic	elogis	location, scale
Lognormal	elnorm	meanlog, sdlog
Lognormal (alternative parameterization)	elnorm.alt	mean, cv
Three-Parameter Lognormal	elnorm3	meanlog, sdlog, threshold
Negative Binomial	enbinom	size, prob
Normal	enorm	mean, sd
Poisson	epois	lambda
Uniform	eunif	min, max
Weibull	eweibull	shape, scale

Table 6.1: Functions to estimate probability distribution parameters.

Distribution Name	Estimation Function	Parameter(s)
Zero-Modified Lognormal (Delta)	`ezmlnorm`	`meanlog, sdlog, p.zero`
Zero-Modified Lognormal (Delta) (alternative parameterization)	`ezmlnorm.alt`	`mean, cv, p.zero`
Zero-Modified Normal	`ezmnorm`	`mean, sd, p.zero`

Table 6.1 (continued): Functions to estimate probability distribution parameters.

Table 6.2 lists the functions available in ENVIRONMENTALSTATS for S-PLUS to estimate quantiles for selected probability distributions. In fact, once you have estimated the parameters of a distribution, you can easily estimate the quantiles of the distribution by using the relevant q*abb* function and plugging in the estimated parameters. The functions listed in **Table 6.2**, however, also allow you to construct confidence intervals for the quantiles.

Distribution Name	Function to Estimate Quantiles
Lognormal	`eqlnorm`
Normal	`eqnorm`
Nonparametric	`eqnpar`
Poisson	`eqpois`

Table 6.2: Functions for estimating and constructing confidence intervals for quantiles.

The S-PLUS function `quantile` allows you estimate the quantiles of a distribution nonparametrically; that is, you don't have to assume any sort of probability distribution. The ENVIRONMENTALSTATS for S-PLUS function `eqnpar` uses the `quantile` function to estimate quantiles, and also allows you to construct confidence intervals for quantiles.

The rest of this chapter gives a brief discussion of the theory of estimating distribution parameters and quantiles and constructing confidence intervals, and shows you how to use the estimating functions in ENVIRONMENTALSTATS for S-PLUS.

Methods for Estimating Distribution Parameters

The three most frequently used methods for estimating the parameters of probability distributions are: the method of moments estimator (MME), the maximum likelihood estimator (MLE), and the minimum variance unbiased estimator (MVUE). More recently, *L*-moment estimators have been introduced and used in the hydrological literature. Often these methods produces slightly different formulas for the estimators, but sometimes they produce the same formula. This section briefly explains the ideas behind MMEs, MLEs, and MVUEs. *L*-moments are explained in the help file **Glossary: *L*-moments.**

Note: Throughout this section, we'll assume that X denotes a random variable and x denotes the actual value of this random variable based on a sample. Thus, x_1, x_2, ..., x_n denote n observations from a particular probability distribution.

Method of Moment Estimators

Method of moment estimators (MMEs) are based on the simple idea of equating the sample moments based on the data with the moments of the probability distribution (see the help file **Glossary: Moments**).

The rth Moment of a Distribution

The ***rth* moment** of a random variable X is often denoted by:

$$\mu'_r = E\left(X^r\right) \tag{6.1}$$

(e.g., Evans et al., 1993, p.13). The first moment of X is called the ***mean*** of X and is usually denoted simply by μ. That is:

$$\mu = \mu'_1 = E(X) \tag{6.2}$$

The formula for the ***rth* sample moment** is:

$$\hat{\mu}'_r = \frac{1}{n}\sum_{i=1}^{n} x_i^r \tag{6.3}$$

Thus, the method of moments estimator of the mean of a distribution is:

$$\hat{\mu}_{mme} = \frac{1}{n}\sum_{i=1}^{n} x_i = \bar{x} \tag{6.4}$$

The quantity

$$\bar{x} = \frac{1}{n} \sum_{i=1}^{n} x_i \tag{6.5}$$

is called the ***sample mean***.

The rth Central Moment of a Distribution

The ***r^th central moment*** of a random variable X is often denoted by:

$$\mu_r = E\left[(X - \mu)^r\right] \tag{6.6}$$

(e.g., Evans et al., 1993, p.13).

The second central moment of X is called the ***variance*** of X and is usually denoted by σ^2. That is:

$$\sigma^2 = \mu_2 = E\left[(X - \mu)^2\right] \tag{6.7}$$

The square root of the second central moment is called the ***standard deviation*** of X and is usually denoted by σ.

The formula for the ***r^th sample central moment*** is:

$$\hat{\mu}_r = \frac{1}{n} \sum_{i=1}^{n} (x_i - \bar{x})^r \tag{6.8}$$

Thus, the method of moments estimator of the variance of a distribution is:

$$\hat{\sigma}^2_{mme} = \frac{1}{n} \sum_{i=1}^{n} (x_i - \bar{x})^2 = s^2_m \tag{6.9}$$

and the method of moments estimator of the standard deviation of a distribution is:

$$\hat{\sigma}_{mme} = \sqrt{\frac{1}{n} \sum_{i=1}^{n} (x_i - \bar{x})^2} = s_m \tag{6.10}$$

The quantity

$$s_m^2 = \frac{1}{n} \sum_{i=1}^{n} \left(x_i - \bar{x}\right)^2 \qquad (6.11)$$

is sometimes called the ***sample variance***, but more often the term "sample variance" refers to the unbiased estimate of variance shown in **Equation (6.20)** below.

Similarly, the quantity s_m is sometimes called the ***sample standard deviation***, but more often the term "sample standard deviation" refers to the standard deviation based on the unbiased estimator of variance.

Example 6.1: Method of Moments Estimators for the Parameters of a Normal Distribution

A normal distribution is characterized by its mean (μ) and variance (σ^2) or standard deviation (σ). The method of moments estimators for the mean, variance, and standard deviation are given by

$$\hat{\mu}_{mme} = \bar{x}$$

$$\hat{\sigma}^2_{mme} = s_m^2 \qquad (6.12)$$

$$\hat{\sigma}_{mme} = s_m$$

(See the help file for `enorm`.)

Example 6.2: Method of Moments Estimator for a Binomial Proportion

A binomial distribution is characterized by two parameters: n, the number of trials, and p, the probability of success for each trial. The single random variable X denotes the number of successes in n trials. The parameter n is set in advance, so the only parameter to estimate is p. As we saw in Chapter 5, if X has a B(n,p) distribution, then the mean of X is

$$E(X) = \mu = np \qquad (6.13)$$

so the method of moments estimator of p is

$$\hat{p}_{mme} = \frac{x}{n} \qquad (6.14)$$

The random variable x can be thought of as the sum of n independent B(1,p) random variables. That is

$$x = \sum_{i=1}^{n} x_i \tag{6.15}$$

where x_i denotes the outcome for the ith trial (1=success, 0=failure). Using this notation, the method of moments estimator of p can be written as

$$\hat{p}_{mme} = \frac{x}{n} = \frac{\sum_{i=1}^{n} x_i}{n} = \bar{x} \tag{6.16}$$

See the help file for `ebinom` for more information.

Maximum Likelihood Estimators

A maximum likelihood estimator (MLE) is based on the simple idea of using the parameter that yields the largest probability of seeing that particular outcome. For example, if you flip a (possibly weighted) coin 1000 times and record 900 heads and 100 tails, does it seem more likely that the probability of a head is 0.5 or 0.9? If in fact the probability of a head is 0.5, the likelihood of seeing exactly 900 heads out of 1000 flips is

```
> dbinom(x=900, size=1000, p=0.5)
[1] 5.958936e-162
```

On the other hand, if the probability of a head is 0.9, the likelihood of seeing exactly 900 heads out of 1000 flips is

```
> dbinom(x=900, size=1000, p=0.9)
[1] 0.04201679
```

Figure 6.1 and **Figure 6.2** show the probability densities of a B(1000, 0.5) and a B(1000, 0.9), respectively. You can see from these figures that getting something close to 900 heads out of 1000 flips is much more likely for the B(1000, 0.9) distribution.

```
> pdfplot("binom", list(size=1000, prob=0.5),
+   left=0.001, right=0.001)
```

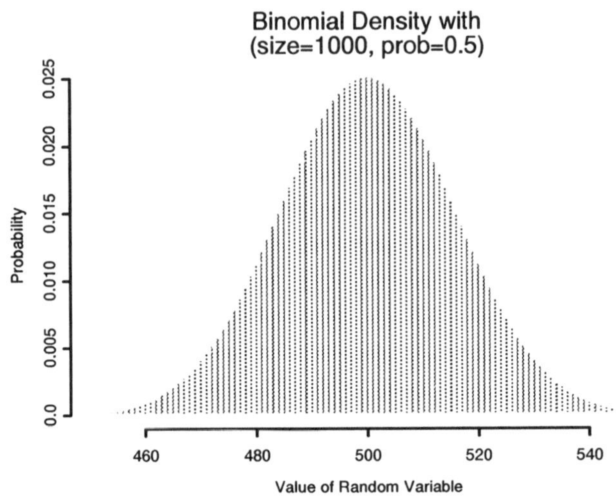

Figure 6.1: Probability density of a B(1000, 0.5) random variable.

```
> pdfplot("binom", list(size=1000, prob=0.9),
+   left=0.001, right=0.001)
```

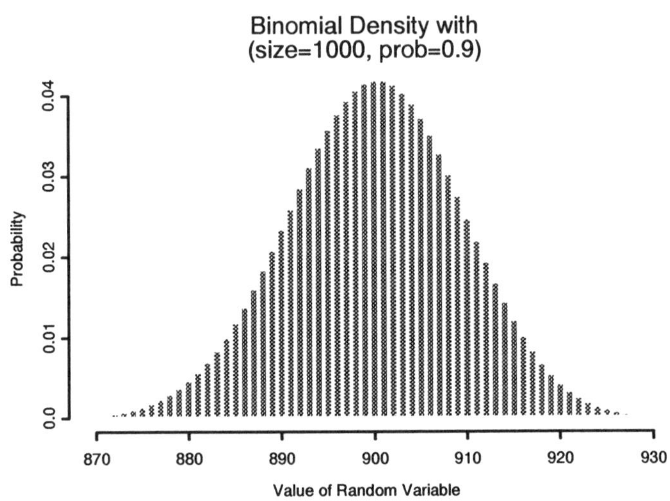

Figure 6.2: Probability density of a B(1000, 0.9) random variable.

Example 6.3: Maximum Likelihood Estimators of the Parameters of a Normal Distribution

The maximum likelihood estimators for the mean, variance, and standard deviation of a normal distribution are given by:

$$\begin{aligned} \hat{\mu}_{mle} &= \bar{x} \\ \hat{\sigma}^2_{mle} &= s^2_m \\ \hat{\sigma}_{mle} &= s_m \end{aligned} \tag{6.17}$$

Comparing **Equation (6.17)** to **Equation (6.12)**, you can see that for a normal distribution the method of moments estimators and maximum likelihood estimators are exactly the same (see the help file for `enorm`).

Example 6.4: Maximum Likelihood Estimator of a Binomial Proportion

The maximum likelihood estimator of the parameter p of a binomial distribution is given by:

$$\hat{p}_{mle} = \frac{x}{n} = \bar{x} \tag{6.18}$$

Comparing **Equation (6.18)** with **Equations (6.14)** and **(6.16)**, you can see that for a binomial distribution the method of moments estimator and the maximum likelihood estimator is the same (see the help file for `ebinom`).

Minimum Variance Unbiased Estimators

If the average value of an estimator equals the population parameter, the estimator is said to be ***unbiased.*** If the average value of the estimator is less than the population parameter, the estimator is said to be negatively biased. If the average value of the estimator is greater than the population parameter, the estimator is said to be positively biased. An estimator with negative or positive bias is called ***biased.*** (See the help file **Glossary: Accuracy, Bias, Precision, and Variability**.)

Unbiased Estimators of the Mean and Variance

The sample mean is an unbiased estimator of the population mean. That is

$$E(\bar{x}) = E(X) = \mu \tag{6.19}$$

An unbiased estimator of the population variance is given by:

$$s^2 = \frac{1}{n-1} \sum_{i=1}^{n} \left(x_i - \bar{x}\right)^2 \tag{6.20}$$

The quantity s^2 is often called the ***sample variance***. The square root of this quantity is often called the ***sample standard deviation*** and is denoted s.

> **Note:** The sample standard deviation s is *not* an unbiased estimator of the population standard deviation σ.

The unbiased estimator of the variance is always larger than the method of moments estimator of variance shown in **Equation (6.11)**:

$$s^2 = \frac{n}{n-1} s_m^2 \tag{6.21}$$

Definition of Minimum Variance Unbiased Estimator

For a particular probability distribution, there may be several possible estimators of a parameter that are unbiased. A ***minimum variance unbiased estimator*** (MVUE) has the smallest variance in the class of all unbiased estimators. The technical details of deriving MVUEs are discussed in basic text books on mathematical statistics.

Example 6.5: MVUEs for the Mean and Variance of a Normal Distribution

The minimum variance unbiased estimators for the mean and variance of a normal distribution are given by the sample mean and variance defined in **Equation (6.5)** and **Equation (6.20)**. That is,

$$\hat{\mu}_{mvue} = \bar{x} \tag{6.22}$$

$$\hat{\sigma}^2_{mvue} = s^2$$

Example 6.6: MVUE of a Binomial Proportion

The MVUE of the parameter p of a binomial distribution is exactly the same as the method of moments estimator and the MLE (see the help file for `ebinom`):

$$\hat{p}_{mvue} = \frac{x}{n} = \bar{x} \tag{6.23}$$

Comparing Different Estimators

Given that there are several ways to estimate the parameters of a probability distribution, how should you decide which method to use? One criterion that is commonly used to compare estimators is the ***mean squared error*** (MSE) (see, for example, Gilliom and Helsel, 1986). The mean squared error of an estimator of some parameter θ is defined as:

$$MSE\left(\hat{\theta}\right) = E\left[\left(\hat{\theta} - \theta\right)^2\right] = Var\left(\hat{\theta}\right) + \left[Bias\left(\hat{\theta}\right)\right]^2 \qquad \textbf{(6.24)}$$

That is, the mean squared error is the average of the square of the distance between the estimator and the parameter it is estimating. Note that for an unbiased estimator, the MSE is the same as its variance.

In practice, a decision-maker needs an estimate of a parameter and a measure of how much "wiggle" is in that estimate. You can use confidence intervals to quantify the degree of "wiggle" in a point estimate (see below). Given that confidence intervals are really what are needed and used in practice, it makes more sense to compare estimators by how well confidence intervals based on these estimators perform. One possible measure of performance is the mean or median width of the confidence interval.

There is not necessarily a one-to-one correspondence between how well an estimator does based on its MSE vs. how well it does based on measuring the performance of its confidence interval. This is because the confidence interval associated with the estimator depends on not only how closely the estimator "bounces" around the true parameter value, but also on how well the estimator of the variance of the estimator performs. (See the section on Confidence Intervals below for more information on how confidence intervals are constructed.)

Confidence Intervals for Distribution Parameters

A ***confidence interval*** for some population parameter θ is an interval on the real line constructed from a sample of *n* observations so as to contain the parameter θ with some specified probability (1-α)100%, where α is some fraction between 0 and 1 (usually α is less than 0.5). The quantity (1-α)100% is called the ***confidence coefficient*** or ***confidence level*** associated with the confidence interval.

A confidence interval is a random interval; that is, the lower and/or upper bounds are random variables that are computed based on sample statistics. Prior to taking one specific sample, the probability that the confidence interval *will* contain θ is (1-α)100%. Once a specific sample is taken and a particular confidence interval is computed, the particular confidence interval associated with that sample either contains θ or it does not. (You usually don't know whether the confidence interval actually contains θ or not because you usually don't know what θ is, which is why you are constructing the confidence interval in the first place!)

If an experiment is repeated N times, and for each experiment a sample is taken and a $(1-\alpha)100\%$ confidence interval for θ is computed, then the number of confidence intervals that actually contain θ is a binomial random variable with parameters N and $p=1-\alpha$.

Constructing a Confidence Interval for the Mean of a Normal Distribution

The most commonly used confidence interval is for the mean of a normal distribution. The formula for a $(1-\alpha)100\%$ confidence interval for the mean of a normal distribution is:

$$\left[\hat{\mu} - t_{n-1,1-\alpha/2} \frac{\hat{\sigma}}{\sqrt{n}} \; , \; \hat{\mu} - t_{n-1,1-\alpha/2} \frac{\hat{\sigma}}{\sqrt{n}} \right] \tag{6.25}$$

$$= \left[\bar{x} - t_{n-1,1-\alpha/2} \frac{s}{\sqrt{n}} \; , \; \bar{x} - t_{n-1,1-\alpha/2} \frac{s}{\sqrt{n}} \right]$$

where the quantity $t_{\nu,p}$ denotes the pth quantile of the Student's t-distribution with ν degrees of freedom (see the help file for **T**). The reasoning behind this formula is explained in most elementary statistical texts (e.g., Zar, 1996, Chapters 7-9).

Constructing a Confidence Interval for a Distribution Parameter: A General Formula

The formula for the confidence interval for the mean of a normal distribution shown in **Equation (6.25)** is exact because the distribution of the sample mean for a normal distribution is itself normal, and because of the relationship between the normal distribution and Student's t-distribution. Formulas for exact confidence intervals are also available for other distributions such as the binomial, exponential, and Poisson distributions (see the help files for `ebinom`, `eexp`, and `epois`).

Often, however, a confidence interval for a parameter is based on a "normal approximation." That is, the distribution of the estimator is assumed to be approximately normal even though the estimator is a function of observations from some distribution other than a normal distribution. (Under certain conditions, which usually hold in practice, the distribution of any maximum likelihood estimator is asymptotically normal.)

In this case, the general formula for an approximate $(1-\alpha)100\%$ confidence interval for the parameter, say θ, is given by:

$$\left[\hat{\theta} - z_{1-\alpha/2}\,\hat{\sigma}_{\hat{\theta}}\ ,\ \hat{\theta} + z_{1-\alpha/2}\,\hat{\sigma}_{\hat{\theta}}\right] \qquad (6.26)$$

where

$$\hat{\sigma}_{\hat{\theta}}$$

denotes an estimate of the standard deviation of

$$\hat{\theta}$$

and z_p denotes the pth quantile of the standard normal distribution. A slightly more conservative formula for the confidence interval replaces the quantile from the normal distribution with a quantile from the Student's t-distribution:

$$\left[\hat{\theta} - t_{\nu,1-\alpha/2}\,\hat{\sigma}_{\hat{\theta}}\ ,\ \hat{\theta} + t_{\nu,1-\alpha/2}\,\hat{\sigma}_{\hat{\theta}}\right] \qquad (6.27)$$

where ν denotes the degrees of freedom associated with the estimate of θ. This confidence interval will always be wider than the one shown in **Equation (6.26)**. The difference in width diminishes as the sample size increases.

Estimating Parameters for Important Distributions in Environmental Statistics

All of the functions in ENVIRONMENTALSTATS for S-PLUS for estimating distribution parameters have help files that explain the methods available for estimating the parameters and optionally constructing confidence intervals. This section gives examples for the normal, lognormal, and binomial distributions using data sets from regulatory guidance documents.

Normal Distribution

In EnvironmentalStats for S-Plus, there are two ways you can estimate the mean of a normal distribution and construct a confidence interval for the mean: use the function `enorm` or use the S-PLUS function `t.test` to perform a t-test since one of the outputs of `t.test` is the confidence interval for the mean (we'll talk more about hypothesis tests in **Chapter 8: Hypothesis Tests & Statistical Models**). The function `enorm` computes an estimate and confidence interval for the one-sample problem. The function `t.test` performs the t-test and computes a confidence interval for the one-sample, two-sample, or paired-sample problem.

Example 6.7: Estimating Mean Aldicarb Level

The guidance document **Statistical Analysis of Ground-Water Monitoring Data at RCRA Facilities** (USEPA, 1989b) contains an example on pages 6-3 to 6-4 dealing with aldicarb concentrations (ppb) at three compliance wells sampled over four months. The maximum concentration limit (MCL) for this example is 7 ppb. A two-sided 98% confidence interval for the mean aldicarb concentration is constructed for each of the three wells. The data are stored in the data frame `epa.89b.aldicarb1.df` (see the help file **Datasets: USEPA (1989b)**).

```
> epa.89b.aldicarb1.df
   Aldicarb Month Well
 1     19.9     1    1
 2     29.6     2    1
 3     18.7     3    1
 4     24.2     4    1
 5     23.7     1    2
 6     21.9     2    2
 7     26.9     3    2
 8     26.1     4    2
 9      5.6     1    3
10      3.3     2    3
11      2.3     3    3
12      6.9     4    3

> attach(epa.89b.aldicarb1.df)

> objects(2)
[1] "Aldicarb" "Month"    "Well"
```

Before we construct the confidence intervals, let's look at the mean levels at each well:

```
> well.means <- sapply(split(Aldicarb, Well), mean)

> well.means
    1     2     3
 23.1 24.65 4.525

> dotplot(names(well.means) ~ well.means, xlim = c(0, 25),
+    xlab = "Mean Aldicarb Concentration (ppb)",
+    ylab = "Well Number")
```

The results are shown in **Figure 6.3**. It is obvious that the concentrations in Wells 1 and 2 are much higher than in Well 3.

Now let's compute the confidence interval for the first well using the `enorm` function.

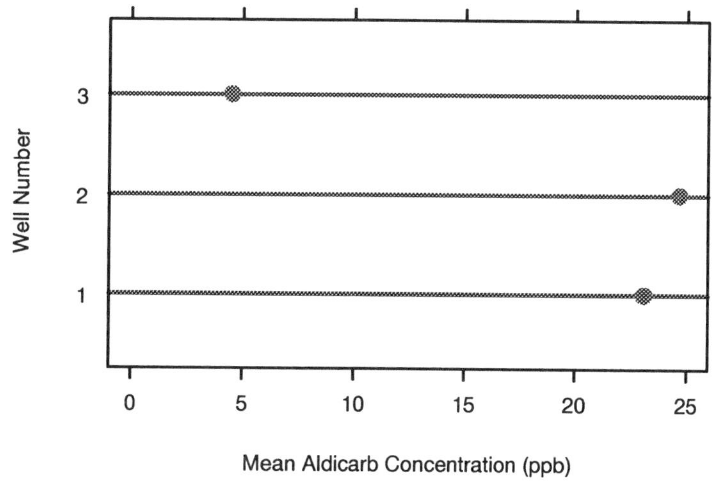

Figure 6.3: Dotchart for aldicarb data.

```
> enorm(Aldicarb[Well==1], ci=T, conf.level=0.98)

Results of Distribution Parameter Estimation
--------------------------------------------

Assumed Distribution:            Normal

Estimated Parameter(s):          mean = 23.1
                                 sd   =  4.93491

Estimation Method:               mvue

Data:                            Aldicarb[Well == 1]

Sample Size:                     4

Confidence Interval for:         mean

Confidence Interval Method:      Exact

Confidence Interval Type:        two-sided

Confidence Level:                98%

Confidence Interval:             LCL = 11.89602
                                 UCL = 34.30398
```

Since this confidence interval does not contain the MCL of 7, Well 1 is out of compliance.

For the above example, the question is whether the concentration of aldicarb is significantly greater than the MCL of 7 ppb at α=0.01. In this case, it is really more appropriate to construct a one-sided lower 99% confidence interval:

```
> enorm(Aldicarb[Well==1], ci=T,
+ ci.type="lower", conf.level=0.99)

Results of Distribution Parameter Estimation
--------------------------------------------

Assumed Distribution:            Normal

Estimated Parameter(s):          mean = 23.1
                                 sd   =  4.93491

Estimation Method:               mvue

Data:                            Aldicarb[Well == 1]

Sample Size:                     4

Confidence Interval for:         mean

Confidence Interval Method:      Exact

Confidence Interval Type:        lower

Confidence Level:                99%

Confidence Interval:             LCL = 11.89602
                                 UCL =      Inf
```

In this case, the lower bound of this one-sided confidence interval does not fall below the MCL of 7, so you may conclude that the well is out of compliance. Note that the lower bound of the one-sided 99% confidence interval is exactly the same as the lower bound of the two-sided 98% confidence interval.

Lognormal Distribution

The lognormal distribution can be characterized by the mean (μ) and standard deviation (σ) of the log-transformed random variable, or by the mean (θ) and coefficient of variation (τ) of the original random variable (see the help file **Lognormal** and the section on the lognormal distribution in the last chapter). If you are interested in characterizing the log-transformed distribution, then simply take the logarithms of the original data and treat them like they come from a normal distribution: use the sample mean to estimate μ, and use the formula in **Equation (6.25)** to construct a confidence interval for μ (see the help file for `elnorm`). If, on the other hand, you want to estimate θ and construct a confidence interval for θ, there are several methods to choose from, all of which are documented in the help file for `elnorm.alt`.

Example 6.8: Estimating the Mean of a Lognormal Distribution and Constructing a Confidence Interval Based on Land's Method

The guidance document **Supplemental Guidance to RAGS: Calculating the Concentration Term** (USEPA, 1992d) contains an example of chromium data (mg/kg) which is assumed to come from a lognormal distribution. The example shows how to compute the 95% upper confidence limit for the mean of the distribution (θ) using Land's (1972) method (see the help file for `elnorm.alt`).

The data for this example are stored in the vector `epa.92d.chromium.vec` (see the help file **Datasets: USEPA (1992d)**).

```
> epa.92d.chromium.vec
 [1]   10   13   20   36   41   59   67  110  110
[10]  136  140  160  200  230 1300
```

Here we'll compute the upper confidence limit using Land's (1972) method and three others.

> **Warning:** Land's (1972) method is exact but is extremely sensitive to the assumption that the data come from a lognormal distribution. Minor deviations from this distribution can result in extremely large values for the upper confidence limit.

```
> elnorm.alt(epa.92d.chromium.vec, ci=T, ci.type="upper")

Results of Distribution Parameter Estimation
--------------------------------------------

Assumed Distribution:            Lognormal

Estimated Parameter(s):          mean = 159.8552
                                 cv   =   1.493994

Estimation Method:               mvue

Data:                            epa.92d.chromium.vec

Sample Size:                     15

Confidence Interval for:         mean

Confidence Interval Method:      Land

Confidence Interval Type:        upper

Confidence Level:                95%

Confidence Interval:             LCL =   0
                                 UCL = 496.6282
```

So the estimated mean is about 160, but the upper confidence limit is about 500! Compare the results based on the other three methods:

```
> elnorm.alt(epa.92d.chromium.vec, ci = T,
+    ci.type = "upper",
+    ci.method="parkin")$interval$limits
 LCL UCL
   0 230

> elnorm.alt(epa.92d.chromium.vec, ci = T,
+    ci.type = "upper", ci.method="cox")$interval$limits
 LCL     UCL
   0 365.527

> elnorm.alt(epa.92d.chromium.vec, ci = T,
+    ci.type = "upper",
+    ci.method="normal.approx")$interval$limits
 LCL     UCL
   0 261.424
```

Binomial Distribution

In EnvironmentalStats for S-Plus, there are two ways you can estimate the parameter of a binomial distribution and construct a confidence interval: use the function `ebinom` or use the S-PLUS function `prop.test` to perform an approximate chi-square test since one of the outputs of `prop.test` is the confidence interval for the binomial proportion (we'll talk more about hypothesis tests in **Chapter 8: Hypothesis Tests & Statistical Models**). The function `ebinom` computes an estimate and confidence interval for the one-sample problem. The function `prop.test` performs the chi-square test and computes a confidence interval for the one-sample and two-sample problem.

Example 6.9: Comparing the Proportion of Detects Between Background and Compliance Wells

The guidance document **Statistical Analysis of Ground-Water Monitoring Data at RCRA Facilities** (USEPA, 1989b) contains an example on pages 8-3 to 8-7 dealing with cadmium concentrations (μg/L) at a background well and some compliance wells. Several of the observations are simply reported as "below the detection limit" (BDL). The example in the guidance document compares the proportion of detected observations at the background well with the proportion of detected observations at the compliance wells. Here, we'll simply compute the proportion of detected values at each kind of well and construct confidence intervals for the probability of detection. The data are stored in the data frame `epa.89b.cadmium.df` (see the help file **Datasets: USEPA (1989b)**).

```
> epa.89b.cadmium.df
   Cadmium.orig Cadmium Censored  Well.type
 1          0.1   0.100        F Background
 2         0.12   0.120        F Background
 3          BDL   0.000        T Background
 4         0.26   0.260        F Background
 .
85          BDL   0.000        T Compliance
86          BDL   0.000        T Compliance
87          BDL   0.000        T Compliance
88          BDL   0.000        T Compliance

> attach(epa.89b.cadmium.df)

> objects(2)
[1] "Cadmium"       "Cadmium.orig" "Censored"
[4] "Well.type"

> sapply(split(!Censored, Well.type), mean)
 Background Compliance
  0.3333333      0.375
```

We see that 33% of the observations at the background well are detects and 37.5 % at the compliance wells are detects. Now let's construct a 95 % confidence interval for the probability of a detect at the background well:

```
> back.data <- !Censored[Well.type=="Background"]

> ebinom(back.data, ci = T)

Results of Distribution Parameter Estimation
--------------------------------------------

Assumed Distribution:            Binomial

Estimated Parameter(s):          size = 24
                                 prob =   0.3333333

Estimation Method:               mle/mme/mvue for 'prob'

Data:                            back.data

Sample Size:                     24

Confidence Interval for:         prob

Confidence Interval Method:      Exact

Confidence Interval Type:        two-sided

Confidence Level:                95%

Confidence Interval:             LCL = 0.1563023
                                 UCL = 0.5532196
```

So the 95% confidence interval says that the probability of a detect at the background well may be anywhere from about 16% to 55%.

Now let's construct the confidence interval for the compliance wells.

```
> comp.data <- !Censored[Well.type=="Compliance"]
```

```
> ebinom(comp.data, ci = T)

Results of Distribution Parameter Estimation
--------------------------------------------

Assumed Distribution:            Binomial

Estimated Parameter(s):          size = 64
                                 prob =  0.375

Estimation Method:               mle/mme/mvue for 'prob'

Data:                            comp.data

Sample Size:                     64

Confidence Interval for:         prob

Confidence Interval Method:      Exact

Confidence Interval Type:        two-sided

Confidence Level:                95%

Confidence Interval:             LCL = 0.2570489
                                 UCL = 0.5049235
```

So the 95% confidence interval says that the probability of a detect at the compliance wells may be anywhere from about 26% to 50%.

Estimating and Constructing Confidence Intervals for Quantiles

Quantiles or percentiles are sometimes used in environmental standards and regulations (e.g., Berthouex and Brown, 1994, p.65). For example, in order to determine compliance, you may be required to estimate an extreme percentile (e.g., the 99th percentile) for the "background level" distribution, and then compare observations at compliance wells or areas to this upper percentile. Or you may be required to compute an upper confidence limit for an extreme percentile and compare this confidence limit with a maximum concentration limit (MCL) or alternate concentration limit (ACL).

Estimating quantiles is fairly straightforward. Once you have estimated the parameters of a distribution, you can easily estimate the quantiles of the distribution by using the relevant `q`*abb* function and plugging in the estimated parameters. Also, the S-PLUS function `quantile` allows you estimate the quantiles of a distribution nonparametrically; that is, you don't have to assume any sort of probability distribution.

A very important detail to consider when dealing with standards based on percentiles (quantiles) is the amount of variability ("wiggle") associated with the estimated percentile. Berthouex and Brown (1994, p.65) note that for a normal (and therefore a lognormal) distribution, the 50th percentile (i.e., median) can be estimated with greater precision than any other percentile, and the precision decreases rapidly for larger or smaller percentiles. Estimates of extreme percentiles are very imprecise, even for data sets with large sample sizes.

Considering these facts, it is very important to quantify the precision of any estimate of an extreme percentile. The ENVIRONMENTALSTATS for S-PLUS functions `eqlnorm`, `eqnorm`, and `eqpois` not only estimate quantiles for the lognormal, normal, and Poisson distributions (respectively), they also allow you to construct confidence intervals for quantiles. The ENVIRONMENTALSTATS for S-PLUS function `eqnpar` uses the S-PLUS `quantile` function to estimate quantiles, and also allows you to construct nonparametric confidence intervals for quantiles.

Confidence intervals for quantiles are based on the idea of ***tolerance intervals***. Tolerance intervals and prediction intervals are discussed in the next chapter.

The rest of this section discusses the ENVIRONMENTALSTATS for S-PLUS functions for estimating quantiles for the normal, lognormal, and Poisson distributions, as well as constructing nonparametric estimates of quantiles.

Normal Distribution

The methods for estimating quantiles and constructing confidence intervals for quantiles for the normal distribution are detailed in the help file for `eqnorm`.

Example 6.10: Using the 95th Percentile from Background as a Standard

The guidance document **Statistical Analysis of Ground-Water Monitoring Data at RCRA Facilities** (USEPA, 1989b) contains an example on page 5-23 of using a standard based on the estimated 95th percentile from lead concentrations (ppm) at two "background" wells. The actual standard used is the one-sided upper confidence limit for the 95th percentile, based on a confidence level of 95%. The data for this example are stored in the data frame `epa.89b.lead.df`.

First lets look at the background well data, estimate the 95th percentile, and compute a one-sided upper confidence limit for this percentile:

```
> epa.89b.lead.df
    Lead Month Well  Well.type
 1  58.0     1    1 Background
 2  54.1     2    1 Background
 3  30.0     3    1 Background
 .
22 183.1     2    6 Compliance
23 198.3     3    6 Compliance
24 160.8     4    6 Compliance

> attach(epa.89b.lead.df)

> objects(2)
[1] "Lead"      "Month"      "Well"       "Well.type"

> back.dat <- Lead[Well.type=="Background"]

> eqnorm(back.dat, p = 0.95, ci = T, ci.type = "upper")

Results of Distribution Parameter Estimation
--------------------------------------------

Assumed Distribution:            Normal

Estimated Parameter(s):          mean = 51.3875
                                 sd   = 16.27061

Estimation Method:               mvue

Estimated Quantile(s):           95'th %ile = 78.15027

Quantile Estimation Method:      qmle

Data:                            back.dat

Sample Size:                     8

Confidence Interval for:         95'th %ile

Confidence Interval Method:      Exact

Confidence Interval Type:        upper

Confidence Level:                95%

Confidence Interval:             LCL = -Inf
                                 UCL = 103.2467
```

So the 95th percentile is estimated as about 78 ppm, and the upper confidence limit for the 95th percentile is about 103 ppm. This large difference between the estimated 95th percentile and the upper confidence limit is due to the very small sample size (n=8).

Now let's check to see what percentage of the observations from the compliance wells are over the standard:

```
> mean(Lead[Well.type == "Compliance"] > 103.2)
[1] 0.5
```

So half of the observations from the compliance wells are above the standard.

> **Note:** Even if there was no difference in the distribution of lead concentrations at the background and compliance wells, we would expect somewhere around 5% of the observations at the compliance wells to be above the standard. Actually, we could expect to see even a bit more than 5% because the way we constructed the confidence interval for the 95th percentile, the upper limit should be above the 95th percentile about 95% of the time (had we repeated this experiment many, many times). The other 5% of the time the upper limits will be *below* the 95th percentile. We don't know whether our particular upper limit is above or below the true 95th percentile.

Lognormal Distribution

The methods for estimating quantiles and constructing confidence intervals for quantiles for the lognormal distribution are detailed in the help file for `eqlnorm`.

Example 6.11: Comparing the Upper Confidence Limit for the 95th Percentile to an ACL

The guidance document **Statistical Analysis of Ground-Water Monitoring Data at RCRA Facilities, Addendum to Interim Final Guidance** (USEPA, 1992c) contains an example on pages 52-54 in which chrysene concentrations (ppb) are compared to an alternate concentration limit (ACL) of 80 ppb. In this example, the document talks about a 95% confidence level associated with an upper tolerance limit (i.e., upper confidence limit for a percentile), but in fact the example incorrectly uses β-expectation tolerance intervals instead of β-content confidence intervals (see the next chapter).

Here, we'll compute the 95% upper confidence limit for the 95th percentile of the concentration at the first well and compare it to the ACL of 80 ppb. The data for this example are stored in the data frame `epa.92c.chrysene.df` (see the help **file Datasets: USEPA (1992c)**).

```
> epa.92c.chrysene.df
   Chrysene Month Well
 1     19.7     1    1
 2     39.2     2    1
 3      7.8     3    1
 .
18     30.5     2    5
19     15.0     3    5
20     23.4     4    5

> attach(epa.92c.chrysene.df)

> Well.1.dat <- Chrysene[Well==1]

> eqlnorm(Well.1.dat, p=.95, ci=T, ci.type="upper")

Results of Distribution Parameter Estimation
--------------------------------------------

Assumed Distribution:            Lognormal

Estimated Parameter(s):          meanlog = 2.813216
                                 sdlog   = 0.6845029

Estimation Method:               mvue

Estimated Quantile(s):           95'th %ile = 51.37354

Quantile Estimation Method:      qmle

Data:                            Well.1.dat

Sample Size:                     4

Confidence Interval for:         95'th %ile

Confidence Interval Method:      Exact

Confidence Interval Type:        upper

Confidence Level:                95%

Confidence Interval:             LCL =   0
                                 UCL = 563.5261
```

So the 95th percentile is estimated as about 51 ppb, but the 95% upper confidence limit for this percentile is a whopping 564 ppb! This is because of the ridiculously small sample size (n=4).

Poisson Distribution

The methods for estimating quantiles and constructing confidence intervals for quantiles for the Poisson distribution are detailed in the help file for eqpois.

Example 6.12: Computing the Upper Confidence Limit for the 95th Percentile

The guidance document **Statistical Analysis of Ground-Water Monitoring Data at RCRA Facilities, Addendum to Interim Final Guidance** (USEPA, 1992c) contains an example on page 40 in which a 95% upper confidence limit is computed for the 95th percentile based on 36 benzene concentrations (ppb) at six background wells. The benzene concentrations are modeled as having come from a Poisson distribution. Of the 36 observations, 33 are reported as "<2" (i.e., below the detection limit). In the example in the guidance document, these observations are set to half of the detection limit in order to compute the confidence limit for the percentile.

> **Note: Example 4.17** shows that these data do not fit a Poisson distribution very well.

The data for this example are stored in epa.92c.benzene1.df (see the help file **Datasets: USEPA (1992c)**).

```
> epa.92c.benzene1.df
   Benzene.orig Benzene Censored Month Well
 1           <2       2        T     1    1
 2           <2       2        T     2    1
 3           <2       2        T     3    1
 .
34           <2       2        T     4    6
35           10      10        F     5    6
36           <2       2        T     6    6

> attach(epa.92c.benzene1.df)

> objects(2)
[1] "Benzene"      "Benzene.orig" "Censored"
[4] "Month"        "Well"

> dat <- Benzene

> dat[Censored] <- 1
```

```
> eqpois(dat, p=0.95, ci=T, ci.type="upper")

Results of Distribution Parameter Estimation
--------------------------------------------

Assumed Distribution:          Poisson

Estimated Parameter(s):        lambda = 1.944444

Estimation Method:             mle/mme/mvue

Estimated Quantile(s):         95'th %ile = 4

Quantile Estimation Method:    Quantile(s) Based on
                               mle/mme/mvue Estimator

Data:                          dat

Sample Size:                   36

Confidence Interval for:       95'th %ile

Confidence Interval Method:    Exact

Confidence Interval Type:      upper

Confidence Level:              95%

Confidence Interval:           LCL = 0
                               UCL = 5
```

So the 95% upper confidence limit for the 95th percentile is 5 ppb.

Nonparametric Estimation of Quantiles and Construction of Confidence Intervals

The nonparametric methods for estimating quantiles and constructing confidence intervals for quantiles are detailed in the help file for `eqnpar`. The basic idea is that a nonparametric estimate of a quantile is a linear interpolation between the order statistics in the sample, and a confidence interval for a quantile is constructed using the order statistics.

Example 6.13: Computing the Upper Confidence Limit for the 95th Percentile Nonparametrically

The guidance document **Statistical Analysis of Ground-Water Monitoring Data at RCRA Facilities, Addendum to Interim Final Guidance** (USEPA, 1992c) contains an example on pages 55-56 in which a nonparametric upper confidence limit is computed for the 95th percentile based on copper concentrations (ppb) at three background wells. The data for the background wells contain several observations that are reported as "<5" (below-detection-limit). Because we are constructing a nonparametric upper confidence limit and there are enough uncensored observations, the non-detects are not a problem. The data for this example are stored in the data frame `epa.92c.copper2.df` (see the help file **Datasets: USEPA (1992c)**).

```
> epa.92c.copper2.df
   Copper.orig Copper Censored Month Well  Well.type
 1         <5     5.0        T     1    1 Background
 2         <5     5.0        T     2    1 Background
 3         7.5    7.5        F     3    1 Background
 .
38         <5     5.0        T     6    5 Compliance
39         5.6    5.6        F     7    5 Compliance
40         <5     5.0        T     8    5 Compliance

> attach(epa.92c.copper2.df)

> objects(2)
[1] "Censored"     "Copper"      "Copper.orig"
[4] "Month"        "Well"        "Well.type"

> back.dat <- Copper[Well.type=="Background"]

> eqnpar(back.dat, p = .95, ci = T, lb = 0,
+   ci.type = "upper", approx.conf.level = .95)

Results of Distribution Parameter Estimation
--------------------------------------------

Assumed Distribution:            None

Estimated Quantile(s):           95'th %ile = 7.925

Quantile Estimation Method:      Nonparametric

Data:                            back.dat

Sample Size:                     24

Confidence Interval for:         95'th %ile

Confidence Interval Method:      exact
```

```
Confidence Interval Type:      upper

Confidence Level:              71%

Confidence Limit Rank(s):      24

Confidence Interval:           LCL = 0
                               UCL = 9.2
```

In this example, we see that there are 24 observations at the background wells. If we use the largest value as the upper confidence limit for the 95th percentile, this yields a 71% confidence interval for the percentile. This is the best we can do with only 24 observations.

Summary

- **Table 6.1** shows all of the functions in ENVIRONMENTALSTATS for S-PLUS for estimating probability distribution parameters. Some of these functions allow you to construct confidence intervals as well. The form of the names of these functions is e*abb*, where *abb* denotes the abbreviation of the distribution name (see column 2 of **Table 5.1**).

- The three most commonly used kinds of estimators are the method of moments estimator (MME), maximum likelihood estimator (MLE), and the minimum variance unbiased estimator (MVUE). *L*-moment estimators have also been introduced and used in the hydrological literature.

- In the past, estimators have been compared based on their mean square error (MSE). It makes more sense to compare estimators based on how well confidence intervals based on these estimators perform.

- A ***confidence interval*** for a distribution parameter is constructed from a sample data set in such a way that if a new sample were taken many, many times and a confidence interval constructed for each new sample, the confidence interval would contain the parameter $(1-\alpha)100\%$ of the time. The quantity $(1-\alpha)100\%$ is called the ***confidence level*** of the confidence interval.

- Quantiles are sometimes used in environmental standards and regulations. Estimates of extreme quantiles (e.g., 95th or 99th percentile) are extremely imprecise, even for large data sets. It is important to quantify the precision associated with an estimated quantile.

- **Table 6.2** lists four functions in ENVIRONMENTALSTATS for S-PLUS for estimating quantiles of probability distributions and constructing confidence intervals for these quantiles.

CHAPTER 7 :

Prediction and Tolerance Intervals

Any activity that requires constant monitoring over time and the comparison of new values to "background" or "standard" values creates a decision problem: if the new values greatly exceed the background values, has a change really occurred, or have the true underlying concentrations stayed the same and this is just a "chance" event? Statistical tests are used as objective tools to decide whether a change has occurred (although the choice of Type I error level and acceptable power are subjective decisions; see **Chapter 8: Hypothesis Tests and Statistical Models**).

For a monitoring program that involves numerous tests over time, figuring out how to balance the *overall* Type I error with the power of detecting a change is not a trivial problem, but it is also a problem that has been dealt with for a long time in the statistical literature under the heading of "multiple comparisons" (see the help file **Glossary: Multiple Comparisons**).

Prediction intervals and tolerance intervals are two tools that you can use to attempt to solve the multiple comparisons problem. This chapter briefly describes these tools and illustrates the use of functions in ENVIRONMENTALSTATS for S-PLUS that you can use to create prediction and tolerance intervals.

Overview

Table 7.1 lists the functions available in ENVIRONMENTALSTATS for S-PLUS for computing prediction intervals and associated quantities (see the help file **Prediction and Tolerance Intervals**). The form of the names of these functions is `pred.int.`*`abb`*, where *`abb`* denotes the abbreviation of the distribution name (see column 2 of **Table 5.1**).

Distribution Name	Function for Prediction Interval	Description
Lognormal	`pred.int.lnorm`	Prediction interval for the next k observations or geometric means from a lognormal distribution.
Normal	`pred.int.norm`	Prediction interval for the next k observations or means from a normal distribution.
Nonparametric	`pred.int.npar`	Nonparametric prediction interval for the next k of K observations.
	`pred.int.npar.n`	Required sample size for a nonparametric prediction interval for the next k of K observations with a specified confidence level.
Poisson	`pred.int.pois`	Prediction interval for the next k observations or sums from a Poisson distribution.

Table 7.1: Functions for prediction intervals.

Table 7.2 lists the functions available in ENVIRONMENTALSTATS for S-PLUS for computing tolerance intervals and associated quantities (see the help file **Prediction and Tolerance Intervals**). The form of the names of these functions is `tol.int.`*abb*, where *abb* denotes the abbreviation of the distribution name (see column 2 of **Table 5.1**).

Davis and McNichols (1987, 1994b), and Gibbons (1987a, 1987b, 1990, 1991a, 1991b, 1994), among others, have suggested using prediction and tolerance intervals for analyzing data from groundwater monitoring programs.

The rest of this chapter gives a brief discussion of the theory of prediction intervals and tolerance intervals, and shows you how to use the functions in ENVIRONMENTALSTATS for S-PLUS for constructing these kinds of intervals using examples from regulatory guidance documents.

Distribution Name	Function for Tolerance Interval	Description
Lognormal	`tol.int.lnorm`	Tolerance interval for a lognormal distribution.

Normal	`tol.int.norm`
	
	`tol.int.norm.K`
		Tolerance interval for a normal distribution.
		
		Compute the constant *K* for a normal tolerance interval.

Nonparametric	`tol.int.npar`
	
	`tol.int.npar.n`
		Nonparametric tolerance interval.
		
		Required sample size for a nonparametric tolerance interval with a specified coverage or confidence level.

Poisson	`tol.int.pois`	Tolerance interval for a Poisson distribution.

Table 7.2: Functions for tolerance intervals.

Prediction Intervals

A ***prediction interval*** for some population is an interval on the real line constructed so that it will contain k future observations or averages from that population with some specified probability $(1-\alpha)100\%$, where α is some fraction between 0 and 1 (usually α is less than 0.5), and k is some pre-specified positive integer. The quantity $(1-\alpha)100\%$ is called the ***confidence coefficient*** or ***confidence level*** associated with the prediction interval.

The basic idea is to assume a particular probability distribution (e.g., normal, lognormal, etc.) for some process generating data (e.g., quarterly observations of chemical concentrations in groundwater), compute sample statistics from a baseline sample, and then use these sample statistics to construct a prediction interval, assuming the distribution of the data does not change in the future.

For example, if X denotes a random variable from some population, a true (1-α)100% two-sided prediction interval for the next k=1 observation of X is given by:

$$\left[x_{\alpha/2} \ , \ x_{1-\alpha/2} \right] \qquad (7.1)$$

where x_p denotes the pth quantile of the distribution of X. Usually the true distribution of X is unknown, so the values of the prediction limits have to be estimated based on estimating the parameters of the distribution of X.

For the usual case when the exact distribution of X is unknown, a prediction interval is thus a random interval; that is, the lower and upper bounds are random variables computed based on sample statistics in the baseline sample. Prior to taking one specific baseline sample, the probability that the prediction interval *will* contain the next k observations is (1-α)100%. Once a specific baseline sample is taken and the prediction interval based on that sample is computed, the probability that that prediction interval will contain the next k observations is not necessarily (1-α)100%, but it should be close to this value.

If an experiment is performed N times, and for each experiment

1. A sample is taken and a (1-α)100% prediction interval for k=1 future observation is computed, and

2. One future observation is generated and compared to the prediction interval,

then the number of prediction intervals that actually contain the future observation generated in step 2 above is a binomial random variable with parameters N and p=1-α.

If, on the other hand, only one baseline sample is taken and only one prediction interval for k=1 future observation is computed, then the number of future observations out of a total of N future observations that will be contained in that one prediction interval is a binomial random variable with parameters N and p=1-α*, where α* depends on the true population parameters and the computed bounds of the prediction interval.

The rest of this section explains how to construct prediction intervals for the lognormal, normal, and Poisson distributions, and also how to construct nonparametric prediction intervals.

Normal and Lognormal Distributions

The details of how to construct prediction intervals for the normal and lognormal distributions are discussed in the help files for `pred.int.norm` and `pred.int.lnorm`, repectively.

Example 7.1: Constructing a Prediction Interval for *k*=1 Future Observation for a Normal Distribution

Let's generate a set of 20 observations from a standard normal distribution and generate a two-sided prediction interval for *k*=1 future observation with associated confidence level 95% (see the help file for `pred.int.norm`):

First set the random seed so you can reproduce this example, then generate the 20 observations:

```
> set.seed(284)

> dat <- rnorm(20, mean=0, sd=1)
```

Now construct the prediction interval:

```
> pred.int.norm(dat)

Results of Distribution Parameter Estimation
--------------------------------------------

Assumed Distribution:           Normal

Estimated Parameter(s):         mean = 0.03816664
                                sd   = 1.042196

Estimation Method:              mvue

Data:                           dat

Sample Size:                    20

Prediction Interval Method:     Exact

Prediction Interval Type:       two-sided

Confidence Level:               95%

Prediction Interval:            LPL = -2.197044
                                UPL =  2.273377
```

Note that the true two-sided 95% prediction interval is given by:

```
> true.interval <- c(LPL = qnorm(.025, mean=0, sd=1),
+    UPL = qnorm(.975, mean=0, sd=1))

> true.interval
       LPL      UPL
 -1.959964 1.959964
```

Also note that the true probability that the computed prediction interval will contain a future observation from a standard normal distribution is given by:

```
> pnorm(2.273377, mean=0, sd=1) -
+    pnorm(-2.197044, mean=0, sd=1)
[1] 0.9744896
```

Example 7.2: Constructing a Prediction Interval for *k*=2 Future Observations for a Normal Distribution

The guidance document **Statistical Analysis of Ground-Water Monitoring Data at RCRA Facilities, Addendum to Interim Final Guidance** (USEPA, 1992c) contains an example on pages 56-59 in which an upper one-sided prediction limit is computed for the next *k*=2 means (each mean based on four observations) based on benzene concentrations (ppb) at a background well.. The data for this example are stored in the data frame `epa.92c.benzene2.df` (see the help file **Datasets: USEPA (1992c)**).

First look at the data and put the background data into an object.

```
> epa.92c.benzene2.df
   Benzene Month  Well.type
 1    12.6     1 Background
 2    30.8     1 Background
 3    52.0     1 Background
 .
18     3.8     5 Compliance
19     2.6     5 Compliance
20    51.9     5 Compliance

> attach(epa.92c.benzene2.df)

> objects(2)
[1] "Benzene"   "Month"     "Well.type"

> back.dat <- Benzene[Well.type == "Background"]
```

Now create the one-sided upper prediction interval. Note that because we are assuming a normal distribution, the lower limit is set to `-Inf` instead of 0.

```
> pred.int.norm(back.dat, k = 2, n.mean = 4,
+    pi.type = "upper")
```

```
Results of Distribution Parameter Estimation
--------------------------------------------

Assumed Distribution:            Normal

Estimated Parameter(s):          mean = 27.51667
                                 sd   = 17.10119

Estimation Method:               mvue

Data:                            back.dat

Sample Size:                     12

Prediction Interval Method:      Bonferroni

Prediction Interval Type:        upper

Confidence Level:                95%

Number of Future Averages:       2

Sample Size for Averages:        4

Prediction Interval:             LPL = -Inf
                                 UPL = 49.24782
```

Now compute the means for the two months of data at the compliance well:

```
> index <- Well.type == "Compliance"

> comp.means <-
+   sapply(split(Benzene[index], Month[index]), mean)

> comp.means
  1  2  3     4      5
 NA NA NA 33.95 26.475
```

Since both means are less than the upper prediction limit of 49.2 ppb, there is no evidence of contamination.

Poisson Distribution

The details of how to construct prediction intervals for the Poisson distribution are discussed in the help file for `pred.int.pois`.

Example 7.3: Constructing a Prediction Interval for the next *k*=1 Sum of Four Observations from a Poisson Distribution

The guidance document **Statistical Analysis of Ground-Water Monitoring Data at RCRA Facilities, Addendum to Interim Final Guidance** (USEPA, 1992c) contains an example on pages 36-38 in which a 99% upper prediction limit is computed for the next *k*=1 sum of 4 observations, based on 36 benzene concentrations (ppb) at six background wells. The benzene concentrations are modeled as having come from a Poisson distribution. Of the 36 observations, 33 are reported as "<2" (i.e., below the detection limit). In the example in the guidance document, these observations are set to half of the detection limit in order to compute the prediction limit.

> **Note: Example 4.17** shows that these data do not fit a Poisson distribution very well.

The data for this example are stored in `epa.92c.benzene1.df` (see the help file **Datasets: USEPA (1992c)**).

```
> epa.92c.benzene1.df
   Benzene.orig Benzene Censored Month Well
 1           <2       2        T     1    1
 2           <2       2        T     2    1
 3           <2       2        T     3    1
 .
34           <2       2        T     4    6
35           10      10        F     5    6
36           <2       2        T     6    6

> attach(epa.92c.benzene1.df)

> objects(2)
[1] "Benzene"       "Benzene.orig" "Censored"
[4] "Month"         "Well"

> dat <- Benzene

> dat[Censored] <- 1

> pred.int.pois(dat, n.sum=4, meth="conditional.approx",
+   pi.type="upper", conf=0.99, round=F)
```

```
Results of Distribution Parameter Estimation
--------------------------------------------

Assumed Distribution:            Poisson

Estimated Parameter(s):          lambda = 1.944444

Estimation Method:               mle/mme/mvue

Data:                            dat

Sample Size:                     36

Prediction Interval Method:      conditional.approx

Prediction Interval Type:        upper

Confidence Level:                99%

Number of Future Sums:           1

Sample Size for Sums:            4

Prediction Interval:             LPL =  0
                                 UPL = 15.28175
```

So if the sum of the next observations from the four downgradient wells is more than 15.3 ppb, you might conclude that at least one of them is out of compliance.

Nonparametric Prediction Intervals

The details of constructing a nonparametric prediction interval are discussed in the help files for `pred.int.npar` and `pred.int.npar.n`.

Example 7.4: Constructing a Nonparametric Prediction Interval for the next *K*=2 Observations.

The guidance document **Statistical Analysis of Ground-Water Monitoring Data at RCRA Facilities, Addendum to Interim Final Guidance** (USEPA, 1992c) contains an example on page 60 in which a nonparametric upper prediction limit is computed for the next K=2 observations, based on 18 arsenic concentrations (ppb) at three background wells. Of the 18 background observations, 9 are reported as "<5" (i.e., below the detection limit). Because we are constructing an upper nonparametric prediction limit and there are enough uncensored observations, the nondetects are not a problem.

The data for this example are stored in `epa.92c.arsenic2.df` (see the help file **Datasets: USEPA (1992c)**).

```
> epa.92c.arsenic2.df
   Arsenic.orig Arsenic Censored Month Well  Well.type
 1          <5      5.0        T     1    1 Background
 2          <5      5.0        T     2    1 Background
 3           8      8.0        F     3    1 Background
 .
22                   NA        F     4    4 Compliance
23           8      8.0        F     5    4 Compliance
24          14     14.0        F     6    4 Compliance

> attach(epa.92c.arsenic2.df)

> objects(2)
[1] "Arsenic"      "Arsenic.orig" "Censored"
[4] "Month"        "Well"         "Well.type"

> back.dat <- Arsenic[Well.type == "Background"]

> pred.int.npar(back.dat, K=2, lb=0, pi.type="upper")

Results of Distribution Parameter Estimation
--------------------------------------------

Assumed Distribution:             None

Data:                             back.dat

Sample Size:                      18

Prediction Interval Method:       Exact

Prediction Interval Type:         upper

Confidence Level:                 90%

Prediction Limit Rank(s):         18

Number of Future Observations:    2

Prediction Interval:              LPL =  0
                                  UPL = 12
```

So using the maximum value as the upper prediction limit based on a sample size of 18 yields a 90% upper prediction interval.

Let's use the function `pred.int.npar.n` to determine how many observations we would need at the background wells to obtain a 95% one-sided upper prediction interval:

```
> pred.int.npar.n(K=2, pi.type = "upper",
+    conf.level = 0.95)
[1] 38
```

So we need at least 38 observations to construct a 95% upper prediction interval for the next two observations.

Tolerance Intervals

A ***tolerance interval*** for some population is an interval on the real line constructed so as to contain β100% of the population (i.e., β100% of all future observations), where $0 < \beta < 1$ (usually β is bigger than 0.5). The quantity β100% is called the ***coverage***.

The basic idea is to assume a particular probability distribution (e.g., normal, lognormal, etc.) for some process generating data (e.g., quarterly observations of chemical concentrations in groundwater), compute sample statistics from a baseline sample, and then use these sample statistics to construct a tolerance interval, assuming the distribution of the data does not change in the future.

For example, if X denotes a random variable from some population, a true β100% two-sided tolerance interval is given by:

$$\left[x_{1-\beta/2} \,,\; x_{\beta/2} \right] \tag{7.2}$$

where x_p denotes the pth quantile of the distribution of X. Usually the true distribution of X is unknown, so the values of the tolerance limits have to be estimated based on estimating the parameters of the distribution of X.

For the usual case when the exact distribution of X is unknown, a tolerance interval is thus a random interval; that is, the lower and upper bounds are random variables computed based on sample statistics in the baseline sample. Given this uncertainty in the bounds, there are two ways to construct tolerance intervals (Guttman, 1970):

- A ***β-content tolerance interval*** with ***confidence level*** (1-α)100% is constructed so that it contains *at least* β100% of the population (i.e., the coverage is at least β100%) with probability (1-α)100%.

- A ***β-expectation tolerance interval*** is constructed so that it contains *on average* β100% of the population (i.e., the average coverage is β100%).

A β-expectation tolerance interval with coverage β100% is equivalent to a prediction interval for one future observation with associated confidence level β100%. Note that there is no explicit confidence level associated with a β-expectation tolerance interval. If a β-expectation tolerance interval is treated as a β-content tolerance interval, the confidence level associated with this tolerance interval is usually around 50% (e.g., Guttman, 1970, Table 4.2, p.76). Thus, a β-content tolerance interval with coverage β100% will usually be wider than a β-expectation tolerance interval with the same coverage if the confidence level associated with the β-content tolerance interval is more than 50%.

It can be shown (e.g., Conover, 1980, pp.119-121) that an upper confidence interval for the pth quantile with confidence level 100(1-α)% is equivalent to an upper β-content tolerance interval with coverage 100*p*% and confidence level 100(1-α)%. Also, a lower confidence interval for the pth quantile with confidence level 100(1-α)% is equivalent to a lower β-content tolerance interval with coverage 100(1-*p*)% and confidence level 100(1-α)%.

The rest of this section explains how to construct tolerance intervals for the lognormal, normal, and Poisson distributions, and also how to construct nonparametric tolerance intervals.

Normal and Lognormal Distributions

The details of how to construct tolerance intervals for the normal and lognormal distributions are discussed in the help files for `tol.int.norm` and `tol.int.lnorm`, repectively.

Example 7.5: Constructing a β-Content Tolerance Interval for a Normal Distribution

The guidance document **Statistical Analysis of Ground-Water Monitoring Data at RCRA Facilities** (USEPA, 1989b) contains an example on pages 6-11 to 6-13 dealing with aldicarb concentrations (ppm) at three compliance wells sampled over four months. The alternative concentration limit (ACL) for this example is 50 ppm, which should not be exceeded more than 5% of the time. A one-sided upper tolerance limit with coverage 95% and associated 95% confidence level is constructed for each well and compared to the ACL.

The data are stored in the data frame `epa.89b.aldicarb2.df` (see the help file **Datasets: USEPA (1989b)**). Here, we'll construct the upper tolerance limit for the first well.

```
> epa.89b.aldicarb2.df
   Aldicarb Month Well
 1     19.9     1    1
 2     29.6     2    1
 3     18.7     3    1
 .
10     23.3     2    3
11     22.3     3    3
12     26.9     4    3
```

```
> attach(epa.89b.aldicarb2.df)

> objects(2)
[1] "Aldicarb" "Month"    "Well"

> tol.int.norm(Aldicarb[Well==1], ti.type="upper")

Results of Distribution Parameter Estimation
--------------------------------------------

Assumed Distribution:             Normal

Estimated Parameter(s):           mean = 23.1
                                  sd   =  4.93491

Estimation Method:                mvue

Data:                             Aldicarb[Well == 1]

Sample Size:                      4

Tolerance Interval Coverage:      95%

Coverage Type:                    content

Tolerance Interval Method:        Exact

Tolerance Interval Type:          upper

Confidence Level:                 95%

Tolerance Interval:               LTL = -Inf
                                  UTL = 48.48456
```

Example 7.6: Constructing a β-Expectation Tolerance Interval for the Lognormal Distribution

The guidance document **Statistical Analysis of Ground-Water Monitoring Data at RCRA Facilities, Addendum to Interim Final Guidance** (USEPA, 1992c) contains an example on pages 52-54 in which chrysene concentrations (ppb) are compared to an alternate concentration limit (ACL) of 80 ppb. In this example, the document talks about a 95% confidence level associated with an upper tolerance limit, but in fact the example uses a β-expectation tolerance interval, not a β-content confidence intervals, so there is no confidence level associated with this tolerance interval.

Here, we'll compute the upper tolerance limit at the first well for the β-expectation tolerance interval with 95% coverage and compare it to the ACL of 80 ppb. The data for this example are assumed to come from a lognormal distribution and are stored in the data frame `epa.92c.chrysene.df` (see the help **file Datasets: USEPA (1992c)**).

```
> epa.92c.chrysene.df
   Chrysene Month Well
 1     19.7     1    1
 2     39.2     2    1
 3      7.8     3    1
 .
18     30.5     2    5
19     15.0     3    5
20     23.4     4    5

> attach(epa.92c.chrysene.df)

> objects(2)
[1] "Chrysene" "Month"    "Well"

> Well.1.dat <- Chrysene[Well==1]

> tol.int.lnorm(Well.1.dat, cov.type = "expectation",
+   ti.type = "upper")

Results of Distribution Parameter Estimation
--------------------------------------------

Assumed Distribution:            Lognormal

Estimated Parameter(s):          meanlog = 2.813216
                                 sdlog   = 0.6845029

Estimation Method:               mvue

Data:                            Well.1.dat

Sample Size:                     4

Tolerance Interval Coverage:     95%

Coverage Type:                   expectation

Tolerance Interval Method:       Exact

Tolerance Interval Type:         upper

Tolerance Interval:              LTL =   0
                                 UTL = 100.911

> log(100.911)
[1] 4.614239
```

The example in the guidance document computes the tolerance limit based on the log-transformed data and gets a value of 4.61 on the log-scale. We computed an upper tolerance limits of 100.9 ppb, so Well 1 has exceeded the ACL of 80 ppb.

Poisson Distribution

The details of how to construct tolerance intervals for the Poisson distribution are discussed in the help file for `tol.int.pois`.

Example 7.7: Constructing a β-Content Tolerance Interval for a Poisson Distribution

The guidance document **Statistical Analysis of Ground-Water Monitoring Data at RCRA Facilities, Addendum to Interim Final Guidance** (USEPA, 1992c) contains an example on page 40 in which an upper tolerance limit is computed for a β-content tolerance interval with 95% coverage and an associated 95% confidence level, based on 36 benzene concentrations (ppb) at six background wells. The benzene concentrations are modeled as having come from a Poisson distribution. Of the 36 observations, 33 are reported as "<2" (i.e., below the detection limit). In the example in the guidance document, these observations are set to half of the detection limit in order to compute the tolerance limit.

> **Note: Example 4.17** shows that these data do not fit a Poisson distribution very well.

The data for this example are stored in `epa.92c.benzene1.df` (see the help file **Datasets: USEPA (1992c)**).

```
> epa.92c.benzene1.df
   Benzene.orig Benzene Censored Month Well
 1           <2       2        T     1    1
 2           <2       2        T     2    1
 3           <2       2        T     3    1
 .
34           <2       2        T     4    6
35           10      10        F     5    6
36           <2       2        T     6    6

> attach(epa.92c.benzene1.df)

> objects(2)
[1] "Benzene"        "Benzene.orig"  "Censored"
[4] "Month"          "Well"

> dat <- Benzene

> dat[Censored] <- 1

> tol.int.pois(dat, ti.type="upper")
```

```
Results of Distribution Parameter Estimation
--------------------------------------------

Assumed Distribution:            Poisson

Estimated Parameter(s):          lambda = 1.944444

Estimation Method:               mle/mme/mvue

Data:                            dat

Sample Size:                     36

Tolerance Interval Coverage:     95%

Coverage Type:                   content

Tolerance Interval Method:       Zacks

Tolerance Interval Type:         upper

Confidence Level:                95%

Tolerance Interval:              LTL = 0
                                 UTL = 5
```

So the upper tolerance limit is 5 ppb.

Nonparametric Tolerance Intervals

The details of how to construct nonparametric tolerance intervals are discussed in the help files for `tol.int.npar` and `tol.int.npar.n`.

Example 7.8: Constructing a Nonparametric β-Content Tolerance Interval

The guidance document **Statistical Analysis of Ground-Water Monitoring Data at RCRA Facilities, Addendum to Interim Final Guidance** (USEPA, 1992c) contains an example on pages 55-56 in which a nonparametric upper tolerance limit is computed for a β-content tolerance interval with attempted coverage of 95% and an associated 95% confidence level, based on copper concentrations (ppb) at three background wells. The data for the background wells contain several observations that are reported as "<5" (below-detection-limit). Because we are constructing a nonparametric upper tolerance limit and there are enough uncensored observations, the non-detects are not a problem. The data for this example are stored in the data frame `epa.92c.copper2.df` (see the help file **Datasets: USEPA (1992c)**).

```
> epa.92c.copper2.df
   Copper.orig Copper Censored Month Well  Well.type
 1         <5      5.0        T     1    1 Background
 2         <5      5.0        T     2    1 Background
 3         7.5     7.5        F     3    1 Background
 .
38         <5      5.0        T     6    5 Compliance
39         5.6     5.6        F     7    5 Compliance
40         <5      5.0        T     8    5 Compliance

> attach(epa.92c.copper2.df)

> objects(2)
[1] "Censored"    "Copper"      "Copper.orig"
[4] "Month"       "Well"        "Well.type"

> back.dat <- Copper[Well.type=="Background"]

> tol.int.npar(back.dat, conf.level = 0.95, lb = 0,
+   ti.type = "upper")

Results of Distribution Parameter Estimation
--------------------------------------------

Assumed Distribution:            None

Data:                            back.dat

Sample Size:                     24

Tolerance Interval Coverage:     88%

Coverage Type:                   content

Tolerance Interval Method:       Exact

Tolerance Interval Type:         upper

Confidence Level:                95%

Tolerance Limit Rank(s):         24

Tolerance Interval:              LTL = 0
                                 UTL = 9.2
```

So using the maximum value of 9.2 out of the 24 observations gives us a tolerance interval with an associated confidence level of 95% but a coverage of only 88% instead of 95%. You need to increase the sample size if you want to increase the coverage, or else decrease the confidence level associated with the tolerance interval.

Let's use the function `tol.int.npar.n` to determine the minimum required sample size to achieve 95% coverage with a 95% confidence level:

```
> tol.int.npar.n(coverage = 0.95, ti.type = "upper",
+    conf.level = 0.95)
[1] 59
```

Finally, let's see what the associated confidence level is using the copper data if we set the coverage to 95%:

```
> tol.int.npar(back.dat, coverage = 0.95, lb = 0,
+    ti.type = "upper")

Results of Distribution Parameter Estimation
--------------------------------------------

Assumed Distribution:             None

Data:                             back.dat

Sample Size:                      24

Tolerance Interval Coverage:      95%

Coverage Type:                    content

Tolerance Interval Method:        Exact

Tolerance Interval Type:          upper

Confidence Level:                 71%

Tolerance Limit Rank(s):          24

Tolerance Interval:               LTL = 0
                                  UTL = 9.2
```

So with 24 observations, using the maximum value as the upper tolerance limit for a β-content tolerance interval with 95% coverage yields an associated confidence level of 71%.

Summary

- **Table 7.1** lists the functions available in EnvironmentalStats for S-Plus for computing predictions intervals and associated quantities (see the help file **Prediction and Tolerance Intervals**). The form of the names of these functions is `pred.int.`*`abb`*, where *`abb`* denotes the abbreviation of the distribution name (see column 2 of **Table 5.1**).

- A ***prediction interval*** for the next k observations is constructed from a sample data set in such a way that if a new sample were taken many, many times and a prediction interval was constructed for each new sample, the prediction intervals would contain k future observations $(1-\alpha)100\%$ of the time. The quantity $(1-\alpha)100\%$ is called the ***confidence level*** associated with the prediction interval.

- **Table 7.2** lists the functions available in EnvironmentalStats for S-Plus for computing tolerance intervals and associated quantities (see the help file **Prediction and Tolerance Intervals**). The form of the names of these functions is `tol.int.`*`abb`*, where *`abb`* denotes the abbreviation of the distribution name (see column 2 of **Table 5.1**).

- A ***tolerance interval*** is constructed to contain a certain percentage of the population (i.e., a certain percentage of all future observations). A ***β-content tolerance interval*** contains *at least* $\beta100\%$ of the population with confidence level $(1-\alpha)100\%$. A ***β-expectation tolerance interval*** contains *on average* $\beta100\%$ of the population.

- There is a one-to-one correspondence between tolerance intervals and confidence intervals for quantiles or percentiles (see Chapter 6).

- Prediction and tolerance intervals have been suggested as statistical tools to use to overcome the multiple comparisons problem for environmental monitoring programs that require frequent statistical tests for the presence of contamination.

CHAPTER 8 :

Hypothesis Tests & Statistical Models

The heart of classical statistics is hypothesis tests. If you are comparing chemical concentrations between a "background" area and a "potentially contaminated" area, how different do the concentrations in these two areas have to be before you decide that the "potentially contaminated" area is in fact contaminated?

S-PLUS contains several functions for performing classical statistical hypothesis tests, such as t-tests, analysis of variance, linear regression, nonparametric tests, quality control procedures and time series analysis (see the S-PLUS documentation and help files).

ENVIRONMENTALSTATS for S-PLUS contains functions for a few statistical tests that are not included in S-PLUS but that are often or sometimes used in environmental statistics. These tests are the Shapiro-Francia, Shapiro-Wilk, and Probability Plot Correlation Coefficient goodness-of-fit tests; Kendall's seasonal test for trend; and the quantile test for a shift in the tail of the distribution. This chapter briefly describes the theory of hypothesis tests, explains the new tests included in ENVIRONMENTALSTATS for S-PLUS, and gives examples of performing these tests.

Overview

Table 8.1 lists the functions available in S-PLUS and ENVIRONMENTALSTATS for S-PLUS for performing classical univariate hypothesis tests (see the S-Plus help files **Statistical Inference**, **ANOVA Models**, **Regression**, and **Nonparametric Statistics**, and the ENVIRONMENTALSTATS for S-PLUS help files **Goodness-of-Fit Tests** and **Hypothesis Tests**). Functions with one asterisk (*) are available in S-PLUS but have been modified in ENVIRONMENTALSTATS for S-PLUS; functions with two asterisks (**) are available only in ENVIRONMENTALSTATS for S-PLUS.

Test Name	Function(s)	Description

Goodness-of-Fit	`chisq.gof*`
	`ks.gof*`
	`ppcc.evd.gof**`
	`ppcc.norm.gof**`
	`sf.gof**`
	`sw.gof**`
		Chi-Square, Kolmogorov-Smirnov, Probability Plot Correlation Coefficient, Shapiro-Francia, and Shapiro-Wilk goodness-of-fit tests.

Student's t-Test	`t.test`	Student's t-test for one-sample, two-samples, and paired-samples.
Wilcoxon Rank Sum, Signed Rank Sum	`wilcox.test`	Nonparametric test for one-sample, two-samples, and paired-samples.
Two-Sample Linear Rank Test	`two.sample.linear.rank.test**`	Two-Sample linear rank test to detect a shift.
Quantile Test	`quantile.test**`	Quantile test to detect a shift in a portion of the "treated" population.
Two-sample Variance Test	`var.test`	F-test to compare two variances.
Analysis of Variance (ANOVA)	`aov`	One-way, Two-way, and Multi-way fixed effects Analysis of Variance.
Nonparametric One-Way ANOVA	`kruskal.test`	Kruskal-Wallis rank sum test for one-way ANOVA.
Nonparametric Two-Way ANOVA	`friedman.test`	Friedman's rank sum test for unreplicated blocked data.
Binomial Test	`binom.test`	Inference on one binomial proportion.

* Available in S-PLUS but modified in ENVIRONMENTALSTATS for S-PLUS

** Available only in ENVIRONMENTALSTATS for S-PLUS.

Table 8.1: Functions for classical univariate hypothesis tests.

Test Name	Function(s)	Description
Chi-Square Test	`chisq.test`	Chi-Square test for a two-dimensional contingency table.
Fisher's Exact Test	`fisher.test`	Fisher's conditionally exact test for a two-dimensional contingency table.
Proportions Test	`prop.test`	Approximate chi-square test for one or several binomial proportions.
McNemar's Test for Paired Count Data	`mcnemar.test`	McNemar's chi-square test for two-dimensional contingency table.
Mantel-Haenszel Test	`mantelhaen.test`	Mantel-Haenszel chi-square test on a three-dimensional contingency table.
Linear Regression	`lm`	Simple and multiple linear regression.
Multiple Comparisons	`multicomp`	Simultaneous or non-simultaneous confidence bounds for estimable linear combinations of parameters of a fixed effects linear model.
Correlation Test	`cor.test`	Test for zero correlation based on Pearson's product moment, Kendall's tau, or Spearman's rank correlation.
Nonparametric Test for Trend	`kendall.trend.test`** `seasonal.kendall.trend.test`**	Kendall's test for trend; also modified for seasons and optionally serial correlation.

* Available in S-PLUS but modified in ENVIRONMENTALSTATS for S-PLUS
** Available only in ENVIRONMENTALSTATS for S-PLUS.

Table 8.1 (continued): Functions for classical univariate hypothesis tests.

S-PLUS also contains several functions for performing time series analysis, quality control analysis, survival analysis, and multivariate analysis (see the S-PLUS documentation and help files for **Time Series**, **Quality Control**, **Survival Analysis**, and **Multivariate Techniques**).

The rest of this chapter gives a brief discussion of the theory of hypothesis tests, and shows you how to use the new functions in ENVIRONMENTALSTATS for S-PLUS for performing goodness-of fit tests, nonparametric two-sample tests, and nonparametric tests for trend.

Hypothesis Tests

A ***hypothesis test*** or ***significance test*** is a formal mathematical mechanism for objectively making a decision in the face of uncertainty. In order to perform a simple hypothesis test, you must reduce your decision-making problem to a binary choice: Yes/No, On/Off, Equal/Not Equal, Bigger Than/Less than or Equal To, etc. One of the choices is designated the null hypothesis, usually denoted H_0 (H-naught), and the other choice is termed the alternative hypothesis, usually denoted H_a.

A hypothesis test is usually used to answer a question about the value of a population parameter. A ***two-sided hypothesis*** test about a population parameter θ is used to test the null hypothesis

$$H_0: \theta = \theta_0 \tag{8.1}$$

versus the alternative hypothesis

$$H_a: \theta \neq \theta_0 \tag{8.2}$$

A ***lower one-sided hypothesis*** test is used to test the null hypothesis

$$H_0: \theta \geq \theta_0 \tag{8.3}$$

versus the alternative hypothesis

$$H_a: \theta < \theta_0 \tag{8.4}$$

and an ***upper one-sided hypothesis*** test is used to test the null hypothesis

$$H_0: \theta \leq \theta_0 \tag{8.5}$$

versus the alternative hypothesis

$$H_a: \theta > \theta_0 \tag{8.6}$$

Parametric vs. Nonparametric Tests

Hypotheses tests are usually based on a ***test statistic***, say T, which is computed from a random sample from the population. For a ***parametric test***, T is usually some estimator of θ and the distribution of T under the null hypothesis depends on the distribution of the population (e.g., normal, lognormal, Poisson, etc.). For a ***nonparametric*** or ***distribution-free test***, T is usually based on the ranks of the data in the random sample, and the distribution of T under the null hypothesis does not depend on the distribution of the population.

For example, for a two-sample t-test, the test statistic is a scaled version of the difference between the two sample means, and both populations are assumed to be normally distributed. For the Wilcoxon rank sum test, the test statistic is the sum of the ranks in the first sample, and the distribution of this statistic, under the null hypothesis, does not depend on the distribution of the two populations.

Type I and Type II Errors (Significance Level and Power)

As stated above, a hypothesis test involves using a particular summary statistic (the test statistic) based on data collected from an experiment or set of observations (e.g., a sample mean or proportion) to determine which choice to make. A test statistic is a ***random*** quantity; if you repeat the experiment or get new observations, you will most likely get a different value for the test statistic. Because you are making your decision based on the value of a random quantity, you will sometimes make the "wrong" choice.

There are two kinds of wrong choices, or errors, associated with a hypothesis test: a Type I error and a Type II error. **Table 8.2** below illustrates the framework of a hypothesis test.

	The Truth	
Your Decision	**H_0 True**	**H_a True**
Do Not Reject H_0	Correct Decision (Probability = 1-α)	**Type II Error** (Probability = β)
Reject H_0	**Type I Error** (Probability = α)	Correct Decision (Probability = 1 - β)

Table 8.2: The framework of a hypothesis test.

If the null hypothesis is true, you can either make the correct decision or you can make a Type I error. If the alternative hypothesis is true, you can either make the correct decision or make a Type II error. The probability of making a Type I error is usually denoted by the Greek letter α (alpha), and the probability of making a Type II error is usually denoted by the Greek letter β (beta).

Note that in the table above, the phrase "Do Not Reject H_0" is used instead of "Decide H_0 is True". This is because in the framework of hypothesis testing, you assume the null hypothesis is true unless you have enough evidence to reject it. If you end up not rejecting H_0 because of the value of your test statistic, you may have unknowingly committed a Type II error; i.e., the alternative hypothesis is really true, but you did not have enough evidence to reject H_0.

The quantity α (the probability of making a Type I error) is called the ***significance level*** or ***α-level*** of the hypothesis test. The quantity 1-β (the probability of correctly rejecting H_0 when it is not true) is called the ***power*** of the hypothesis test. (Sometimes α or 100α% is called the ***false positive rate***, and β or 100β% is called the ***false negative rate***. These terms are more frequently used when the hypothesis test is based on a medical screening test to determine whether a person has a particular disease or medical condition.)

The critical aspect of any hypothesis test is the acceptable values for the probabilities of making a Type I or Type II error. ***The choice of values for α and β is a subjective, policy decision***. The possible choices for α and β are limited by the sample size of the experiment (usually denoted n), the variability inherent in the data (usually denoted σ), and the magnitude of the difference between the null and alternative hypothesis (usually denoted δ or Δ). Conventional choices for α are 0.01, 0.05, and 0.1, but these choices should be made in the context of balancing the cost of a Type I and Type II error. For most hypothesis tests, there is a well-defined relationship between α, β, n and the scaled difference δ/σ (see, for example, the help file for `plot.t.test.design`).

P-Values

When you perform a hypothesis test, you usually compute a quantity called the ***p-value***. The p-value is the probability of seeing a test statistic as extreme or more extreme than the one you observed, assuming the null hypothesis is true. If the p-value is less than or equal to the specified value of α (the Type I Error level), you reject the null hypothesis. If the p-value is greater than α, you do not reject the null hypothesis.

The Relationship Between Hypothesis Tests and Confidence Intervals

Consider the null hypothesis shown in **Equation (8.1)**, where θ is some population parameter of interest (e.g., mean, proportion, etc.). There is a one-to-one relationship between hypothesis tests concerning θ and a confidence interval for this parameter. A $(1-\alpha)100\%$ confidence interval for θ consists of all possible values of θ that are associated with not rejecting the null hypothesis at significance level α.

Goodness-of-Fit Tests

Goodness-of-fit tests are used to determine whether a particular data set appears to have come from a specific probability distribution. Sometimes, you specify not only the distribution (e.g., normal, lognormal, Poisson, etc.), but also the parameters associated with the distribution as well. More often, however, you are interested in simply testing whether the data appear to come from a probability distribution without specifying the exact values of the parameters of the distribution.

S-PLUS comes with two built-in functions for performing goodness-of-fit tests: the function `chisq.gof` performs a chi-square goodness of fit test, and the function `ks.gof` performs the Kolmogorov-Smirnov goodness-of-fit test. ENVIRONMENTALSTATS for S-PLUS includes special methods of printing and plotting the results of goodness-of-fit tests, so both `chisq.gof` and `ks.gof` have been slightly modified in ENVIRONMENTALSTATS for S-PLUS.

ENVIRONMENTALSTATS for S-PLUS also includes four other goodness-of-fit tests: `sw.gof` performs the Shapiro-Wilk goodness-of-fit test, `sf.gof` performs the Shapiro-Francia goodness-of-fit test, `ppcc.norm.gof` performs the probability plot correlation coefficient goodness-of-fit test for a normal distribution, and `ppcc.evd.gof` performs the probability plot correlation coefficient goodness-of-fit test for the extreme value distribution.

Most of the time goodness-of-fit tests are used to assess how well a data set appears to come from a normal or lognormal distribution, because classical parametric hypothesis tests assume the data come from a normal distribution. The assumption of normality, however is not as important as the assumption of independent observations and equal variance (e.g., Millard, 1985).

For large sample sizes, almost any goodness-of-fit test will reject the null hypothesis, since "real" data are never distributed according to any theoretical distribution (Conover, 1980, p.367). For most cases, however, the distribution of "real" data is close enough to some theoretical distribution that fairly accurate results may be provided by assuming that particular theoretical distribution. Similarly, for small sample sizes, most goodness-of-fit tests will not have much power of detecting deviations from the hypothesized distribution unless the deviations are quite extreme.

Example 8.1: Testing the Goodness-of-Fit to a Normal Distribution

The guidance document **Statistical Analysis of Ground-Water Monitoring Data at RCRA Facilities, Addendum to Interim Final Guidance** (USEPA, 1992c) contains an example on pages 6-19 involving a set of nickel concentrations (ppb) at four monitoring wells. The nickel concentrations are assumed to come from a normal distribution and several different goodness-of-fit tests are performed to assess the validity of this assumption. The data for this example are stored in the data frame `epa.92c.nickel1.df` (see the help file **Datasets: USEPA (1992c)**).

First look at the data: create a histogram and normal Q-Q plot.

```
> epa.92c.nickel1.df
   Nickel Month Well
 1   58.8     1    1
 2    1.0     2    1
 3  262.0     3    1
 4   56.0     4    1
 5    8.7     5    1
 6   19.0     1    2
 7   81.5     2    2
 8  331.0     3    2
 9   14.0     4    2
10   64.4     5    2
11   39.0     1    3
12  151.0     2    3
13   27.0     3    3
14   21.4     4    3
15  578.0     5    3
16    3.1     1    4
17  942.0     2    4
18   85.6     3    4
19   10.0     4    4
20  637.0     5    4

> attach(epa.92c.nickel1.df)

> objects(2)
[1] "Month"  "Nickel" "Well"
```

```
> hist(Nickel, xlab="Nickel (ppb)", ylab="Frequency")
```

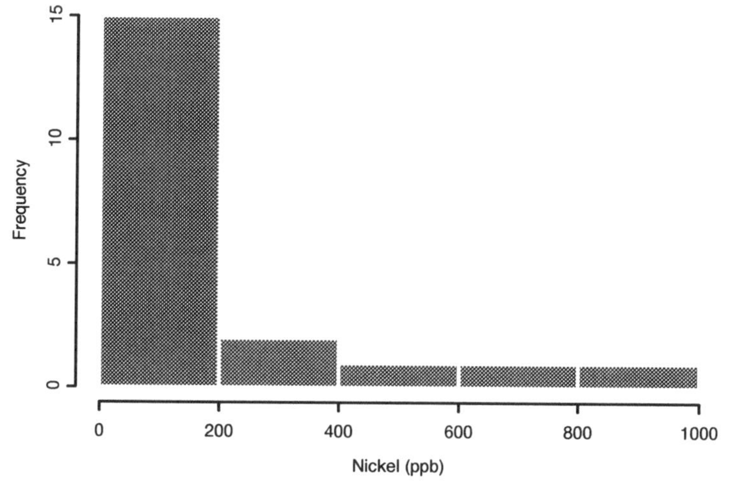

Figure 8.1: Histogram of the nickel data.

```
> qqplot(Nickel)
```

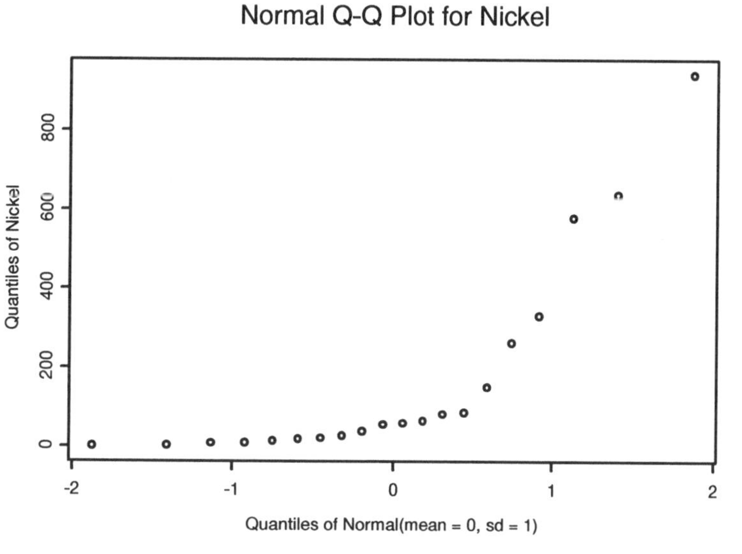

Figure 8.2: Normal Q-Q plot for the nickel data.

The data certainly do not appear to come from a normal distribution; they are much too skewed. A lognormal distribution may provide a better fit (see **Example 8.3** below).

Now perform the Shapiro-Wilk goodness-of-fit test.

```
> sw.gof(Nickel)

Results of Goodness-of-Fit Test
-------------------------------

Test Method:                    Shapiro-Wilk GOF

Hypothesized Distribution:      Normal

Estimated Parameter(s):         mean = 169.525
                                sd   = 259.7175

Estimation Method:              mvue

Data:                           Nickel

Sample Size:                    20

Test Statistic:                 W = 0.6785233

Test Statistic Parameter:       n = 20

P-value:                        2.157839e-5

Alternative Hypothesis:         True cdf does not equal the
                                Normal Distribution.
```

The extremely small p-value indicates that a normal distribution does not fit these data very well. The Shapiro-Francia test (`sf.gof`) and probability plot correlation coefficient test (`ppcc.norm.gof`) yield similar results:

Example 8.2: Plotting the Results of a Goodness-of-Fit Test

ENVIRONMENTALSTATS for S-PLUS includes a special method for plotting the results of goodness-of-fit tests that allows you to show the observed and fitted distribution, observed and fitted cumulative distribution, Q-Q plots, and test results. First, let's just look at the observed and fitted distribution and the observed and fitted cumulative distribution.

```
> sw.gof.norm <- sw.gof(Nickel)

> plot(sw.gof.norm)

Make a plot selection (or 0 to exit):

1: plot: All
2: plot: Observed Distribution Overlaid with Fitted
   Distribution
3: plot: Observed CDF Overlaid with Fitted CDF
4: plot: Observed Quantiles vs. Fitted Quantiles
   (Q-Q Plot)
5: plot: Tukey Mean-Difference Q-Q Plot
6: plot: Results of GOF Test
Selection: 2

Make a plot selection (or 0 to exit):

1: plot: All
2: plot: Observed Distribution Overlaid with Fitted
   Distribution
3: plot: Observed CDF Overlaid with Fitted CDF
4: plot: Observed Quantiles vs. Fitted Quantiles
   (Q-Q Plot)
5: plot: Tukey Mean-Difference Q-Q Plot
6: plot: Results of GOF Test
Selection: 3
```

The results of choosing plot types 2 and 3 are shown in **Figure 8.3** and **Figure 8.4** below. It is obvious that a normal distribution does not fit these data very well.

Now let's call `plot.gof.summary` to summarize the results of the goodness-of-fit test (see **Figure 8.5**). The function `plot.gof.summary` creates four plots on one page: the observed distribution overlaid with the fitted distribution, the observed cdf overlaid with the fitted cdf, a Q-Q plot, and the results of the goodness-of-fit test. All of the plots show that the distribution of the nickel data is skewed to the right compared to a normal distribution.

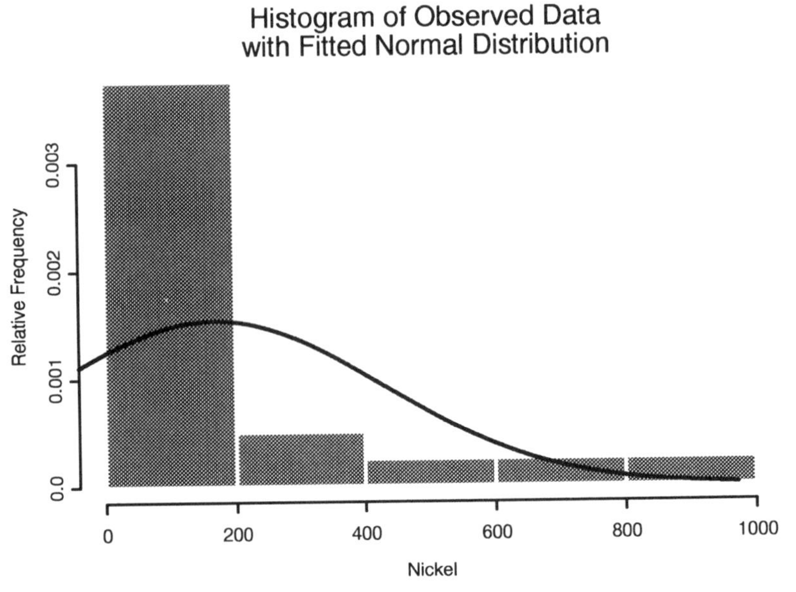

Figure 8.3: Observed and fitted distribution for nickel data.

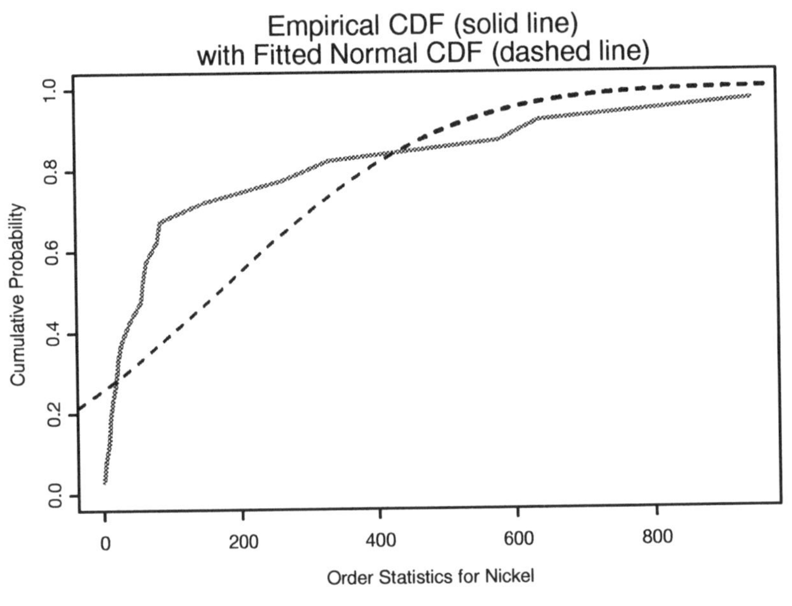

Figure 8.4: Observed and fitted cumulative distribution for the nickel data.

```
> plot.gof.summary(sw.gof.norm)
```

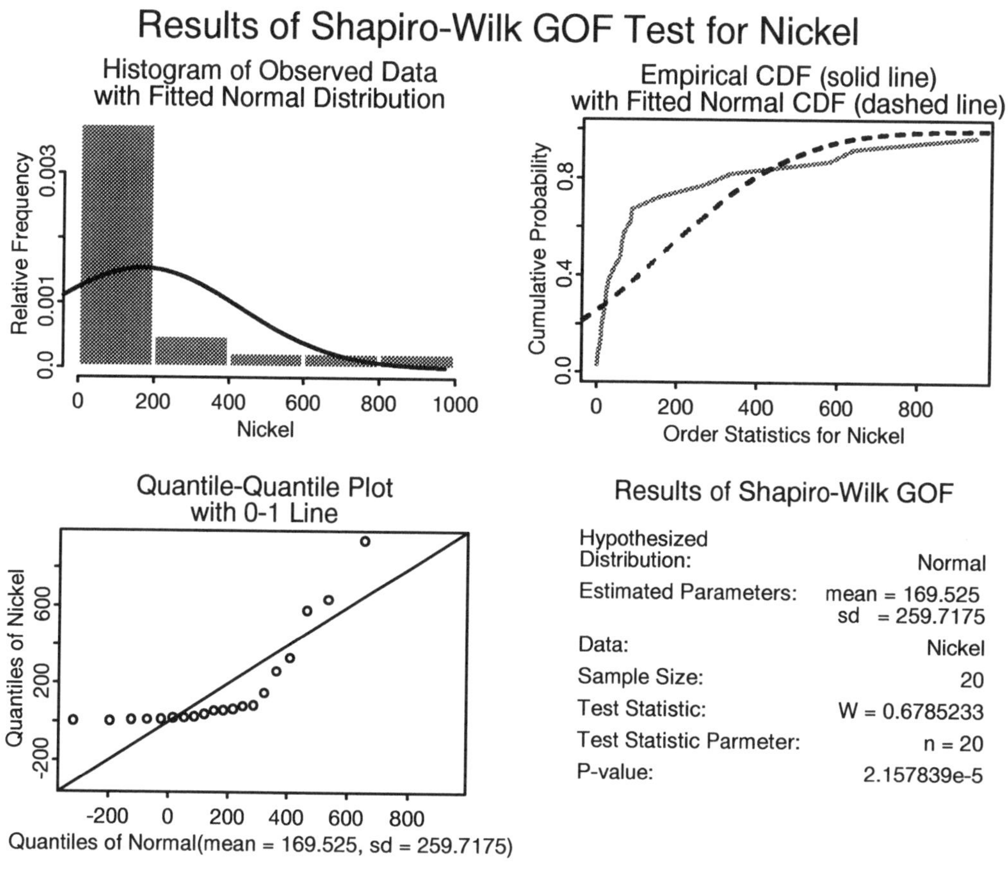

Figure 8.5: Summary of Shapiro-Wilk normal goodness-of-fit test for the nickel data.

Example 8.3: Testing the Goodness-of-Fit to a Lognormal Distribution

We'll repeat the last example, but assume the data come from a lognormal distribution instead.

```
> sw.gof(Nickel, distribution = "lnorm")

Results of Goodness-of-Fit Test
-------------------------------

Test Method:                    Shapiro-Wilk GOF

Hypothesized Distribution:      Lognormal

Estimated Parameter(s):         meanlog = 3.918529
                                sdlog   = 1.801404
```

```
Estimation Method:              mvue

Data:                           Nickel

Sample Size:                    20

Test Statistic:                 W = 0.9788851

Test Statistic Parameter:       n = 20

P-value:                        0.9189022

Alternative Hypothesis:         True cdf does not equal the
                                Lognormal Distribution.
```

```
> sw.gof.lnorm <- sw.gof(Nickel, distribution = "lnorm")

> plot.gof.summary(sw.gof.lnorm, digits = 2)
```

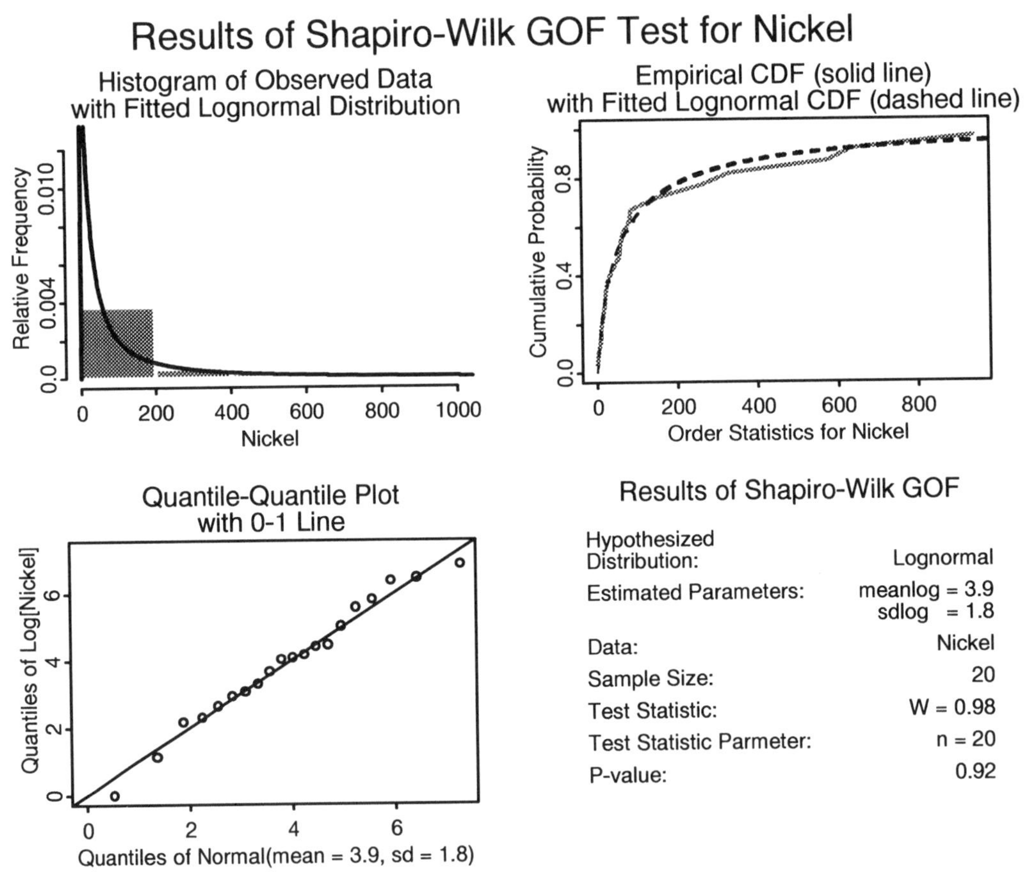

Figure 8.6: Summary of Shapiro-Wilk lognormal goodness-of-fit test for the nickel data.

Given the large p-value and no evidence of large deviations in the summary plots, it appears that a lognormal distribution is an adequate fit for these data.

Nonparametric Two-Sample Tests

ENVIRONMENTALSTATS for S-PLUS includes two different functions for comparing two samples without assuming a particular form for the underlying probability distribution. A two-sample linear rank test is used to detect a shift in location between two populations. The quantile test is used to detect a shift in a portion of one of the populations (i.e., a shift in the tail). Each of these tests is discussed below.

Two-Sample Linear Rank Tests

The function `two.sample.linear.rank.test` let's you perform a two-sample linear rank test. Two sample linear rank tests compare two samples by combining the samples and ranking the observations, then summing up some function of the ranks (the ***scores***) for the first sample (Hettmansperger, 1984, Chapter 3). The Wilcoxon rank sum test is a special kind of two-sample linear rank test in which the function of the ranks is the identity function, so you just sum up the raw ranks of the first sample (see the S-PLUS help file `wilcox.test`). Other possible functions to use include the inverse of the normal (Gaussian) cumulative distribution function (normal scores test), the sign of the quantity [rank minus the mean of the ranks] (Mood's median test), and a form of the inverse of the rank (Savage scores test).

Based on the theory of locally most powerful tests, certain functions are best for certain underlying probability distributions (see the help file for `two.sample.linear.rank.test`). General two-sample linear rank tests have been included in ENVIRONMENTALSTATS for S-PLUS mainly to give a foundation for two-sample linear rank tests for censored data (see **Chapter 9: Censored Data**). In practice, for most data sets there should be little difference in the results between two-sample linear rank tests based on different scores.

Example 8.4: Comparing TcCB Concentrations at Two Different Sites

The guidance document **Statistical Methods for Evaluating the Attainment of Cleanup Standards, Volume 3: Reference-Based Standards for Soils and Solid Media** (USEPA, 1994b) contains an example on pages 6.22-6.26 involving measurements of 1,2,3,4-Tetrachlorobenzene (TcCB) concentrations (ppb) in the soil at a Reference area and a Cleanup area. There are 47 measurements from the Reference area and 77 from the Cleanup area. The question of interest is whether there is any residual contamination left in the Cleanup area.

The data for this example are stored in the data frame `epa.94b.tccb.df` (see the help file **Datasets: USEPA (1994b)**). One of the 77 observations in the Cleanup area is recorded as ND (not detected). For the purposes of this example we'll set this to the smallest detected value in the Cleanup area (0.09 ppb).

First, look at the data:

```
> epa.94b.tccb.df
    TcCB.orig   TcCB Censored       Area
  1      0.22   0.22        F Reference
  2      0.23   0.23        F Reference
  3      0.26   0.26        F Reference
  .
122     18.40  18.40        F Cleanup
123     51.97  51.97        F Cleanup
124    168.64 168.64        F Cleanup

> attach(epa.94b.tccb.df)

> objects(2)
[1] "Area"      "Censored"  "TcCB"      "TcCB.orig"

> levels(Area)
[1] "Cleanup"   "Reference"

> boxplot(split(TcCB, Area), ylab = "TcCB (ppb)")
```

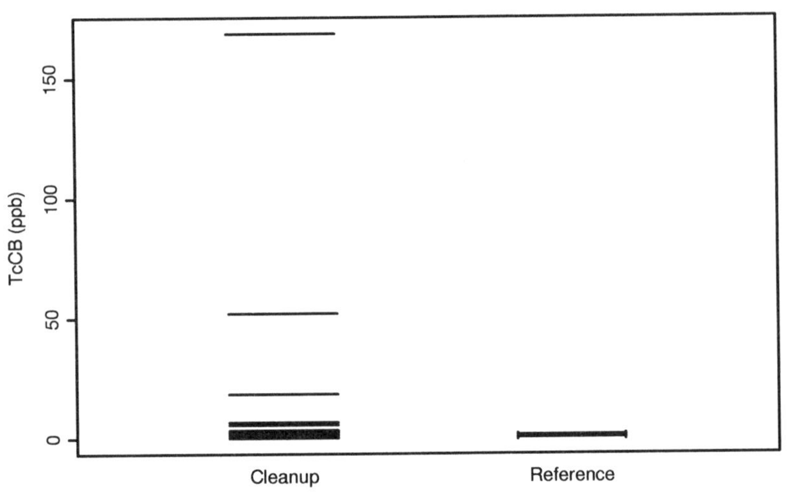

Figure 8.7: Boxplots of the TcCB data.

These data are extremely skewed. Let's look at the log-transformed data.

```
> boxplot(split(log(TcCB), Area),
+    ylab = "log [ TcCB (ppb) ]")
```

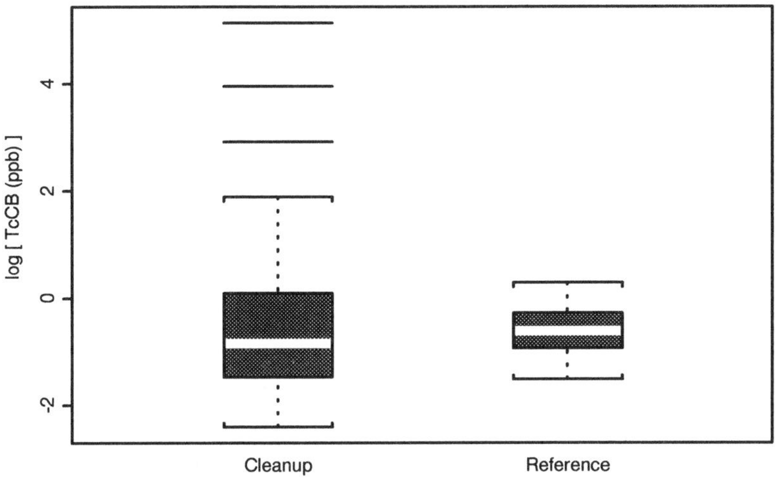

Figure 8.8: Boxplots for log-transformed TcCB data.

It looks like most of the observations in the Cleanup area are of a similar magnitude to the observations in the Reference area, but there appear to be some extreme values in the Cleanup area.

Now let's compare the two areas with the Wilcoxon rank sum test and Mood's median test. Because we are only interested in whether the Cleanup area concentrations are larger than the Reference area concentrations, we'll perform a one-sided test.

```
> summary(Area)
 Cleanup Reference
      77        47

> ref <- TcCB[Area == "Reference"]

> cleanup <- TcCB[Area == "Cleanup"]
```

By default, the function `two.sample.linear.rank.test` performs the Wilcoxon rank sum test. By setting the argument `test="moods.median"` , we can also perform Mood's median test.

```
> two.sample.linear.rank.test(cleanup, ref,
+   alternative = "greater")

Results of Hypothesis Test
--------------------------

Null Hypothesis:              Fx(t) = Fy(t)

Alternative Hypothesis:       Fx(t) > Fy(t)
                              for at least one t

Test Name:                    Two-Sample Linear Rank Test:
                              Wilcoxon Rank Sum Test
                              Based on Normal Approximation

Data:                         x = cleanup
                              y = ref

Sample Sizes:                 nx = 77
                              ny = 47

Test Statistic:               z = 1.171872

P-value:                      0.1206242

> two.sample.linear.rank.test(cleanup, ref,
+   alternative = "greater", test = "moods.median")

Results of Hypothesis Test
--------------------------

Null Hypothesis:              Fx(t) = Fy(t)

Alternative Hypothesis:       Fx(t) > Fy(t)
                              for at least one t

Test Name:                    Two-Sample Linear Rank Test:
                              Mood's Median Test
                              Based on Normal Approximation

Data:                         x = cleanup
                              y = ref

Sample Sizes:                 nx = 77
                              ny = 47

Test Statistic:               z = 1.659208

P-value:                      0.04853697
```

Note that in this case Mood's median test is significant at the 0.05 level, whereas the Wilcoxon rank test is not even significant at the 0.10 level.

The Quantile Test

The quantile test is used to detect a shift in a portion of one of the populations (i.e., a shift in the tail) (Johnson et al., 1987). That is, rather than testing for the whole distribution of the "treated" population being shifted relative to the "control" population, the quantile test compares the upper tails of the two distributions. This kind of test seems well-suited for testing for residual contamination at a cleaned up area.

The ENVIRONMENTALSTATS for S-PLUS function `quantile.test` allows you to perform the quantile test. In order to perform this test, you have to decide on a rank r. Once r is set, you rank the observations in the combined samples, and then compute k, the number of observations from the "treated" group with ranks greater than or equal to r. If k is "too large", the test rejects the null hypothesis that the two populations are the same in favor of the alternative hypothesis that a portion of the "treated" group is shifted to the right of the "control" group.

The best choice for r depends on the sample sizes from each group and how small you want the p-value to be. The function `quantile.test.p.value` can help you decide on the value of r.

Example 8.5: Using the Quantile Test to Compare the Reference and Treatment Areas

The guidance document **Statistical Methods for Evaluating the Attainment of Cleanup Standards, Volume 3: Reference-Based Standards for Soils and Solid Media** (USEPA, 1994b) contains an example on pages 7.23-7.24 in which the quantile test is applied to the TcCB data discussed above. We'll reproduce the example here.

First let's determine what value we want to use for r assuming we set the significance level to 0.01.

```
> p.vals <- quantile.test.p.value(m=47, n=77,
+    r=c(rep(8,3), rep(9,3), rep(10,3)),
+    k=c(6,7,8,7,8,9,8,9,10))

> round(p.vals, 3)
[1] 0.355 0.122 0.019 0.264 0.081 0.011 0.193
[8] 0.053 0.007
```

Using r=9, we'll get a p-value of 0.011 if k=9. Now let's perform the test.

```
> quantile.test(ref, cleanup, target.r = 9)

Results of Hypothesis Test
--------------------------

Null Hypothesis:               e = 0

Alternative Hypothesis:        0 < e <= 1, where
                               Fy(t) = (1-e)*Fx(t) + e*Fz(t),
                               Fz(t) <= Fx(t) for all t,
                               and Fx != Fz

Test Name:                     Quantile Test

Data:                          x = ref
                               y = cleanup

Sample Sizes:                  nx = 47
                               ny = 77

Test Statistics:               k (# y obs of r largest) = 9
                               r                        = 9

Test Statistic Parameters:     m           = 47
                               n           = 77
                               quantile.ub =  0.928

P-value:                       0.01136926
```

So the quantile test definitely picks up the shift in the tail of the distribution of the TcCB concentrations in the Cleanup area (see **Figure 8.8**).

Nonparametric Tests for Trend

One way to test for trend over time in a series of measurements is to use simple linear regression and use time as the predictor variable. This method works well if the trend is linear, the observations are not serially correlated, and if the distribution of the observations is normal (Gaussian). Nonparametric tests for trend allow you to test for trend without having to assume the trend in linear (only monotonic) and without having to make an assumption about the distribution of the observations.

Nonparametric tests for trend are based on nonparametric estimates of correlation between two random variables, say X and Y, where X is some measure of time. Nonparametric estimates of correlation such as Spearman's rho and Kendall's tau use the ranks of the data to assess the strength of the relationship between X and Y. Spearman's rho is simply the Pearson product moment correlation based on the ranks (Conover, 1980, p.252). Kendall's tau is based on the number of concordant pairs minus the number of discordant pairs, where a pair (x_1, y_1) and (x_2, y_2) is concordant if $x_1 > x_2$ implies $y_1 > y_2$, or $x_1 < x_2$ implies $y_1 < y_2$ (Conover, 1980, p.256).

In the case of testing for trend, the variable x is assumed to denote time and to increase sequentially. In this case, Kendall's tau is based on looking simply at the number of times $y_i < y_j$ minus the number of times $y_i > y_j$, where $i < j$.

The ENVIRONMENTALSTATS for S-PLUS function `kendall.trend.test` allows you to test for trend using Kendall's tau statistic. It also returns a nonparametric estimate of slope, which is relevant only if you want to assume the trend is linear. See the help file for this function for more information.

Hirsch et al. (1982) introduced a modification of Kendall's test for trend that allows for seasonality in observations collected over time. They call this test the ***seasonal Kendall test***. Their test is appropriate for testing for trend in each season when the trend is always in the same direction across all seasons. Van Belle and Hughes (1984) introduced a heterogeneity test for trend which is appropriate for testing for trend in any direction in any season. Hirsch and Slack (1984) proposed an extension to the seasonal Kendall test that allows for serial dependence in the observations. The ENVIRONMENTALSTATS for S-PLUS function `seasonal.kendall.trend.test` includes all of these tests, as well as an extension of the van Belle-Hughes heterogeneity test to the case of serial dependence. See the help file for this function for more information.

Example 8.6: Testing for Monotonic Trend Using Simulated Data

In this example, we'll generate 25 observations from a model with an exponential trend and test the null hypothesis of no trend over time. Note that in this case the estimates of slope and intercept only indicate the linear part of the trend because the true underlying model is exponential, not linear.

First we'll generate the data and look at the true and observed relationships:

```
> set.seed(250)

> errors <- rnorm(25, sd=.25)

> x <- 1:25

> mu.y <- exp(x/25)

> y <- mu.y * exp(errors)

> plot(x, mu.y, main="True Model")

> plot(x, y, main="Observed Data")
```

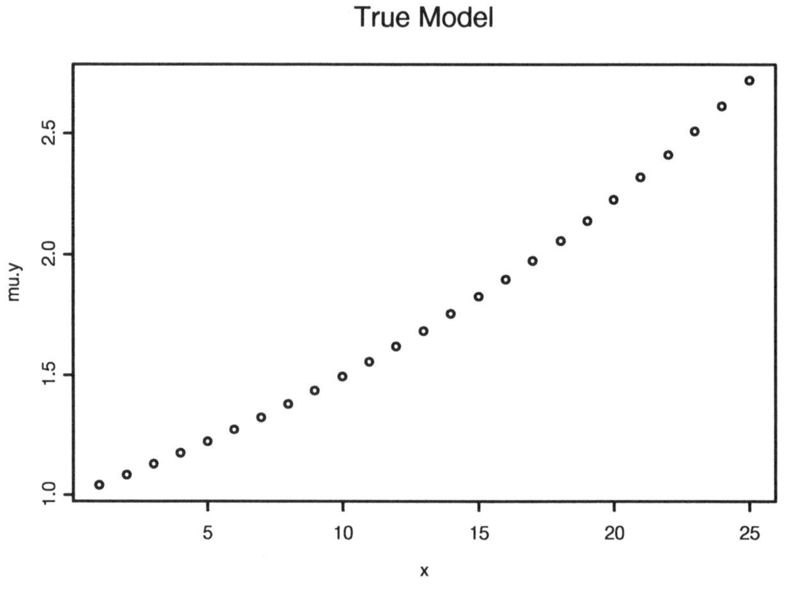

Figure 8.9: True relationship between *X* and mean value of *Y*.

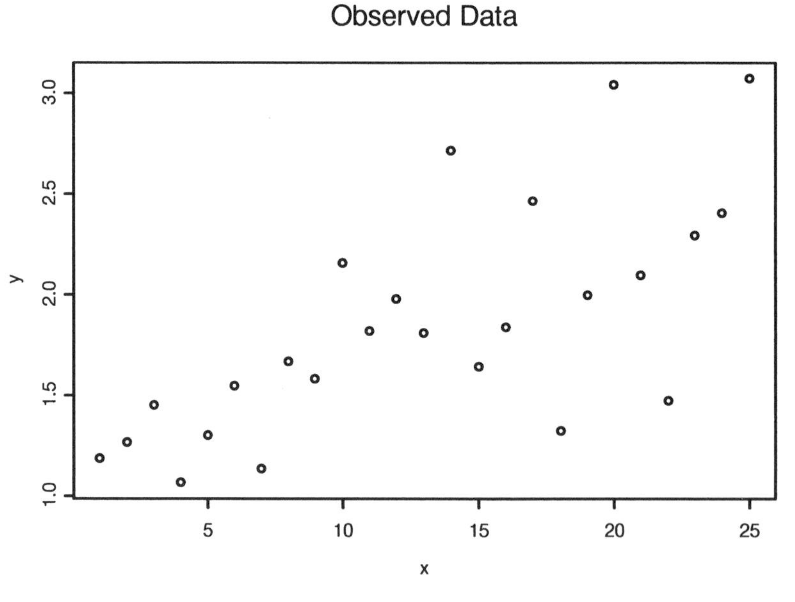

Figure 8.10: Observed relationship between *X* and *Y*.

Now let's perform the test for trend:

```
> kendall.trend.test(y, x)

Results of Hypothesis Test
--------------------------

Null Hypothesis:               tau = 0

Alternative Hypothesis:        True tau is not
                               equal to 0

Test Name:                     Kendall's Test for Trend
                               (with continuity correction)

Estimated Parameter(s):        tau       = 0.5733333
                               slope     = 0.0522136
                               intercept = 1.130756

Estimation Method:             slope:      Theil/Sen
                                           Estimator
                               intercept:  Conover's
                                           Estimator

Data:                          y = y
                               x = x

Sample Size:                   25

Test Statistic:                z = 3.9937

P-value:                       0.00006505027

Confidence Interval for:       slope

Confidence Interval Method:    Gilbert's Modification
                               of Theil/Sen Method

Confidence Interval Type:      two-sided

Confidence Level:              95%

Confidence Interval:           LCL = 0.03538303
                               UCL = 0.07796641
```

The small p-value indicates significant trend.

Example 8.7: Testing for Trend Within Two Seasons Using Simulated Data

In this example, we'll generate observations for 25 years and 2 seasons from a model with an exponential trend in each season, and test the null hypothesis of no trend over time. Note that in this case the estimates of slope and intercept should be ignored because the true underlying model is exponential, not linear.

First we'll generate the data and look at the true and observed relationships:

```
> set.seed(250)

> errors <- matrix(rnorm(50, sd=.25), 25, 2)

> x <- matrix(1:25, 25, 2)

> season.effect <- cbind(rep(0,25), rep(1,25))

> mu.y <- exp(x/25 + season.effect)

> y <- mu.y * exp(errors)

> matplot(x, mu.y, xlab="Time",
+   ylab="True Mean for Each Season", main="True Model")

> matplot(x, y, xlab="Time",
+   ylab="Observations for Each Season",
+   main="Observed Data")
```

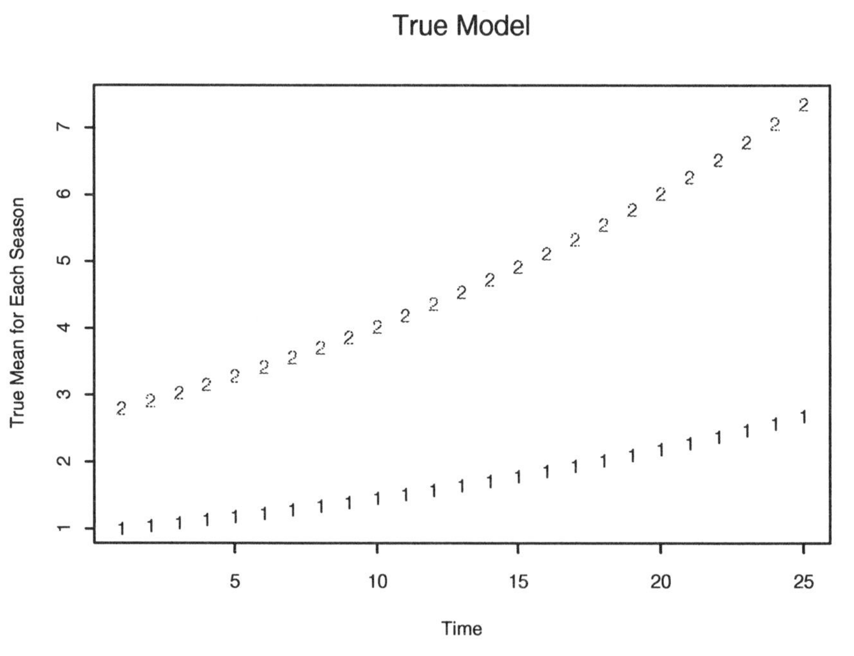

Figure 8.11: True relationship between X and mean value of Y for each season.

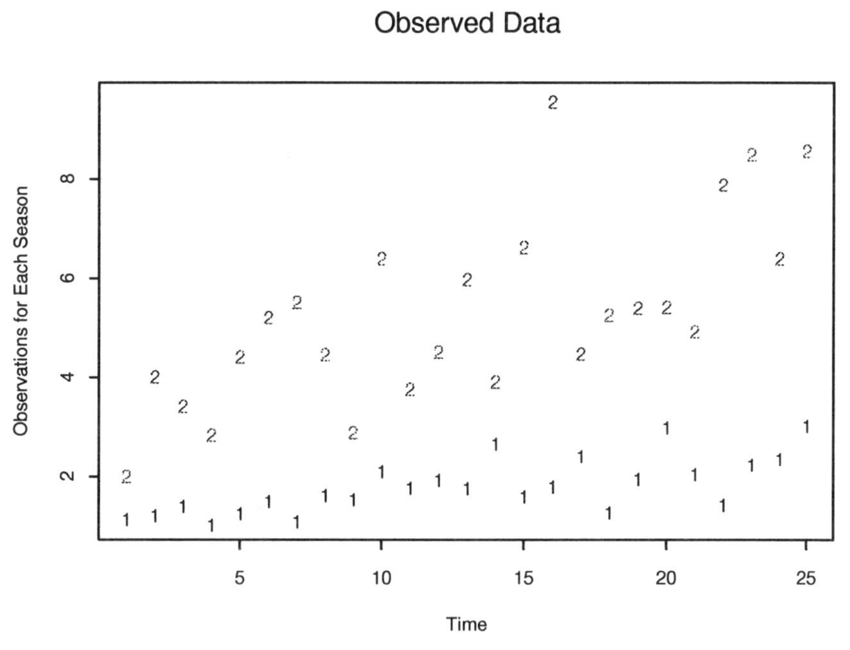

Figure 8.12: Observed relationship between X and Y for each season.

Now let's perform the seasonal test for trend:

```
> seasonal.kendall.trend.test(y)

Results of Hypothesis Test
--------------------------

Null Hypothesis:             All 2 values of tau = 0

Alternative Hypothesis:      The seasonal taus are not
                             all equal (Chi-Square
                             Heterogeneity Test)

                             At least one seasonal tau != 0
                             and all non-zero tau's have
                             the same sign (z Trend Test)

Test Name:                   Seasonal Kendall Test
                             for Trend (with
                             continuity correction)

Estimated Parameter(s):      tau       = 0.5633333
                             slope     = 0.08008812
                             intercept = 2.028826
```

```
Estimation Method:           tau:        Weighted Average
                                         of Seasonal
                                         Estimates
                             slope:      Hirsch et al.'s
                                         Modification of
                                         Thiel/Sen
                                         Estimator
                             intercept:  Median of
                                         Seasonal Estimates

Data:                        y = y

Sample Sizes:                1     = 25
                             2     = 25
                             Total = 50

Test Statistics:             Chi-Square (Het) = 0.009818182
                             z (Trend)        = 5.565371833

Test Statistic Parameter:    df = 1

P-values:                    Chi-Square (Het) = 0.92
                             z (Trend)        = 2.616e-008

Confidence Interval for:     slope

Confidence Interval Method:    Gilbert's Modification of
                               Theil/Sen Method

Confidence Interval Type:    two-sided

Confidence Level:            95%

Confidence Interval:         LCL = 0.0519801
                             UCL = 0.122521
```

The chi-square test for heterogeneity in trends between seasons yields a p-value of 0.92, but the z-test for overall trend yields a p-value of 3 x 10^{-8}. You can also look at seasonal estimates of slope and intercept (see the help file for `seasonal.kendall.trend.test` for more information).

Summary

- **Table 8.1** lists the functions available in S-PLUS and ENVIRONMENTALSTATS for S-PLUS for performing classical univariate hypothesis tests.

- **Table 8.2** illustrates the framework of a hypothesis test. There are two kinds of errors. The Greek letter α denotes the probability of a Type I error and the Greek letter β denotes the probability of a Type II error. The probability of committing a Type I error is called the ***significance level*** of the test. The quantity $1-\beta$ is called the ***power*** of the test.

- The choices for α and β are ***subjective, policy decisions*** that should be made by balancing the cost of making each kind of error.

- There is a one-to-one correspondence between confidence intervals and hypothesis tests.

- Goodness-of-fit tests let you assess how well a data set appears to have come from a particular probability distribution. S-PLUS includes the chi-square (`chisq.gof`) and Kolmogorov-Smirnov (`ks.gof`) goodness-of-fit tests. ENVIRONMENTALSTATS for S-PLUS includes functions for performing the Shapiro-Wilk (`sw.gof`), Shapiro-Francia (`sf.gof`), and probability plot correlation coefficient (`ppcc.norm.gof`, `ppcc.evd.gof`) goodness-of-fit tests, as well as functions for plotting the results of goodness-of-fit tests.

- Two-sample linear rank tests (`two.sample.linear.rank.test`) let you test for a shift between two populations. The quantile test (`quantile.test`) lets you test for a shift in a portion of one of the populations.

- Nonparametric tests for trend (`kendall.trend.test` and `seasonal.kendall.trend.test`) allow you to test for a monotonic (not necessarily linear) trend without having to make an assumption about the distribution of the observations.

CHAPTER 9 :

Censored Data

So far, almost all of the data we have dealt with in our examples have been complete. For the few data sets containing censored observations (reported as "below-detection-limit"), we've arbitrarily set the censored observation(s) to the detection limit. Usually, however, you need a more objective way of dealing with nondetects.

ENVIRONMENTALSTATS for S-PLUS includes several functions for dealing with data sets that contain censored observations, including functions for constructing empirical cdf plots and Q-Q plots, estimating distribution parameters and quantiles, constructing confidence intervals, constructing prediction and tolerance intervals, performing goodness-of-fit tests, and comparing two groups. Furthermore, standard nonparametric tests can often be used with censored data since these tests depend only on the ranks of the data, not the actual data values.

This chapter briefly describes the terminology associated with censored data, then discusses the techniques for dealing with censored data that are available in ENVIRONMENTALSTATS for S-PLUS.

Overview

Table 9.1 lists the functions available in ENVIRONMENTALSTATS for S-PLUS for dealing with censored data.

There are also methods for estimating quantiles and computing approximate prediction limits and tolerance limits that are not listed in **Table 9.1**. See the help file **Censored Data** for more information.

The rest of this chapter briefly describes the terminology associated with censored data, then shows you how to use the functions in ENVIRONMENTALSTATS for S-PLUS for dealing with censored data.

Technique/Test	Function(s)	Description
Empirical CDF Plot	`ecdfplot.censored`	Plot empirical CDF plot.
	`cdf.compare.censored`	Plot empirical CDF plot with theoretical CDF plot or another empirical CDF plot.
Q-Q Plot	`qqplot.censored`	Quantile-quantile plot for censored data.
Estimate Parameters of a Lognormal Distribution	`elnorm.singly.censored`	Singly censored data.
	`elnorm.alt.singly.censored`	Singly censored data, alternative parameterization.
	`elnorm.multiply.censored`	Multiply censored data.
	`elnorm.alt.multiply.censored`	Multiply censored data, alternative parameterization.
Estimate Parameters of a Normal Distribution	`enorm.singly.censored`	Singly censored data.
	`enorm.multiply.censored`	Multiply censored data.
Estimate Parameters of a Poisson Distribution	`epois.singly.censored`	Singly censored data.
	`epois.multiply.censored`	Multiply censored data.

Goodness-of-Fit Tests	`ppcc.norm.singly.censored.gof`
	`sf.singly.censored.gof`
	`sw.singly.censored.gof`
		Probability Plot Correlation Coefficient, Shapiro-Francia, and Shapiro-Wilk goodness-of-fit tests for singly censored data.

Two-Sample Linear Rank Test	`two.sample.linear.rank.test.censored`	Two-Sample linear rank test to detect a shift for multiply censored data.

Table 9.1: Functions for dealing with censored data.

Terminology

A sample of data contains ***censored observations*** if some of the observations are reported only as being below or above some censoring level. In environmental data analysis, censored data sets are common, with values being reported as "less than the detection limit" (e.g., USEPA, 1992c, p.25; Porter et al., 1988). Statistical methods for dealing with censored data sets have a long history in the field of survival analysis and life testing (see the help file **References: Censored Data -- General References**).

Classification of Censored Data

There are four major ways to classify censored data: truncated vs. censored, left vs. right vs. double, single vs. multiple (progressive), and censored Type I vs. censored Type II (Cohen, 1991, pp. 3-5). Most environmental data sets with nondetect values are either ***Type I left singly censored*** or ***Type I left multiply (progressively) censored***.

Truncated vs. Censored

A sample is ***left truncated*** (also called truncated on the left) if only values above a known truncation point T are reported or used. The number of observations falling below T and their actual values are unknown or ignored. A sample is ***right truncated*** (also called truncated on the right) if only values below a known truncation point T are reported or used.

A sample of N observations is ***left singly censored*** (also called singly censored on the left) if c observations are known only to fall below a known censoring level T, while the remaining n ($n=N-c$) uncensored observations falling above T are fully measured and reported. A sample of N observations is ***right singly censored*** (also called singly censored on the right) if c observations are known only to fall above a known censoring level T, while the remaining n ($n=N-c$) uncensored observations falling below T are fully measured and reported.

Left vs. Right vs. Double

The definitions of left truncated, right truncated, left censored, and right censored are given above. A sample is ***doubly truncated*** if it is truncated both on the left and the right. A sample is ***doubly censored*** if it is censored both on the left and the right.

Single vs. Multiple (Progressive)

A sample is ***singly truncated*** (e.g., singly left truncated) if there is only one truncation point T. A sample is ***singly censored*** (e.g., singly left censored) if there is only one censoring level T.

A sample is ***multiply truncated*** or ***progressively truncated*** (e.g., multiply left truncated) if there are several truncation points $T_1, T_2, ..., T_p$, where $T_1 < T_2 < ... < T_p$. A sample is ***multiply censored*** or ***progressively censored*** (e.g., multiply left censored) if there are several censoring levels $T_1, T_2, ..., T_p$, where $T_1 < T_2 < ... < T_p$.

Censored Type I vs. Censored Type II

A censored sample has been subjected to ***Type I censoring*** if the censoring level(s) is(are) known in advance, so that given a fixed sample size N, the number of censored observations c (and hence the number of uncensored observations n) is a random outcome. Type I censored samples are sometimes called time-censored samples (Nelson, 1982, p.248).

A censored sample has been subjected to ***Type II censoring*** if the sample size (N) and number of censored observations (c; and hence the number of uncensored observations n) are fixed in advance, so that the censoring level(s) are random outcomes. Type II censored samples are sometimes called failure-censored samples (Nelson, 1982, p.248).

Examples of Different Kinds of Censored Data Sets

Example 9.1: Type I Left Singly Censored

USEPA (1992c, p.30) contains a data set where concentrations of zinc (ppb) are measured at 5 background wells over eight separate sampling periods. Of the 40 values reported, 20 are reported as "<7". See the description of `epa.92c.zinc.df` in the help file **Datasets: USEPA (1992c)** for more information.

Example 9.2: Type I Left Multiply (Progressively) Censored

Millard and Deverel (1988) present an example of copper concentrations that include values below one of 4 detection limits (see the help file for `millard.deverel.88.df` for more information). Helsel and Cohn (1988) present an example of 56 silver concentrations (mg/L) that include 34 values below one of 12 detection limits (see the help file for `helsel.cohn.88.silver.df` for more information).

Example 9.3: Type I Left Singly Truncated

If, in **Example 9.1** above, the zinc concentrations reported as "<7" were eliminated from the sample completely, leaving only the 20 uncensored observations and no indication of how many observations were excluded, this would constitute a type I left singly truncated sample.

Example 9.4: Type I Right Singly Censored

This kind of sample frequently occurs in life testing experiments. For example, a set of N electronic components may be put "on-line" all at the same time. During the experiment, the time until failure is noted for each component. After a given time period T (e.g., 10 days) the experiment is terminated, even though not all of the components have failed by this time. For the components that have not failed, it is known only that their time until failure is greater than T.

Example 9.5: Type II Right Singly Censored

If we use the same set up as in **Example 9.4**, but terminate the experiment after *n* of the *N* components have failed, this produces a type II right singly censored sample.

Example 9.6: Type I Right Multiply (Progressively) Censored

This kind of sample frequently occurs in survival studies. A number of people are enrolled over the course of the entry period (e.g., 1 year) in a study of a new drug therapy whose purpose is to prolong the life of terminally ill patients. Individuals are assigned to the treatment or control group, and are periodically seen until the study ends or until they die (or for some other reason stop coming for check ups.) At the end of the study, information is available on survival times: when or if individuals died during the study. Since not everyone entered the study at the same time, there may be several people who are still alive at the end of the study, but who are censored at different censoring levels.

Defining the Detection Limit

Much confusion and controversy surrounds the term ***detection limit***. There are, in fact, several terms, concepts, and equations associated with determining the presence or absence and concentration of a chemical in a physical sample.

The need for the concept of a detection limit stems from the fact that there is ***bias*** and ***variability*** in any instrument used to measure chemical concentrations in a physical sample. Thus, an instrument may show no concentration when the chemical is present, or it may show a positive concentration when the chemical is absent. Furthermore, the bias and variability may be different for different concentrations. The field of instrument calibration deals with statistical methods that are used to measure and account for the bias and variability in instruments used to measure chemical concentrations. Keith (1991, Chapter 10) gives a good introduction and explanation of the definitions and concepts associated with a detection limit. Currie (1995) gives a recent comprehensive set of definitions associated with defining detection limits.

USEPA (1992b, pp. 2-25, G-4) defines the ***detection limit*** as the lowest concentration level which can be determined to be different from a blank.. Kaiser (1965), Glasser et al. (1981), and Gibbons (1994, p. 96) define the ***method detection limit*** (MDL) as the minimum concentration of a substance that can be identified, measured, and reported with 99% confidence that the analyte concentration is greater than zero. This is the definition adopted in the Code of Federal Regulations, Title 40, Part 136, Appendix B.

Currie (1968) defines three different limits: a ***decision limit***, a ***detection limit***, and a ***determination limit***. USEPA (1987c) defines a ***practical quantitation limit*** (PQL) as “the lowest level achievable by good laboratories within specified limits during routine laboratory operating conditions.” Gibbons (1994, pp.122-123) points out that this is a vague definition and has been operationally defined by the USEPA in three different ways.

Gibbons (1995) criticizes the methods used to define traditional detection limit estimators, but, as one of the reviewers in this article points out, his calculated detection limits are an order of magnitude less than the range of data used to produce them.

Rocke and Lorenzato (1995) introduced a new model for measurement error in analytical chemistry that involves two variance components: a constant variance component (additive) for low concentrations, and a constant coefficient of variation component (multiplicative) for larger concentrations. This model may help provide some resolution in defining the meaning of the term ***detection limit***.

See the references in the help file **References: Censored Data -- Detection Limit** for more information on, and definitions of, the above concepts.

Commonly Used Data Qualifiers

Table 9.2 shows a list of qualifiers commonly used by analytical laboratories to report results. All persons involved in an environmental study should understand the meaning and source of these qualifiers.

Qualifier	Meaning
J	Result is of limited use due to discrepancies in holding times, blank analyses, duplicate analyses, spike analyses, or laboratory contamination problems.
L	(Formerly known as "trace"). Result is of limited use because it is between the instrument/method detection limit and the contract detection/quantitation limit.
N	Result probably acceptable but is just outside the calibration range or the recovery is just outside the specification range.
R	Results are unusable due to discrepancies in analytical technique/protocol, improper calibration, outside calibration range, outside specified recovery windows, or blunder.
U	Reading was below instrument detection limit (inorganics), or method detection limit (organics).

Table 9.2: Commonly used analytical laboratory data qualifiers.

Graphical Assessment of Censored Data

ENVIRONMENTALSTATS for S-PLUS contains functions for creating empirical cumulative distribution function plots and quantile-quantile (Q-Q) plots for censored data. Each of these is discussed and illustrated in this section.

Empirical CDF Plots

The ENVIRONMENTALSTATS for S-PLUS function `ecdfplot.censored` allows you to create an empirical cumulative distribution function plot for singly or multiply censored data. The function `cdf.compare.censored` allows you to compare an empirical cdf based on singly or multiply censored data with a theoretical cdf or with another empirical cdf.

Example 9.7: Creating an Empirical CDF Plot for Multiply Censored Silver Concentrations

The data frame `helsel.cohn.88.silver.df` contains 56 observations of silver concentrations (mg/L), of which 34 are censored below one of 12 distinct censoring levels. Here we'll create an empirical cdf plot, and include the censored observations in the plot.

```
> helsel.cohn.88.silver.df
   Ag.orig    Ag Censored
 1     0.8   0.8        F
 2     0.1   0.1        F
 3     2     2.0        F
 .
54     <1    1.0        T
55     <1    1.0        T
56     <0.1  0.1        T

> attach(helsel.cohn.88.silver.df)

> names(helsel.cohn.88.silver.df)
[1] "Ag.orig"  "Ag"       "Censored"

> ecdfplot.censored(Ag, Censored,
+    xlab = "Ag (microgram / Liter)",
+    main=paste("Empirical CDF Plot for Silver",
+    "(Method of Michael-Schucany)", sep="\n"),
+    include.cen=T)
```

The plot is shown in **Figure 9.1**. The data look extremely skewed to the right.

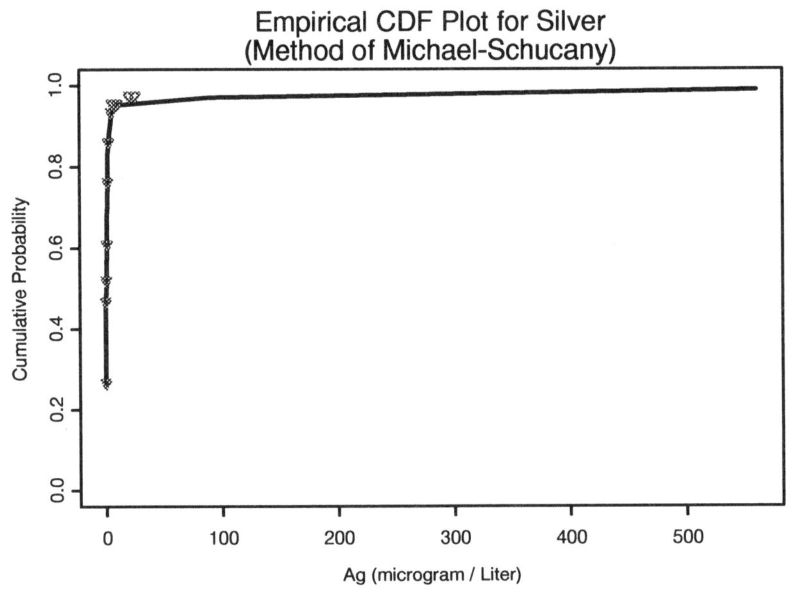

Figure 9.1: Empirical cdf plot of the silver data.

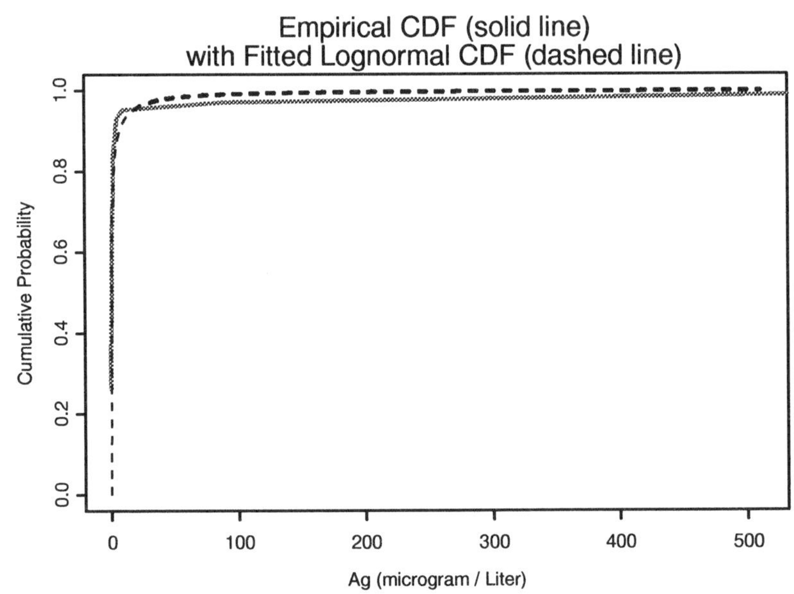

Figure 9.2: Empirical cdf of silver data with fitted lognormal cdf.

Example 9.8: Comparing the Empirical CDF of the Silver Data to a Lognormal CDF

In this example, we'll use the function `cdf.compare.censored` to compare the cdf for the silver data with a cdf from a lognormal distribution. The parameters for the lognormal distribution are estimated from the data (see the next section, **Estimating Distribution Parameters**).

```
> cdf.compare.censored(Ag, Censored,
+    distribution = "lnorm",
+    xlab = "Ag (microgram / Liter)")
```

The plot is shown in **Figure 9.2**. The lognormal distribution appears to provide a good fit to these data. You could also take the logarithm of the data and specify a normal distribution (`dist="norm"`) in the call to `cdf.compare.censored`.

Example 9.9: Comparing the Empirical CDF of Copper and Zinc Concentrations from Two Geological Sites

The data frame `millard.deverel.88.df` contains information on copper and zinc concentrations (μg/L) in shallow ground water from two geological zones (alluvial fan and basin-trough) in the San Joaquin Valley, CA. There are 68 samples from the alluvial fan zone and 50 from the basin-trough zone. Some observations are reported as "<DL" where DL denotes a detection limit. There are multiple detection limits for both the copper and zinc data in each of the geological zones.

Here, we'll compare the empirical cdf's of copper and zinc between the alluvial fan zone and the basin-trough zone. First, we need to separate the data by element (copper vs. zinc) and location (alluvial fan vs. basin-trough).

```
> attach(millard.deverel.88.df)

> names(millard.deverel.88.df)
[1] "Cu.orig"      "Cu"            "Cu.censored"
[4] "Zn.orig"      "Zn"            "Zn.censored"
[7] "Zone"         "Location"

> levels(Zone)
[1] "Alluvial.Fan" "Basin.Trough"

> Cu.AF <- Cu[Zone == "Alluvial.Fan"]

> Cu.AF.cen <- Cu.censored[Zone == "Alluvial.Fan"]

> Cu.BT <- Cu[Zone == "Basin.Trough"]

> Cu.BT.cen <- Cu.censored[Zone == "Basin.Trough"]

> Zn.AF <- Zn[Zone == "Alluvial.Fan"]
```

```
> Zn.AF.cen <- Zn.censored[Zone == "Alluvial.Fan"]

> Zn.BT <- Zn[Zone == "Basin.Trough"]

> Zn.BT.cen <- Zn.censored[Zone == "Basin.Trough"]
```

Now we can call `cdf.compare.censored` to compare the empirical cdf of copper between the alluvial fan and basin-trough zones.

```
> cdf.compare.censored(Cu.AF, Cu.AF.cen, "left",
+   Cu.BT, Cu.BT.cen, "left")
```

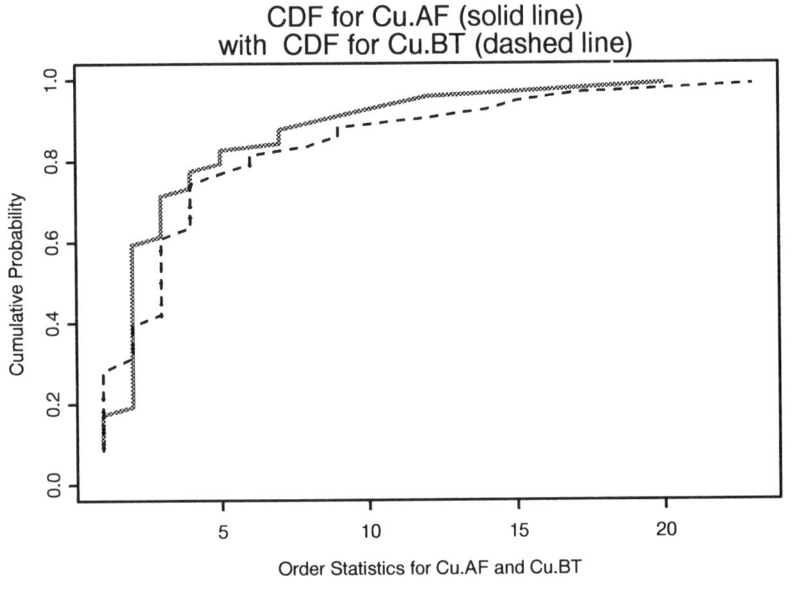

Figure 9.3: Empirical cdf of copper for alluvial fan vs. basin-trough zones.

The results are shown in **Figure 9.3**. We see that in general, the copper concentrations are higher in the basin-trough zone, but this difference between the empirical cdf's may or may not be significant (see the section **Two-Sample Linear Rank Tests** later in this chapter).

Figure 9.4 shows the results of calling `cdf.compare.censored` to compare the empirical cdf of zinc between the alluvial fan and basin-trough zones. Here we see that there are numerous concentrations of zinc in the alluvial fan zone that are much larger than most or all of the concentrations in the basin-trough zone.

```
> cdf.compare.censored(Zn.AF, Zn.AF.cen, "left",
+   Zn.BT, Zn.BT.cen, "left")
```

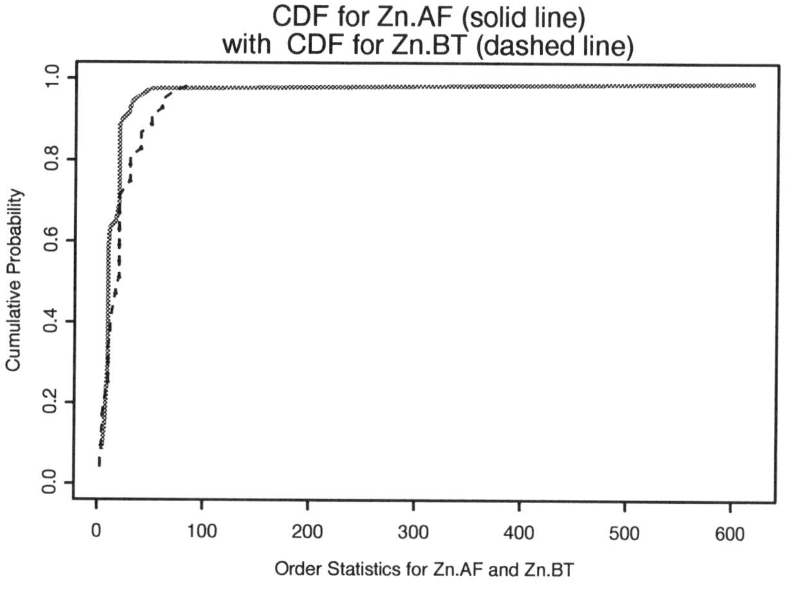

Figure 9.4: Empirical cdf of zinc for alluvial fan vs. basin trough zones.

Quantile-Quantile Plots

The ENVIRONMENTALSTATS for S-PLUS function `qqplot.censored` lets you create a Q-Q plot based on singly or multiply censored data.

Example 9.10: Using Q-Q Plots to Assess the Goodness-of-Fit of the Silver Data to a Lognormal Distribution

Using the same data as in **Example 9.7** and **Example 9.8** (i.e., the data frame `helsel.cohn.88.silver.df`), let's assess how well a lognormal distribution fits these data by constructing a Q-Q plot and a Tukey mean-difference Q-Q plot.

Figure 9.5 shows the result of calling `qqplot.censored` to produce a standard Q-Q plot. **Figure 9.6** shows the result of calling `qqplot.censored` with the argument `plot.type="Tukey"` to produce a Tukey mean-difference Q-Q plot. Both of these Q-Q plots indicate that the lognormal distribution may be an adequate fit to these data, although some systematic deviation may be present.

```
> qqplot.censored(Ag, Censored, dist = "lnorm",
+   prob.meth="h", add.line=T, ylim = c(-6, 6))
```

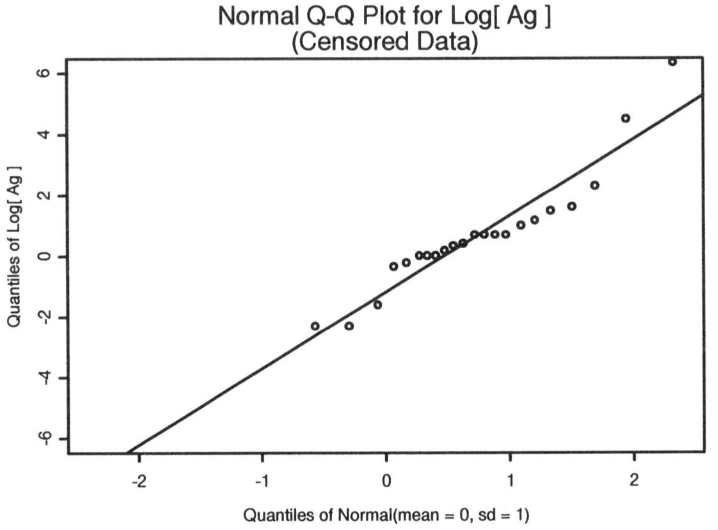

Figure 9.5: Normal Q-Q plot for the log-transformed silver data.

```
> qqplot.censored(Ag, Censored, dist = "lnorm",
+   prob.meth="h", plot.type = "T", estimate.params = T,
+   add.line = T)
```

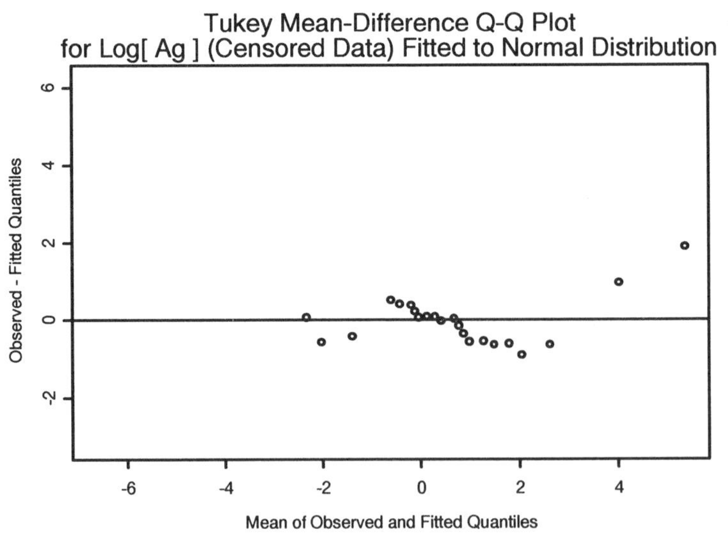

Figure 9.6: Tukey mean-difference Q-Q plot for the log-transformed silver data.

Example 9.11: Censored Data vs. Zero-Modified Distributions

Sometimes nondetects are modeled as "0" values (e.g., Owen and DeRouen, 1980). To assess whether modeling nondetects as "0" values appears to be a good idea, you can compare the results of creating a Q-Q plot with `qqplot.censored` vs. the results of creating a Q-Q plot with the `qqplot` function where you specify a zero-modified distribution.

The guidance document **Statistical Analysis of Ground-Water Monitoring Data at RCRA Facilities, Addendum to Interim Final Guidance** (USEPA, 1992c) contains an example on pages 28-34 involving a set of zinc concentrations (ppb) at five background wells. There are several values reported as "<7". The example compares the Q-Q plot based on assuming the data are censored with the Q-Q plot based on assuming the "<7" values are actually 0's. The data for this example are stored in the data frame `epa.92c.zinc.df` (see the help file **Datasets: USEPA (1992c)**).

```
> epa.92c.zinc.df
    Zinc.orig  Zinc Censored Sample Well
 1      <7      7.00        T      1    1
 2      11.41 11.41        F      2    1
 3      <7      7.00        T      3    1
 .
38      12.35 12.35        F      6    5
39      <7      7.00        T      7    5
40      8.74    8.74        F      8    5

> attach(epa.92c.zinc.df)

> names(epa.92c.zinc.df)
[1] "Zinc.orig" "Zinc"       "Censored"
[4] "Sample"     "Well"
```

First we'll create the normal Q-Q plot and the Tukey mean-difference Q-Q plot assuming the data are censored. **Figure 9.7** and **Figure 9.8** show the result of calling `qqplot.censored`. A systematic deviation from normality is very apparent in the Tukey mean-difference Q-Q plot.

Now create Q-Q plots assuming the "<7" values are actually 0's and so the data come from a zero-modified normal distribution (see the help file **Zero-Modified Normal**).

```
> Zinc.w.0 <- Zinc

> Zinc.w.0[Censored] <- 0
```

Figure 9.9 and **Figure 9.10** show the result of calling `qqplot` with the argument `dist="zmnorm"` to specify a zero-modified normal distribution. Both of these Q-Q plots indicate the zero-modified normal distribution is a good fit to these data.

```
> qqplot.censored(Zinc, Censored, add.line = T)
```

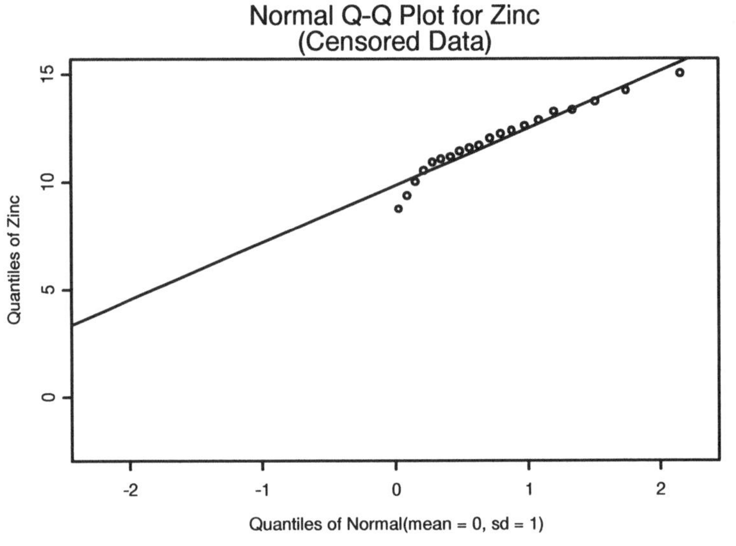

Figure 9.7: Normal Q-Q plot for zinc data with singly censored observations.

```
> qqplot.censored(Zinc, Censored, add.line = T,
+    plot.type = "Tukey", estimate.params = T)
```

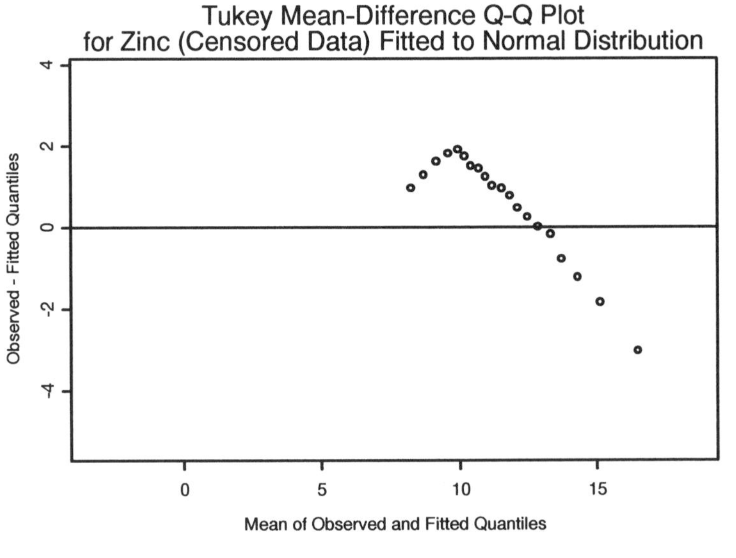

Figure 9.8: Tukey mean-difference Q-Q plot for zinc data with singly censored observations.

```
> qqplot(Zinc.w.0, dist = "zmnorm", estimate.params = T,
+   add.line = T)
```

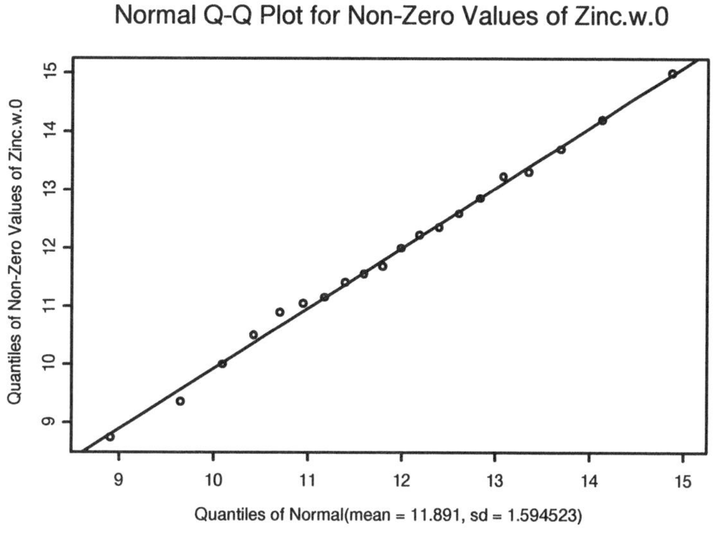

Figure 9.9: Q-Q plot for zinc data assuming a zero-modified normal distribution.

```
> qqplot(Zinc.w.0, dist = "zmnorm", estimate.params = T,
+   add.line = T, plot.type = "Tukey")
```

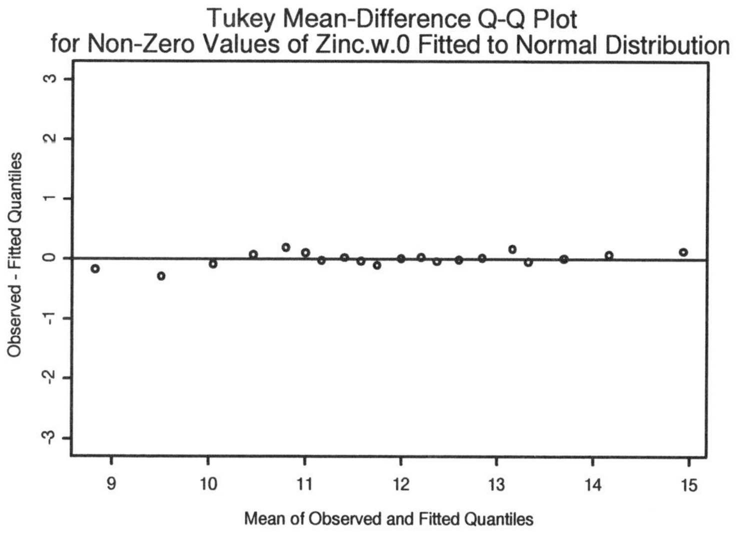

Figure 9.10: Tukey mean-difference Q-Q plot for the zinc data assuming a zero-modified normal distribution.

Estimating Distribution Parameters

ENVIRONMENTALSTATS for S-PLUS contains functions for estimating the distribution parameters of a normal, lognormal, and Poisson distribution, based on singly or multiply censored data (see the help file **Censored Data: Estimating Distribution Parameters and Quantiles**). The usual functions for estimating quantiles for these distributions (i.e., not assuming censored data) can also be used if you supply them with the results of the parameter estimation functions for censored data.

Example 9.12: Estimating the Parameters of a Normal Distribution Based on Singly Censored Data

The guidance document **Statistical Analysis of Ground-Water Monitoring Data at RCRA Facilities, Interim Final Guidance** (USEPA, 1989b) contains an example on pages 8-7 to 8-11 involving a set of sulfate concentrations (mg/L) in which three values are reported as "<1,450". The sulfate concentrations are assumed to come from a normal distribution and the example illustrates how to compute the maximum likelihood estimates of mean and standard deviation. The data for this example are stored in the data frame `epa.89b.sulfate.df` (see the help file **Datasets: USEPA (1989b)**).

```
> epa.89b.sulfate.df
   Sulfate.orig Sulfate Censored
 1         1850    1850        F
 2         1760    1760        F
 3        <1450    1450        T
 .
22         1770    1770        F
23         1790    1790        F
24         1780    1780        F

> attach(epa.89b.sulfate.df)

> names(epa.89b.sulfate.df)
[1] "Sulfate.orig" "Sulfate"      "Censored"
```

```
> enorm.singly.censored(Sulfate, Censored, ci = T)

Results of Distribution Parameter Estimation
Based on Type I Censored Data
--------------------------------------------

Assumed Distribution:            Normal

Censoring Side:                  left

Censoring Level(s):              1450

Estimated Parameter(s):          mean = 1723.995
                                 sd   =  153.6451

Estimation Method:               mle

Data:                            Sulfate

Censoring Variable:              Censored

Sample Size:                     24

Percent Censored:                12.5%

Confidence Interval for:         mean

Assumed Sample Size:             23.10555

Confidence Interval Method:      Normal Approximation

Confidence Interval Type:        two-sided

Confidence Level:                95%

Confidence Interval:             LCL = 1657.556
                                 UCL = 1790.434
```

Example 9.13: Estimating the Parameters of a Lognormal Distribution Based on Multiply Censored Data

Using the same silver data as in **Example 9.7**, **Example 9.8**, and **Example 9.10**, let's assume the data come from a lognormal distribution and estimate the parameters.

```
> attach(helsel.cohn.88.silver.df)

> objects(2)
[1] "Ag"       "Ag.orig"  "Censored"
```

```
> elnorm.multiply.censored(Ag, Censored, ci=T)

Results of Distribution Parameter Estimation
Based on Type I Censored Data
-------------------------------------------

Assumed Distribution:            Lognormal

Censoring Side:                  left

Censoring Level(s):               0.1  0.2  0.3  0.5  1
                                  2    2.5  5    6    10
                                 20   25

Estimated Parameter(s):          meanlog = -1.040572
                                 sdlog   =  2.354847

Estimation Method:               mle

Data:                            Ag

Censoring Variable:              Censored

Sample Size:                     56

Percent Censored:                60.7%

Confidence Interval for:         meanlog

Assumed Sample Size:             22

Confidence Interval Method:      Normal Approximation

Confidence Interval Type:        two-sided

Confidence Level:                95%

Confidence Interval:             LCL = -2.012302
                                 UCL = -0.06884105
```

```
> elnorm.alt.multiply.censored(Ag, Censored, ci=T)

Results of Distribution Parameter Estimation
Based on Type I Censored Data
--------------------------------------------

Assumed Distribution:            Lognormal

Censoring Side:                  left

Censoring Level(s):               0.1  0.2  0.3  0.5  1
                                  2    2.5  5    6    10
                                 20   25

Estimated Parameter(s):          mean =  5.652396
                                 cv   = 15.96972

Estimation Method:               mle

Data:                            Ag

Censoring Variable:              Censored

Sample Size:                     56

Percent Censored:                60.7%

Confidence Interval for:         mean

Assumed Sample Size:             22

Confidence Interval Method:      Normal Approximation
                                 Based on Delta Method

Confidence Interval Type:        two-sided

Confidence Level:                95%

Confidence Interval:             LCL =  0
                                 UCL = 25.93041
```

Goodness-of-Fit Tests

EnvironmentalStats for S-Plus contains three functions for performing goodness-of-fit tests for a normal distribution based on singly censored data: `ppcc.norm.singly.censored.gof` (probability plot correlation coefficient), `sf.singly.censored.gof` (Shapiro-Francia), and `sw.singly.censored.gof` (Shapiro-Wilk) (see the help file **Censored Data: Goodness-of-Fit Tests**).

Example 9.14: Shapiro-Wilk Test for Normality Based on the Singly Censored Zinc Data

Using the same zinc data as in **Example 9.11**, perform a goodness-of-fit test assuming the "<7" values are censored. Then perform a goodness-of-fit test assuming the data are from a zero-modified normal distribution and that the "<7" values are really 0's.

```
> attach(epa.92c.zinc.df)

> zinc.censored.gof <-
+   sw.singly.censored.gof(Zinc, Censored)

> zinc.censored.gof

Results of Goodness-of-Fit Test
Based on Type I Censored Data
-------------------------------

Test Method:                     Shapiro-Wilk GOF
                                 (Singly Censored Data)

Hypothesized Distribution:       Normal

Censoring Side:                  left

Censoring Level(s):              7

Estimated Parameter(s):          mean = 7.632708
                                 sd   = 4.821067

Estimation Method:               mle

Data:                            Zinc

Censoring Variable:              Censored

Sample Size:                     40

Percent Censored:                50%

Test Statistic:                  W = 0.9203649

Test Statistic Parameters:       N     = 40
                                 DELTA =  0.5

P-value:                         0.1668357

Alternative Hypothesis:          True cdf does not equal the
                                 Normal Distribution.
```

In this case, the p-value is not extremely small, yet indicates there may be a deviation from normality. The Q-Q plots in **Example 9.11** showed a systematic lack of fit. Let's look at the plots associated with this goodness-of-fit test.

```
> plot.gof.censored.summary(zinc.censored.gof, digits = 2)
```

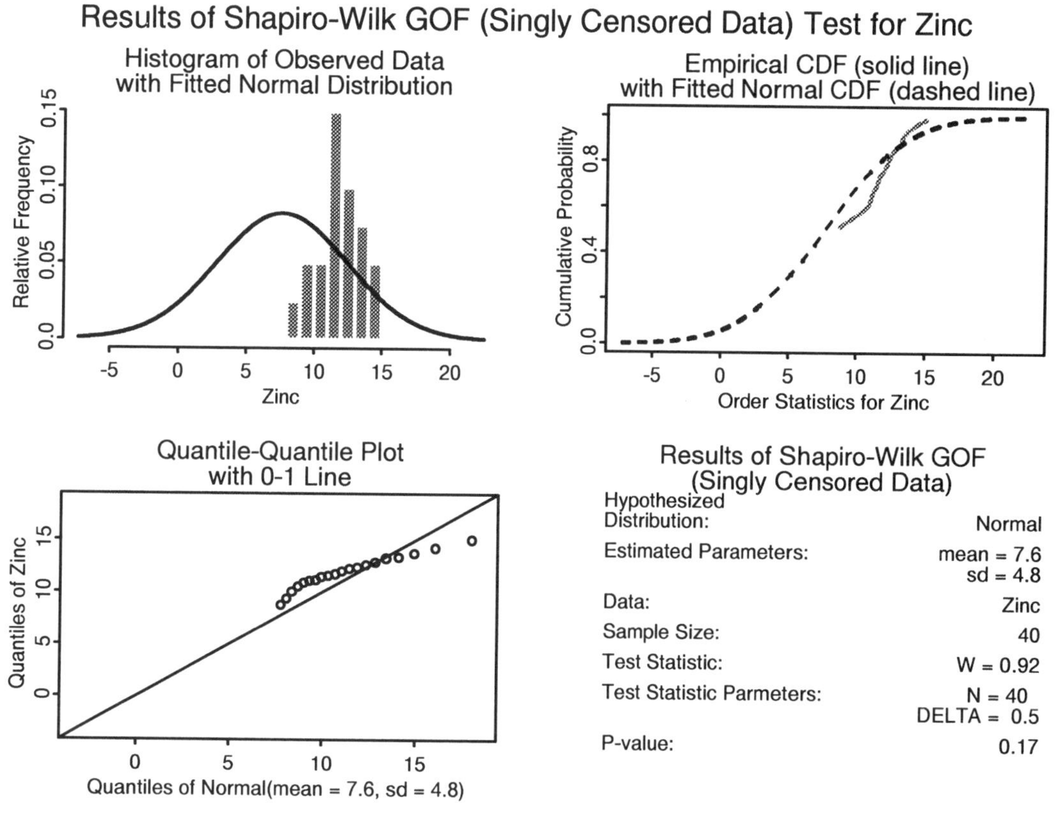

Figure 9.11: Results of Shapiro-Wilk goodness-of-fit test for singly censored zinc data.

Now let's do the goodness-of-fit test assuming the "<7" values are really 0's and that the data comes from a zero-modified normal distribution.

```
> Zinc.w.0 <- Zinc

> Zinc.w.0[Censored] <- 0

> zinc.zmnorm.gof <- sw.gof(Zinc.w.0, dist="zmnorm")
```

```
> zinc.zmnorm.gof

Results of Goodness-of-Fit Test
-------------------------------

Test Method:                     Shapiro-Wilk GOF

Hypothesized Distribution:       Zero-Modified Normal

Estimated Parameter(s):          mean          = 11.891
                                 sd            =  1.594523
                                 p.zero        =  0.5
                                 mean.zmnorm   =  5.9455
                                 sd.zmnorm     =  6.123235

Estimation Method:               mvue

Data:                            Zinc.w.0

Sample Size:                     40

Test Statistic:                  W = 0.9945601

Test Statistic Parameter:        n = 20

P-value:                         0.9999819

Alternative Hypothesis:          True cdf does not equal the
                                 Zero-Modified Normal
                                 Distribution.
```

The p-value here is certainly much larger. Now let's plot the results of this goodness-of-fit test. The plots in **Figure 9.12** show that the zero-modified normal distribution appears to be a good fit to the zinc data.

```
> plot.gof.summary(zinc.zmnorm.gof, digits = 3)
```

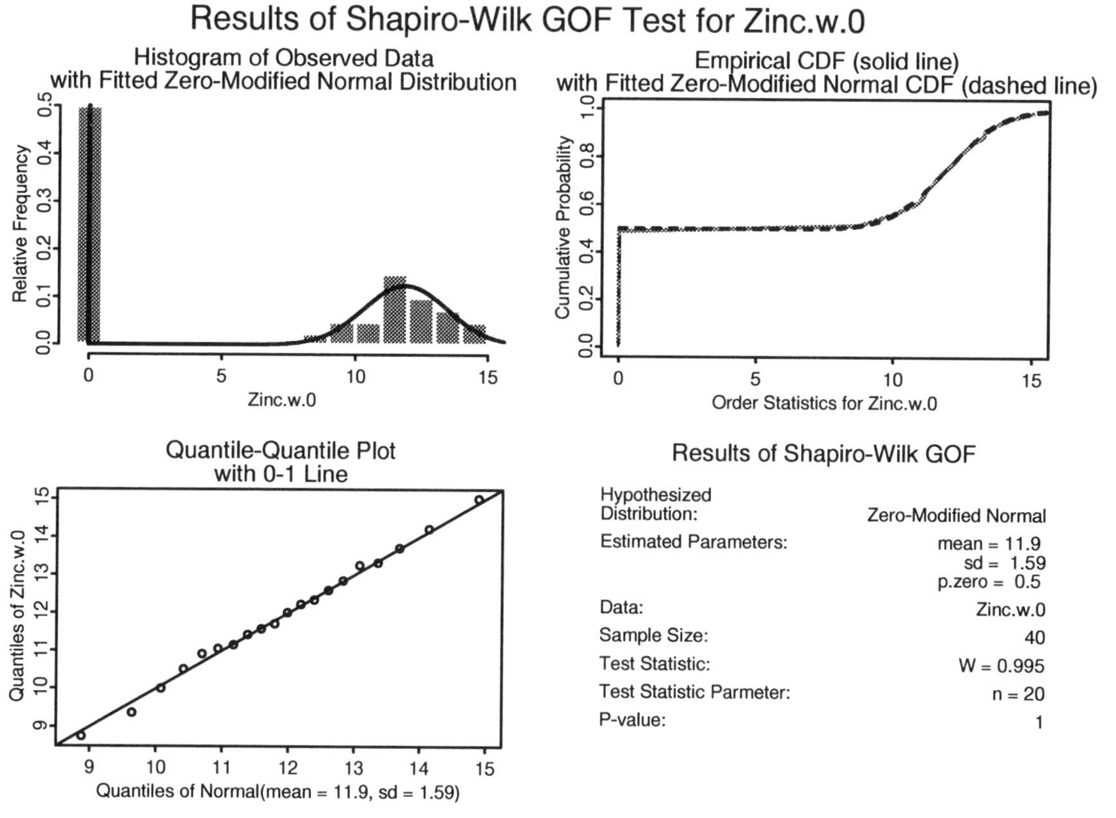

Figure 9.12: Results of Shapiro-Wilk goodness-of-fit test to zinc data assuming zero-modified normal distribution.

Two-Sample Linear Rank Tests

The function `two.sample.linear.rank.test.censored` let's you perform a two-sample linear rank test to compare two groups, based on singly or multiply censored data (see the help file **Censored Data: Hypothesis Tests**).

Example 9.15: Comparing Copper and Zinc Concentrations in the Soil at Two Different Sites Based on Multiply Censored Data

Using the same copper and zinc data as in **Example 9.9**, test to see whether there is a significant difference in concentrations between the alluvial fan and basin-trough zones.

First we'll compare the copper concentrations. Looking at the data, note the large number of tied, uncensored observations:

```
> table(Cu.AF[!Cu.AF.cen])
 1  2 3 4 5 7 8 9 10 11 12 16 20
 5 21 6 3 3 3 1 1  1  1  1  1  1

> table(Cu.BT[!Cu.BT.cen])
 1 2 3 4 5 6 8 9 12 14 15 17 23
 7 4 8 5 1 2 1 2  1  1  1  1  1
```

The large number of tied observations suggests that the hypergeometric variance should be used. In this case, the p-value is 0.2, so there is not enough evidence to say there is a difference in copper concentrations between the two geological zones.

```
> two.sample.linear.rank.test.censored(Cu.AF,
+   Cu.AF.cen, Cu.BT, Cu.BT.cen, var = "h")

Results of Hypothesis Test
Based on Censored Data
--------------------------

Null Hypothesis:                Fx(t) = Fy(t)

Alternative Hypothesis:         Fx(t) != Fy(t) for at
                                least one t

Test Name:                      Two-Sample Linear
                                Rank Test:
                                Normal Scores Test Using
                                Prentice Survival
                                Estimator Based on
                                Prentice and Marek (1979)
                                with Hypergeometric
                                Variance
```

```
Censoring Side:                  left

Data:                            x = Cu.AF
                                 y = Cu.BT

Censoring Variable:              x = Cu.AF.cen
                                 y = Cu.BT.cen

Number NA/NaN/Inf's Removed:     x = 3
                                 y = 1

Sample Sizes:                    nx = 65
                                 ny = 49

Percent Censored:                x = 26.2%
                                 y = 28.6%

Test Statistic:                  z = 1.279016

P-value:                         0.2008913
```

Now let's perform the test for the zinc concentrations.

```
> two.sample.linear.rank.test.censored(Zn.AF,
+  Zn.AF.cen, Zn.BT, Zn.BT.cen, var = "h")

Results of Hypothesis Test
Based on Censored Data
--------------------------

Null Hypothesis:                 Fx(t) = Fy(t)

Alternative Hypothesis:          Fx(t) != Fy(t) for at
                                 least one t

Test Name:                       Two-Sample Linear
                                 Rank Test:
                                 Normal Scores Test Using
                                 Prentice Survival
                                 Estimator Based on
                                 Prentice and Marek (1979)
                                 with Hypergeometric
                                 Variance

Censoring Side:                  left

Data:                            x = Zn.AF
                                 y = Zn.BT

Censoring Variable:              x = Zn.AF.cen
                                 y = Zn.BT.cen
```

```
Number NA/NaN/Inf's Removed:     x = 1
                                 y = 0

Sample Sizes:                    nx = 67
                                 ny = 50

Percent Censored:                x = 23.9%
                                 y =  8  %

Test Statistic:                  z = 2.476636

P-value:                         0.01326269
```

Here, the small p-value indicates there is a significant difference between the zinc concentrations in the two geologic zones.

Summary

- **Table 9.1** lists the functions available in ENVIRONMENTALSTATS for S-PLUS for dealing with censored data. These functions include methods to create empirical cdf plots and Q-Q plots; estimate distribution parameters for the normal, lognormal, and Poisson distributions; perform normal goodness-of-fit tests; and compare two groups based on a two-sample linear rank test.

- Most environmental data sets that include censored observations are Type I left singly censored or Type I left multiply censored data.

- The term "detection limit" may mean different things to different people. It is important to know exactly what meaning is being used when defining and discussing the detection limit.

CHAPTER 10 :

Monte Carlo Simulation and Risk Assessment

Several government agencies are charged with evaluating the potential health and ecological effects of environmental toxicants. Based on their assessments, these agencies set standards for acceptable concentration levels of these toxicants in air, water, soil, food, etc. The process of modeling exposure to a toxicant and predicting health or ecological effects is termed risk assessment.

In the past, estimates of risk were often based solely on setting values of the input variables (e.g., body weight, dose, etc.) to particular point estimates and producing a single point estimate of risk, with little, if any, quantification of the uncertainty associated with the estimated risk. More recently, several practitioners have advocated "probabilistic" risk assessment, in which the input variables are considered random variables, so the result of the risk assessment is a probability distribution for predicted risk or exposure.

This chapter discusses the concepts of Monte Carlo simulation, sensitivity and uncertainty analysis, and risk assessment, and shows you how to use functions in S-PLUS and ENVIRONMENTALSTATS for S-PLUS to perform probabilistic risk assessment.

Overview

Human and ecological risk assessment involves characterizing the exposure to a toxicant for one or several populations, quantifying the relationship between exposure (dose) and health or ecological effects (response), determining the risk (probability) of a health or ecological effect given the observed level(s) of exposure, and characterizing the uncertainty associated with the estimated risk (Hallenbeck, 1993, p.1).

Activity	Function(s)	Description
Generate Random Numbers From a Specified Probability Distribution	`rabb`	Create a vector of random numbers via simple random sampling (see column 2 of **Table 5.1** for a list of possible values of `abb`).
	`simulate.vector`	Create a vector of random numbers via simple random sampling or Latin Hypercube sampling.
Generate Random Numbers from a Multivariate Distribution	`rmvnorm`	Generate random numbers from a multivariate normal distribution.
	`simulate.mv.matrix`	Generate a multivariate matrix of random numbers from specified distributions, based on a specific rank correlation matrix, using either simple random sampling or Latin Hypercube sampling.
Compute a vector of estimates of risk or exposure based on a set of values for the input variables (e.g., the result of Monte Carlo simulation)	`eval.expr`	Evaluate an S-PLUS expression, and optionally specify a matrix, data frame, or list containing the values of variables used in the expression.

Table 10.1: Functions useful for Monte Carlo simulation and probabilistic risk assessment.

Risk assessment is an enormous field of research and practice. The merits and disadvantages of particular exposure models (including fate and transport models) and dose-response models are not discussed in this chapter. Several journals, text books, and web sites are dedicated to risk assessment. Three standard texts include Hallenbeck (1993), Suter (1993), and Rodricks (1992). Other references are listed in the help file **References: Risk Assessment**.

In the past, estimates of risk were often based solely on setting values of the input variables (e.g., body weight, dose, etc.) to particular point estimates and producing a single point estimate of risk, with little, if any, quantification of the uncertainty associated with the estimated risk. More recently, several practitioners have advocated "probabilistic" risk assessment, in which the input variables are considered random variables, so the result of the risk assessment is a probability distribution for predicted risk or exposure.

Usually, the equation describing risk or exposure is so complicated that it is not feasible to determine the output distribution using analytical methods, so the distribution of risk or exposure is derived via Monte Carlo simulation. **Table 10.1** lists several functions available in S-PLUS and ENVIRONMENTALSTATS for S-PLUS that are useful for performing Monte Carlo simulation and probabilistic risk assessment.

This chapter discusses the concepts of Monte Carlo simulation, sensitivity and uncertainty analysis, and risk assessment, and shows you how to use functions in S-PLUS and ENVIRONMENTALSTATS for S-PLUS to perform probabilistic risk assessment.

Mathematical Models and Monte Carlo Simulation

Risk assessment involves several activities, including constructing a mathematical model to relate exposure to a toxicant to the probability of an adverse effect. Before we talk about risk assessment, we'll discuss the concepts and practices associated with constructing and assessing the behavior of mathematical models.

Monte Carlo Simulation

Monte Carlo simulation is a method of investigating the distribution of a random variable by simulating random numbers (Gentle, 1985). Usually, the random variable of interest, say Y, is some function of one or more other random variables:

$$Y = h(\underline{X}) = h(X_1, X_2, \ldots, X_k) \tag{10.1}$$

For example, Y may be an estimate of the median of a population with a Cauchy distribution (see the S-PLUS help file for **Cauchy**), in which case the vector of random variables $\underline{X}$ represents k independent and identically distributed observations from some particular Cauchy distribution. As another example, Y may be the incremental lifetime cancer risk due to ingestion of soil contaminated with benzene (Thompson et al., 1992; Hamed and Bedient, 1997). In this case the random vector $\underline{X}$ may represent observations from several kinds of distributions that characterize exposure and dose-response, such as benzene concentration in the soil, soil ingestion rate, average body weight, the cancer potency factor for benzene, etc. These distributions may or may not be assumed to be independent of one another (Smith et al., 1992; Bukowski et al., 1995).

Sometimes the input variables X_1, X_2, …, X_k are called ***input parameters***. This terminology can be confusing, however, since the input variables are often random variables and therefore have ***distribution parameters*** associated with their probability distributions (e.g., mean and standard deviation for a normally distributed input variable).

Sometimes the distribution of Y in **Equation (10.1)** above can be derived analytically based on statistical theory (Springer, 1979; Slob, 1994). Often, however, the function h is complicated and/or the elements of the random vector $\underline{X}$ involve several kinds of probability distributions, making it difficult or impossible to derive the exact distribution of Y. In this case, Monte Carlo simulation can be used to approximate the distribution of Y. Monte Carlo simulation is often used in risk assessment, specifically in sensitivity and uncertainty analysis.

Monte Carlo simulation involves creating a large number of realizations of the random vector $\underline{X}$, say n, and computing Y for each of the n realizations of $\underline{X}$. The resulting distribution of Y, or some characteristic of this distribution (e.g., the mean), is then assumed to be "close" to the true distribution or distribution characteristic of Y. The adequacy of the approximation depends on a number of factors, including how well the mathematical relationship described in **Equation (10.1)** reflects the true relationship between Y and $\underline{X}$, how well the specified distribution of $\underline{X}$ reflects its true distribution (including any possible dependencies between the individual elements of $\underline{X}$), and how many Monte Carlo samples or trials (n) are created.

Usually, Monte Carlo simulation involves generating random numbers from some specified theoretical probability distribution, such as a normal, lognormal, beta, etc. (see the help file **Probability Distributions and Random Numbers**). When the simulation is done based on an empirical distribution, this is also called ***bootstrapping*** (Efron and Tibshirani, 1993) (see the help files for `remp`, `simulate.vector`, and `simulate.mv.matrix`).

Vose (1996, p.40) claims that the term "Monte Carlo" comes from the code name of an American project on the atom bomb during World War II, and not from the town of the same name in Monaco that is well known for its casinos. This, however, begs the question of the origin of the code name. Gentle (1985, p.612) notes that Monte Carlo simulation was used extensively during World War II and immediately after during work on the development of the atomic bomb. He states that the name "Monte Carlo" dates from that time period and comes from the casino in Monaco with the same name. Hayes (1993, p.114) states that the modern form of Monte Carlo simulation was invented at Los Alamos shortly after World War II by Stanislaw Ulam while he was working on a problem of the diffusion of neutrons, and that the term Monte Carlo method is "named in honor of the well-known generator of random integers (between 0 and 36) in the Mediterranean principality."

References that address the issues of how to properly perform and report the results of a Monte Carlo simulation study include Hoaglin and Andrews (1975), Law and Kelton (1991), Burmaster and Anderson (1994), and Vose (1996).

Example 10.1: Simulating the Distribution of the Sum of Two Normal Random Variables

Suppose *X* and *Y* are two independent standard normal random variables. Then the distribution of *W=X+Y* is normal with a mean of 0 and a variance of 2. Suppose, however, that we don't know how to derive the distribution of *W*. We can use Monte Carlo simulation to investigate the shape of the distribution of *W*, as well as compute characteristics of the distribution (e.g., mean, median, standard deviation, quantiles, etc.)

First we'll set the seed of the random number generator so you can reproduce this example. Then we'll generate 1,000 observations for *X* and *Y*, and then add them together.

```
> set.seed(20)

> X.1000 <- rnorm(1000)

> Y.1000 <- rnorm(1000)

> W.1000 <- X.1000 + Y.1000
```

Now let's look at the empirical distribution of *W* based on the Monte Carlo simulation and compare it to its true distribution (see **Figure 10.1**).

```
> hist(W.1000, nclass = 25, prob = T,
+    xlim = c(-6, 6), ylim = c(0, 0.3),
+    xlab = "W", ylab = "Relative Frequency")

> pdfplot("norm", list(mean = 0, sd = sqrt(2)),
+    add = T)

> title(main = paste(
+    "Empirical and Theoretical Distributions",
+    "(Based on 1,000 Observations)", sep = "\n"))
```

We can also look at some summary statistics and compare them to the true population values.

```
> emp.stats.1000 <- c(mean = mean(W.1000),
+    sd = sd(W.1000),
+    quantile(W.1000,
+    probs = c(0.1, 0.25, 0.75, 0.9)))

> pop.stats <- c(0, sqrt(2),
+    qnorm(c(0.1, 0.25, 0.75, 0.9), sd = sqrt(2)))

> names(pop.stats) <- names(emp.stats.1000)
```

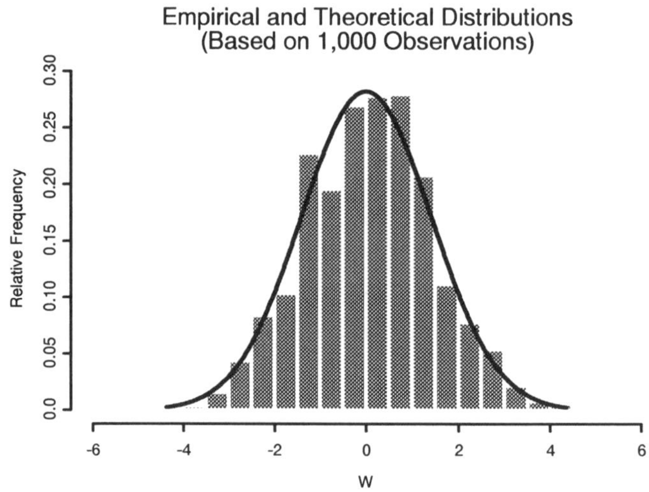

Figure 10.1: Empirical and theoretical distribution of the sum of two N(0,1) random variables.

```
> cbind(Empirical = round(emp.stats.1000, 2),
+    Population = round(pop.stats, 2))
     Empirical Population
mean      0.05       0.00
  sd      1.42       1.41
 10%     -1.73      -1.81
 25%     -0.98      -0.95
 75%      0.98       0.95
 90%      1.83       1.81
```

The simulated distribution is remarkably "close" to the true distribution.

Now let's see what happens when we generate 10,000 values for *X* and *Y*. First we'll generate the values:

```
> X.10000 <- rnorm(10000)

> Y.10000 <- rnorm(10000)

> W.10000 <- X.10000 + Y.10000
```

Now let's look at the empirical distribution of *W* based on the Monte Carlo simulation and compare it to its true distribution:

```
> hist(W.10000, nclass = 100, prob = T,
+    xlim = c(-6, 6), ylim = c(0, 0.3),
+    xlab = "W", ylab = "Relative Frequency")

> pdfplot("norm", list(mean = 0, sd = sqrt(2)),
+    add = T)

> title(main = paste(
+    "Empirical and Theoretical Distributions",
+    "(Based on 10,000 Observations)", sep = "\n"))
```

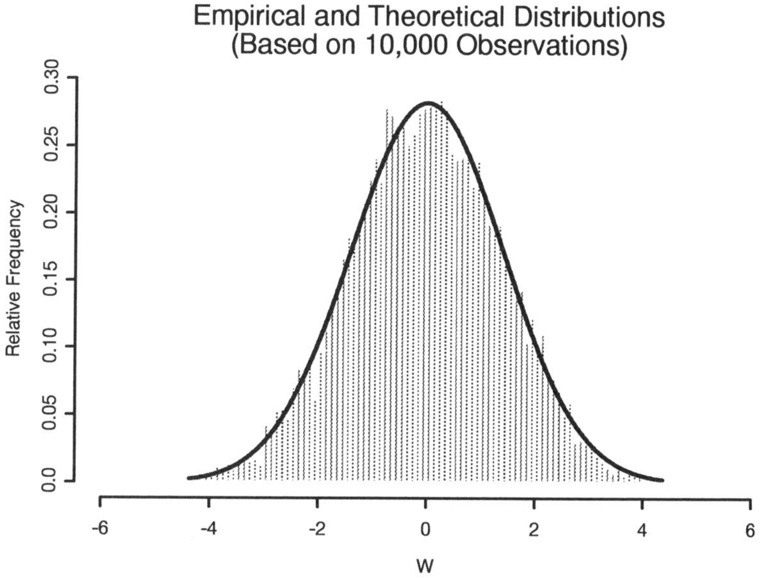

Figure 10.2: Empirical and theoretical distribution of the sum of two N(0,1) random variables.

Now let's look at some summary statistics and compare them to the true population values.

```
> emp.stats.10000 <- c(mean = mean(W.10000),
+    sd = sd(W.10000),
+    quantile(W.10000,
+    probs = c(0.1, 0.25, 0.75, 0.9)))
```

```
> cbind(Emp.1000 = round(emp.stats.1000, 2),
+    Emp.10000 = round(emp.stats.10000, 2),
+    Population = round(pop.stats, 2))
      Emp.1000 Emp.10000 Population
mean      0.05     -0.01       0.00
  sd      1.42      1.41       1.41
 10%     -1.73     -1.80      -1.81
 25%     -0.98     -0.97      -0.95
 75%      0.98      0.96       0.95
 90%      1.83      1.81       1.81
```

As we increase the number of Monte Carlo trials, the simulated distribution tends to get "closer" to the true distribution. This is called the ***Law of Large Numbers***.

Generating Random Numbers

A random number is a realization of a random variable, say X. For many people, the term random number initially conjures up an image of somehow choosing an integer between a specified lower and upper bound (e.g., 1 and 10), where each number is equally likely to be chosen. In that case, the random variable X is a discrete uniform random variable with probability density (mass) function given by:

$$f(x) = \Pr(X = x) = \frac{1}{10}, \quad x = 1, 2, \ldots, 10 \tag{10.2}$$

In general, a random number can be a realization of a random variable from any kind of probability distribution (e.g., uniform, normal, lognormal, gamma, empirical, etc.)

Generating Random Numbers From a Uniform Distribution

The S-PLUS function `runif` generates pseudo-random numbers from a (continuous) uniform distribution (see the S-PLUS help file for **Uniform**). Uniform random number generation in S-PLUS is adapted from Marsaglia et al. (1973), which couples a multiplicative-congruential generator with a feedback shift register. References that discuss generating pseudo-random numbers include Kenedy and Gentle (1980), Rubenstein (1981), Ripley (1987), Law and Kelton (1991), Hayes (1993), and Barry (1996). Venables and Ripley (1997, pp.166-167) explain the S-PLUS pseudo-random number generator in detail.

Pseudo-random number generators start with an initial seed, and then appear to generate random numbers, although these numbers are actually generated by a deterministic mechanism. Each time a set of random numbers is generated, the value of the seed changes. If you start with the same seed, you will get the same sequence of pseudo-random numbers. You can use the S-PLUS function `set.seed` to set the seed of the random number generator.

The period of a random number generator is the number of random numbers that can be generated before the sequence repeats itself. For most starting seeds, the period of the random number generator in S-PLUS is $2^{30} \times 4{,}292{,}868{,}097 \approx 4.6 \times 10^{18}$ (about 4.6 quintillion) (Venables and Ripley, 1997, p.167).

Generating Random Numbers From an Arbitrary Distribution

The S-PLUS and ENVIRONMENTALSTATS for S-PLUS functions of the form `rabb` generate random numbers from several theoretical probability distributions (see column 2 of **Table 5.1** for a list of possible values of *abb*). For example, the function `rnorm` generates random numbers from a normal distribution.

To generate random numbers from a specified probability distribution, most computer software programs use the ***inverse transformation method*** (Rubenstein, 1981, pp.38-43; Law and Kelton, 1991, pp.465-474; Vose, 1996, pp.39-40). Suppose the random variable U has a U[0,1] distribution, that is, a uniform distribution with parameters `min=0` and `max=1` (see the S-PLUS help file for **Uniform**). Let F denote the cumulative distribution function (cdf) of the specified probability distribution. Then the random variable X defined by:

$$X = F^{-1}(U) \qquad \textbf{(10.3)}$$

has the specified distribution, where the quantity F^{-1} denotes the inverse of the cdf function F. Thus, to generate a set of random numbers from any distribution, all you need is a set of random numbers from a U[0,1] distribution and a function that computes the inverse of the cdf function for the specified distribution.

Figure 10.3 illustrates the inverse transformation method for a standard normal distribution, with U=`0.8`. In this case, the random number generated is `qnorm(0.8)=0.8416212`.

```
> cdfplot(xlab = "Random Number (x)",
+   ylab = "Cumulative Frequency [ F(x) ]",
+   main=paste("Example of Inverse Transformation",
+   "Method\nfor Standard Normal Distribution"))

> usr <- par("usr")

> arrows(usr[1], 0.8, qnorm(0.8), 0.8,
+   lwd = 2, col = 2)

> arrows(qnorm(0.8), 0.8, qnorm(0.8), usr[3],
+   lwd = 2, col = 2)
```

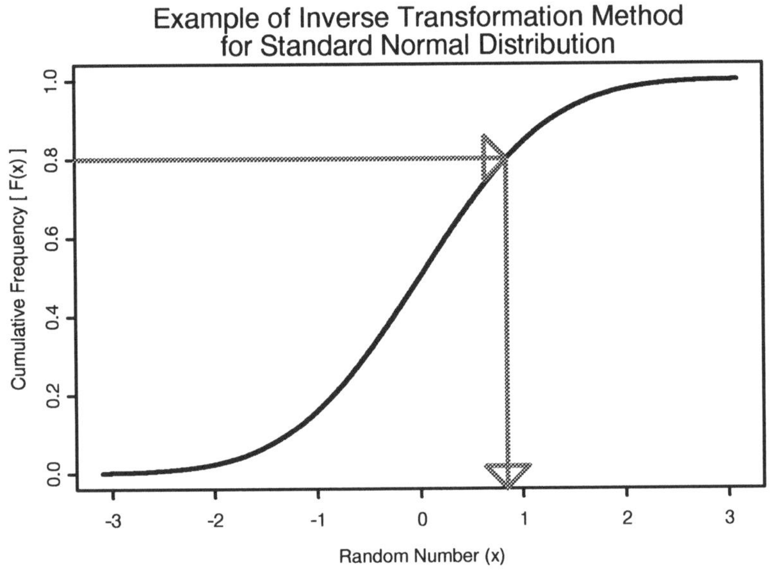

Figure 10.3: Example of inverse transformation method.

Latin Hypercube Sampling

Latin Hypercube sampling, sometimes abbreviated LHS, is a method of sampling from a probability distribution (one random variable) or a joint probability distribution (several random variables) that ensures all portions of the probability distribution are represented in the sample. It was introduced in the published literature by McKay et al. (1979). Other references include Iman and Conover (1980, 1982), and Vose (1996, pp.42-45). Latin Hypercube sampling is an extension of quota sampling, and when applied to the joint distribution of k random variables, can be viewed as a k-dimensional extension of Latin square sampling, thus the name (McKay et al., 1979, p.240).

Latin Hypercube sampling was introduced to overcome the following problem in Monte Carlo simulation based on simple random sampling (SRS). Suppose we want to generate random numbers from a specified distribution. If we use simple random sampling, there is a low probability of getting very many observations in an area of low probability of the distribution. For example, if we generate n observations from the distribution, the probability that none of these observations falls into the upper 98th percentile of the distribution is 0.98^n. So, for example, there is a 13% chance that out of 100 random numbers, none will fall at or above the 98th percentile. If we are interested in reproducing the shape of the distribution, we will need a very large number of observations to ensure that we can adequately characterize the tails of the distribution (Vose, 1996, p.40).

Latin Hypercube sampling was developed in the context of using computer models that required enormous amounts of time to run and for which only a limited number of Monte Carlo simulations could be implemented. In cases where it is fairly easy to generate tens of thousands of Monte Carlo trials, Latin Hypercube sampling may not offer any real advantage.

Algorithm for LHS

Latin Hypercube sampling works as follows for a single probability distribution. If we want to generate *n* random numbers from the distribution, the distribution is divided into *n* intervals of equal probability $1/n$. A random number is then generated from each of these intervals. For *k* independent probability distributions, LHS is applied to each distribution, and the resulting random numbers are matched at random to produce *n* random vectors of dimension *k*.

The three figures below illustrate Latin Hypercube sampling for n=4 based on using the inverse transformation method for generating random numbers, assuming a standard normal distribution. In this case, the interval [0,1] is divided into the four intervals [0, 0.25], [0.25, 0.5], [0.5, 0.75], and [0.75, 1]. Next, a uniform random number is generated within each of these intervals. The four numbers generated are (to two decimal places) 0.04, 0.35, 0.70, and 0.79. Finally, the standard normal random numbers associated with the inverse cumulative distribution function of the four uniform random numbers are computed. Note that we call the function `set.seed` to set the seed of the random number generator so you can reproduce this example.

First let's create the plot of the probability density function and show the intervals of equal probability:

```
> pdfplot(curve.fill = F, xlab = "Random Number (x)",
+   ylab = "Probability Density [ f(x) ]",
+   main =
+   paste("Example of Latin Hypercube Sampling",
+   "with 4 Equal-Probability Intervals", sep = "\n"))

> usr <- par("usr")

> u.bounds <- seq(0.25, 0.75, by = 0.25)

> u.bounds
[1] 0.25 0.50 0.75

> x.bounds <- qnorm(u.bounds)

> segments(x.bounds, usr[3],
+   x.bounds, dnorm(x.bounds), lwd = 2, col = 2)
```

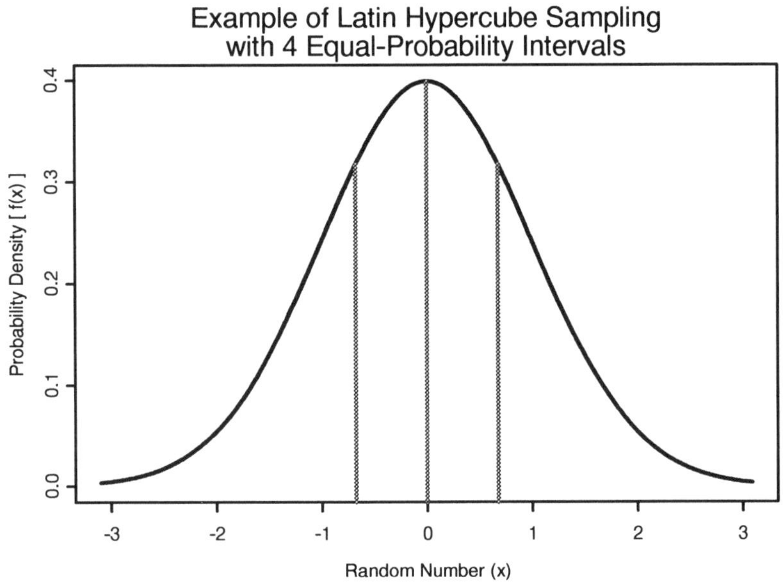

Figure 10.4: Four equal-probability intervals for a N(0,1) distribution.

Now let's create the plot of the cumulative distribution function and show the intervals of equal probability:

```
> cdfplot(xlab = "Random Number (x)",
+   ylab = "Cumulative Frequency [ F(x) ]",
+   main =
+   paste("Example of Latin Hypercube Sampling",
+   "with 4 Equal-Probability Intervals", sep = "\n"))

> usr <- par("usr")

> segments(usr[1], u.bounds, x.bounds, u.bounds,
+   lwd = 2, col = 2)

> segments(x.bounds, u.bounds, x.bounds, usr[3],
+   lwd = 2, col = 2)
```

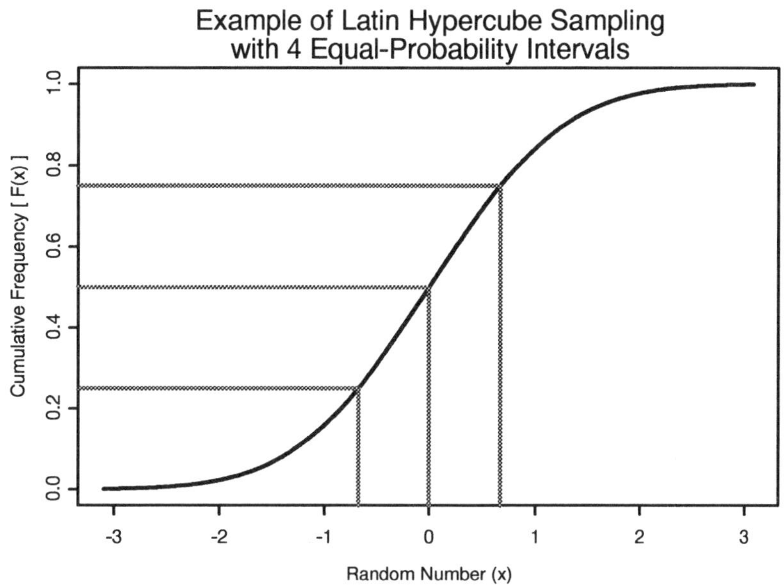

Figure 10.5: Four equal-probability intervals for a N(0,1) distribution.

Now let's generate the four random numbers and add them to the above plot:

```
> set.seed(5)

> u <- runif(4, min = seq(0, 0.75, by = 0.25),
+   max = seq(0.25, 1, by = 0.25))

> u
[1] 0.03777221 0.34740211 0.69535166 0.79149643

> x <- qnorm(u)

> x
[1] -1.7771448 -0.3923438  0.5110776  0.8116245

> arrows(usr[1], u, x, u, col = 3, lty = 2)

> arrows(x, u, x, usr[3], col = 3, lty = 2)
```

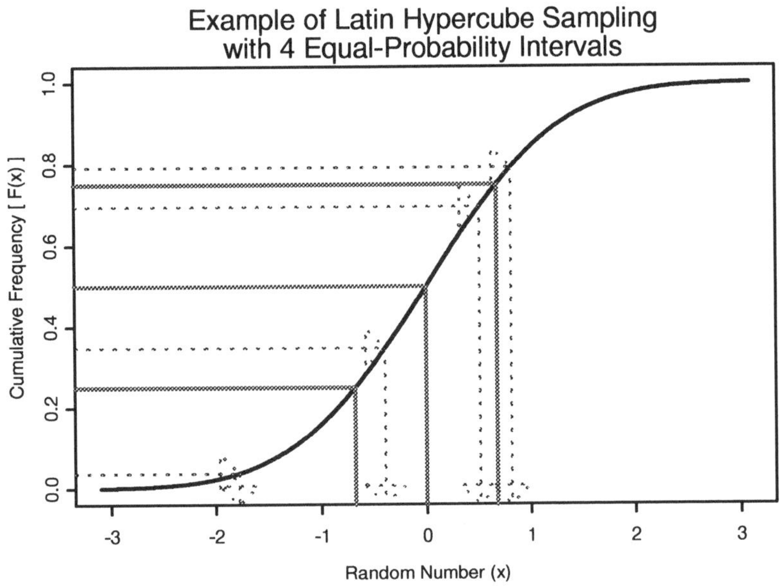

Figure 10.6: Visual explanation of generating 4 random numbers from a N(0,1) distribution using Latin Hypercube sampling.

Properties of LHS

Let Y denote the outcome variable for a Monte Carlo simulation, and suppose Y is a function of k independent random variables as shown in **Equation (10.1)** above. McKay et al. (1979) consider the class of estimators of the form

$$T(Y_1, Y_2, \ldots, Y_n) = \frac{1}{n} \sum_{i=1}^{n} g(Y_i) \tag{10.4}$$

where g is an arbitrary function. This class of estimators includes the mean, the rth sample moment, and the empirical cumulative distribution function. Set

$$\tau = E[T(Y)] \tag{10.5}$$

McKay et al. (1979) show that under LHS, T is an unbiased estimator of τ, and also, if h (from **Equation (10.1)** above) is monotonic in each of its arguments and g is monotonic, then the variance of T under LHS is less than or equal to the variance of T under SRS.

Stein (1987) shows that the variance of the sample mean of Y under LHS is asymptotically less than the variance of the sample mean under simple random sampling whether or not the function h is monotonic in its arguments. Unfortunately, for most cases of LHS, the formula for the true variance of the sample mean is difficult to derive, and thus a good estimate of true variance is not available. Using the usual formula of dividing the sample variance by the sample size will usually overestimate the true variance of the sample mean.

Iman and Conover (1980) and Stein (1987) suggest producing several independent Latin Hypercube samples, say N, and for each Latin Hypercube sample computing the sample mean based on the n observations within that sample. (They call this method ***replicated Latin Hypercube sampling***.) You can then estimate the variance of the sample mean by computing the usual sample variance of these N sample means. Note that this method can be applied to any quantity of interest, such as the median, 95th percentile, etc.

EnvironmentalStats for S-Plus Functions for LHS

The EnvironmentalStats for S-Plus functions `simulate.vector` and `simulate.mv.matrix` allow you to generate random numbers using either simple random sampling or Latin Hypercube sampling. The function `simulate.vector` generates a vector of random numbers from a specified theoretical or empirical probability distribution. The function `simulate.mv.matrix` generates a matrix of random numbers from specified probability distributions based on a specified rank correlation matrix using the algorithm of Iman and Conover (1982).

Example 10.2: Simple Random Sampling vs. Latin Hypercube Sampling

Let's compare the result of SRS and LHS for a standard normal distribution with n=`100`. First let's compare the observed data to the true probability density.

```
> x.srs <- simulate.vector(100, seed = 10)

> x.lhs <- simulate.vector(100, sample.method = "LHS")

> hist(x.srs, prob = T, breaks = seq(-3.5, 3.5, by = 0.5),
+   ylim = c(0, 0.45), xlab = "Random Numbers from SRS",
+   ylab = "Relative Frequency",
+   main = "Results of Simple Random Sampling")

> pdfplot(add = T)

> hist(x.lhs, prob = T, breaks = seq(-3.5, 3.5, by = 0.5),
+   ylim = c(0, 0.45), xlab = "Random Numbers from LHS",
+   ylab = "Relative Frequency",
+   main = "Results of Latin Hypercube Sampling")

> pdfplot(add = T)
```

The results are shown in **Figure 10.7** and **Figure 10.8**.

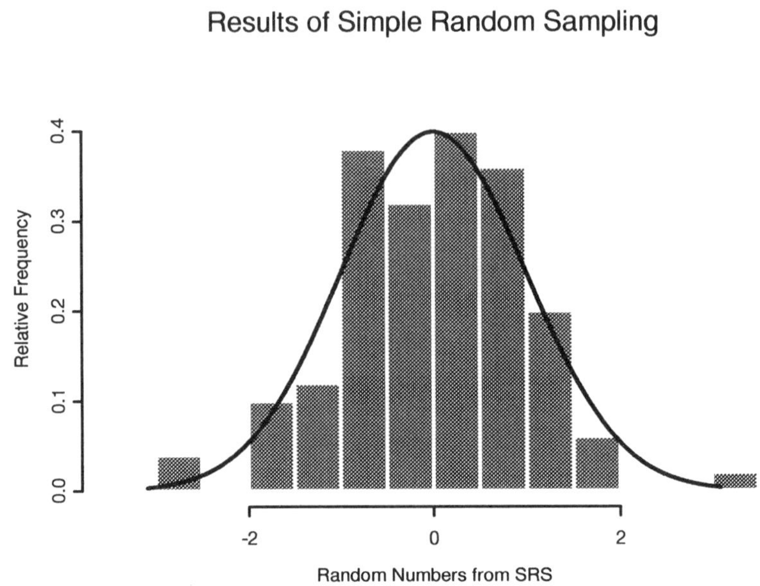

Figure 10.7: Results of simple random sampling from a N(0,1) distribution.

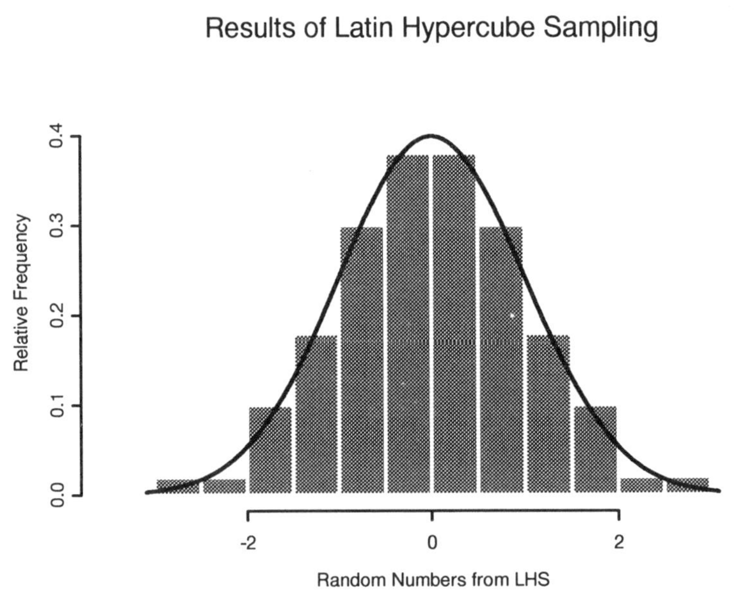

Figure 10.8: Results of Latin Hypercube sampling from a N(0,1) distribution.

Now let's compare the empirical cumulative distribution functions to the true cdf's.

```
> ecdfplot(x.srs,
+    main=paste("Empirical CDF vs. True CDF",
+    "for Simple Random Sampling", sep = "\n"))

> cdfplot(add = T, cdf.col = 2, cdf.lty = 2)

> legend(-2.5, 1, c("Empirical", "True"),
+    lty = 1:2, lwd = 2, col = 1:2)

> ecdfplot(x.lhs, xlim = c(-3, 3),
+    main=paste("Empirical CDF vs. True CDF",
+    "for Latin Hypercube Sampling", sep = "\n"))

> cdfplot(add = T, cdf.col = 2, cdf.lty = 2)

> legend(-2.5, 1, c("Empirical", "True"),
+    lty = 1:2, lwd = 2, col = 1:2)
```

The results are shown in **Figure 10.9** and **Figure 10.10**.

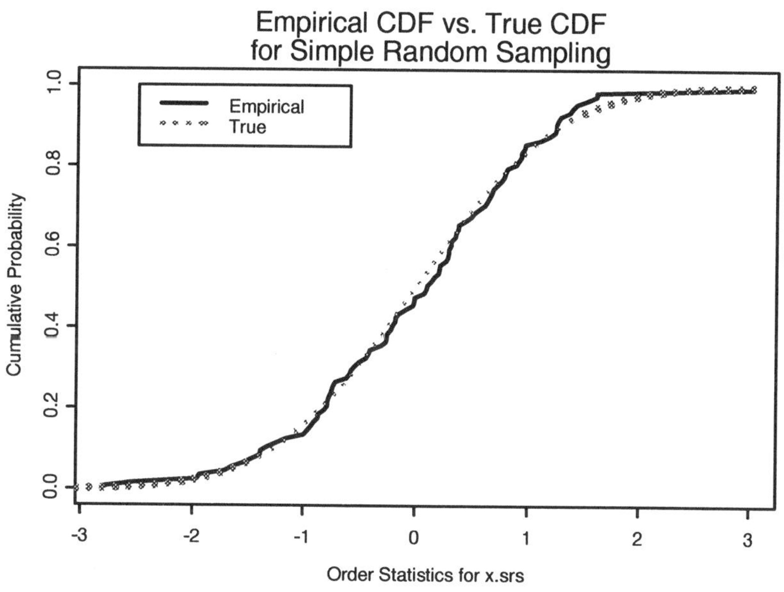

Figure 10.9: Results of simple random sampling from a N(0,1) distribution.

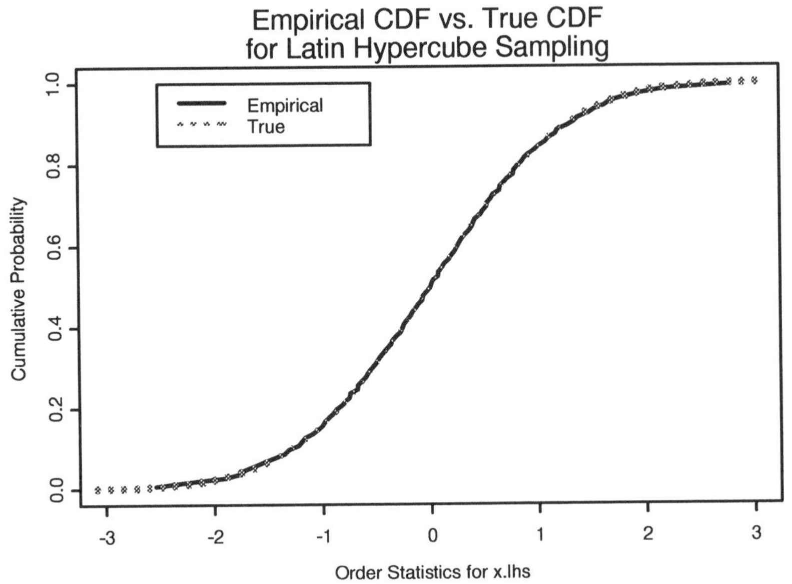

Figure 10.10: Results of Latin Hypercube sampling from a N(0,1) distribution.

Uncertainty and Sensitivity Analysis

Uncertainty analysis and sensitivity analysis are terms used to describe various methods of characterizing the behavior of a complex mathematical/computer model. The model in **Equation (10.1)** above is different from most conventional statistical models, where the form of the model is:

$$Y = h(\underline{X}) + \varepsilon \qquad \textbf{(10.6)}$$

(e.g., linear regression models, generalized linear models, nonlinear regression models, etc.). In **Equation (10.6)**, the vector $\underline{X}$ is assumed to be set or observed at fixed values, and for fixed values of $\underline{X}$ the response variable Y deviates about its mean value according to the distribution of the error term ε. This kind of model is useful when we are interested in the specific relationship between Y and $\underline{X}$, and we want to predict the value of Y for a specified value of $\underline{X}$. Furthermore, this kind of model is fit using paired observations of Y and $\underline{X}$.

In **Equation (10.1)**, Y is assumed to be observed without error, that is, the value of Y is deterministic for a set value of $\underline{X}$. The output variable Y, however, is a random variable because the input variables $X_1, X_2, \ldots, X_k$ are assumed to be a combination of random variables and constant terms. This kind of model is useful when we are interested in describing the distribution of Y taken over all possible (read as "reasonable and realistic") combinations of the input variables. Furthermore, paired observations of Y and $\underline{X}$ are usually not available to validate this kind of model, hence, there is some amount of unquantifiable uncertainty associated with this kind of model.

Uncertainty analysis involves describing the variability or distribution of values of the output variable Y that is due to the collective variation in the input variables $\underline{X}$ (Iman and Helton, 1988, p.72). This description usually involves graphical displays such as histograms, empirical pdf plots, and empirical cdf plots, as well as summary statistics such as the mean, median, standard deviation, coefficient of variation, 95th percentile, etc.

Sensitivity analysis involves determining how the distribution of Y changes with changes in the individual input variables. It is used to identify which input variables contribute the most to the variation or uncertainty in the output variable Y (Iman and Helton, 1988, p.72). Sensitivity analysis is also used in a broader sense to determine how changing the distributions of the input variables and/or their assumed correlations or even changing the form of the model affects the output (Thompson et al., 1992; Smith et al., 1992; Cullen, 1994; Shlyakhter, 1994; Bukowski et al., 1995; Hamed and Bedient, 1997; USEPA, 1997).

Important vs. Sensitive Parameters

Two useful concepts associated with sensitivity analysis are ***important parameters*** (variables) and ***sensitive parameters*** (variables) (Crick et al., 1987 as cited in Hamby, 1994, p.137). Sensitive parameters have a substantial influence on the resulting distribution of the output variable Y, that is, small changes in the value of a sensitive parameter result in substantial changes in Y. Important parameters have some amount of uncertainty and/or variability associated with them and this variability contributes substantially to the resulting variability in the output variable Y.

The figures below illustrate these two concepts for the simple case of three input variables. **Figure 10.11** is an example of an important variable. An important variable is always sensitive. **Figure 10.12** is an example of a variable that is not sensitive, and hence not important. **Figure 10.13** is an example of a variable that is not important. This variable may not be sensitive, like the one in **Figure 10.12**, or it may be sensitive like the one in **Figure 10.11** but it is not important because of its limited variability.

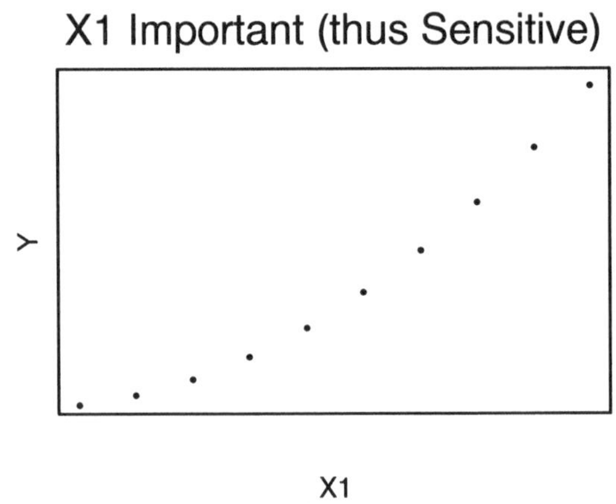

Figure 10.11: An important and thus sensitive variable.

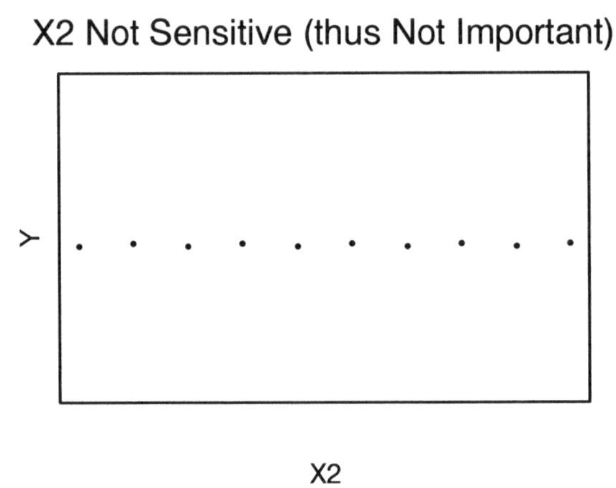

Figure 10.12: A variable that is not sensitive and thus not important.

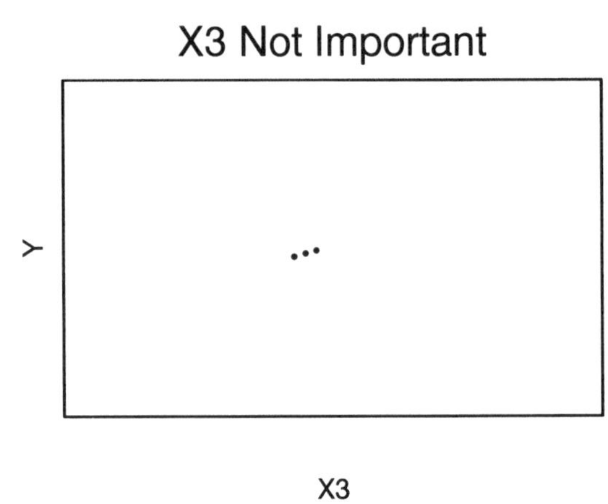

Figure 10.13: A variable that is not important.

Uncertainty vs. Variability

The terms uncertainty and variability have specific meanings in the risk assessment literature that don't necessarily match their meanings in the statistical literature or everyday language. The term ***variability*** refers to the inherent heterogeneity of a particular variable (parameter). For example, there is natural variation in body weight and height between individuals in a given population. The term ***uncertainty*** refers to a lack of knowledge about specific parameters, models, or factors (Morgan and Henrion, 1990; Hattis and Burmaster, 1994; Rowe, 1994; Bogen, 1995; USEPA, 1997, p.10). For example, we may be uncertain about the true distribution of exposure to a toxic chemical in a population (parameter uncertainty due to lack of data, measurement errors, sampling errors, systematic errors, etc.), or we may be uncertain how well our model of incremental lifetime cancer risk reflects reality (model uncertainty due to simplification of the process, mis-specification of the model structure, model misuse, use of inappropriate surrogate variables, etc.), or we may be uncertain about whether a chemical is even present at a site of concern (scenario uncertainty due to descriptive errors, aggregation errors, errors in professional judgment, incomplete analysis, etc.).

We can usually reduce uncertainty through further measurement or study. We cannot reduce variability, since it is inherent in the variable. Note that in the risk assessment literature, measurement error contributes to uncertainty; we can decrease uncertainty by decreasing measurement error. In the statistical literature, measurement error is one component of the variance of a random variable.

Note that a parameter (input variable) may have little or no variability associated with it, yet still have uncertainty associated with it (e.g., the speed of light is constant, but we only know its value to a given number of decimal places).

This meaning of the term ***uncertainty*** should not be confused with the term ***uncertainty analysis*** described above. Uncertainty analysis characterizes the distribution of the output variable Y. The output variable Y varies due to the fact that the input variables are random variables. The distributions of the input random variables reflect both variability (inherent heterogeneity) and uncertainty (lack of knowledge).

Sensitivity Analysis Methods

Sensitivity analysis methods can be classified into three groups: one-at-a-time deviations from a baseline case, factorial design and response surface modeling , and Monte Carlo simulation (Hamby, 1994). Each of these kinds of sensitivity analysis is briefly discussed below. For more detailed information, see the section **Sensitivity Analysis Methods** in the help file **Glossary: Uncertainty and Sensitivity Analysis**. Several studies indicate that using Monte Carlo simulation in conjunction with certain sensitivity measures usually provides the best method of determining sensitivity of the parameters.

One-at-a-Time Deviations From a Baseline Case

These sensitivity analysis methods include differential analysis and measures of change in output to change in input. Differential analysis is simply approximating the variance of the output variable Y at a particular value of the input vector $\underline{X}$ (called the ***baseline case***) by using a first-order Taylor series expansion (Kotz and Johnson, 1985, Volume 8, pp.646-647; Downing et al., 1985; Seiler, 1987; Iman and Helton, 1988; Hamby, 1994, pp.138-139). This approximating equation for the variance of Y is useful for quantifying the proportion of variability in Y that is accounted for by each input variable. Unfortunately, the approximation is usually good only in a small region close to the baseline case, and the relative contribution of each input variable to the variance of Y may differ dramatically for differently chosen baseline cases. Also, differential analysis requires the calculation of partial derivatives, which may or may not be a simple task, depending on the complexity of the input function h (see **Equation (10.1)** above).

Measures of change in output to change in input include the ratio of percent change in Y to percent change in X_i (Hamby, 1994, pp.139-140), the ratio of percent change in Y to change in X_i in units of the standard deviation of X_i (Hamby, 1994, p.140; Finley and Paustenbach, 1994), the percent change in Y as X_i ranges from its minimum to maximum value (Hamby, 1994, p.140), and spider plots, which are plots of Y vs. percent change in X_i, or Y vs. percentiles of X_i (Vose, 1996, p.282-283).

Factorial Design and Response Surface Modeling

The concepts of factorial designs and response surfaces come from the field of experimental design (Box et al., 1978). In the context of sensitivity analysis for computer models, n distinct values of the input vector $\underline{X}$ are chosen (usually reflecting the possible ranges and medians of each of the k input variables), and the model output Y is recorded for each of these input values. Then a multiple linear regression model (a ***response surface***) is fit to these data. The fitted model is called the fitted response surface, and this response surface is used as a replacement for the computer model (Downing et al., 1985; Iman and Helton, 1988; Hamby, 1994, p.140). The sensitivity analysis is based on the fitted response surface. The reason for using a response surface to replace the actual model output is that some computer models are, or used to be, very costly to run, whereas computing output based on a response surface is relatively inexpensive.

One way to rank the importance of the variables in the response surface is to simply compare the magnitudes of the estimated coefficients. The estimated coefficients, however, depend on the units of the predictor variables in the model, so most analysts use standardized regression coefficients (Iman and Helton, 1988; Hamby, 1994, p.146). The standardized regression coefficients are simply the coefficients that are obtained from fitting the response surface model based on the "standardized" output variable and the "standardized" predictor (input) variables. That is, for each variable, each observation is replaced by subtracting the mean (for that variable) from the observation and dividing by the standard deviation (for that variable).

Weisberg (1985, p.186) warns against using standardized coefficients to determine the importance of predictor variables because they depend on the range of the predictor variables. So for example, the variable $X3$ in **Figure 10.13** above may be very important but it has a very limited range. In the context of sensitivity analysis for computer models, however, if the analyst is comfortable with assuming the range of input variables will not change appreciably in the future, then standardized coefficients are an acceptable way to rank the importance of variables.

Iman and Helton (1988) compared uncertainty and sensitivity analysis of several models based on differential analysis, factorial design with a response surface model, and Monte Carlo simulation using Latin Hypercube sampling. The main outcome they looked at for uncertainty analysis was estimating the cumulative distribution function. They found that the models were too mathematically complex to be adequately represented by a response surface. Also, the results of differential analysis gave widely varying results depending on the values chosen for the baseline case. The method based on Monte Carlo simulation gave the best results.

Monte Carlo Simulation

Monte Carlo simulation is used to produce a distribution of Y values based on generating a large number of values of the input vector $\underline{X}$ according to the joint distribution of $\underline{X}$. There are several possible ways to produce a distribution for Y, including varying all of the input parameters (variables) simultaneously, varying one parameter at a time while keeping the others fixed at baseline values, or varying the parameters in one group while keeping the parameters in the other groups at fixed baseline values. This section briefly describes sensitivity methods that can be used with Monte Carlo simulation results.

- **Histograms, Empirical CDF Plots, Percentiles of Output**. A simple graphical way to assess the effect of different input variables or groups of input variables on the distribution of Y is to look at how the histogram and empirical cdf of Y change as you vary one parameter at a time or vary parameters by groups (Thompson et al., 1992). Various quantities such as the mean, median, 95th percentile, etc. can be displayed on these plots as well.

- **Scatterplots**. Another simple graphical way to asses the effect of different input variables on the distribution of Y and their relationship to each other is to look at pairwise scatterplots (see the S-PLUS help files for `pairs`, `xyplot`, and `brush`).

- **Correlations and Partial Correlations**. A quantitative measure of the relationship between Y and an individual input variable X_i is the correlation between these two variables, computed based on varying all parameters simultaneously (Saltelli and Marivoet, 1990; Hamby, 1994, pp.143-144). The Pearson product moment correlation measures the strength of linear association between Y and X_i, while Spearman rank correlation measures the strength of any monotonic relationship between Y and X_i (see the help file for `cor`). Vose (1996, pp.280-282) suggests using "Tornado Charts", which are simply horizontal barcharts displaying the values of the correlations (see the S-PLUS help files for `barchart` and `barplot`).

Individual correlations are hard to interpret when some or most of the input variables are highly related to one another. One way to get around this problem is to look at partial correlation coefficients. See the section **Sensitivity Analysis Methods** in the help file **Glossary: Uncertainty and Sensitivity Analysis Methods** for more information.

- **Change in Output to Change in Input**. Any of these types of measures that are described above under the subsection **One-at-a-Time Deviations from a Baseline Case** can be adapted to the results of a Monte Carlo simulation. Additional measures include relative deviation, in which you vary one parameter at a time and compute the coefficient of variation (CV) of Y for each case (Hamby, 1994, p.142), and relative deviation ratio, in which you vary one parameter at a time and compute the ratio of the CV of Y to the CV of X_i (Hamby, 1994, pp.142-143).

- **Response Surface**. This methodology that was described above in the subsection **Factorial Design and Response Surface Modeling** can be adapted to the results of a Monte Carlo simulation. In this case, use the model input and output to fit a regression equation (possibly stepwise) and then use standardized coefficients to rank the input variables (Iman and Helton, 1988; Saltelli and Marivoet, 1990).

- **Comparing Groupings Within Input Distributions Based on Partitioning the Output Distribution**. One final method of sensitivity analysis that has been used with Monte Carlo simulation is to divide the distribution of the output variable Y into two or more groups, and then to compare the distributions of an input variable that has been split up based on these groupings (Saltelli and Marivoet, 1990; Hamby, 1994, pp.146-150; Vose, 1996, pp.296-298). For example, you could divide the distribution of an input variable X_i into two groups based on whether the values yielded a value of Y below the median of Y or above the median of Y. You could then use any of the Trellis functions (e.g., `bwplot` or `histogram`) to compare the distributions of these two groups, compare these distributions with a goodness-of-fit test (see the help files for `ks.gof` and `cdf.compare`), or compare the means or medians of these distributions with the t-test (using `t.test`) or Wilcoxon rank sum test (using `wilcox.test`). A significant difference between the two distributions is an indication that the input variable is important in determining the distribution of Y.

Uncertainty Analysis Methods

A specific model such as **Equation (10.1)** with specific joint distributions of the input variables leads to a specific distribution of the output variable. The process of describing the distribution of the output variable is called ***uncertainty analysis*** (Iman and Helton, 1988, p.72). This description usually involves graphical displays such as histograms, empirical pdf plots, and empirical cdf plots, as well as summary statistics such as the mean, median, standard deviation, coefficient of variation, 95th percentile, etc.

Sometimes the distribution of Y in **Equation (10.1)** above can be derived analytically based on statistical theory (Springer, 1979; Slob, 1994). For example, if the function h describes a combination of products and ratios, and all of the input variables have a lognormal distribution, then the output variable Y has a lognormal distribution as well, since products and ratios of lognormal random variables have lognormal distributions. Many risk models, however, include several kinds of distributions for the input variables, and some risk models are not easily described in a closed algebraic form. In these cases, the exact distribution of Y can be difficult or almost impossible to derive analytically.

This rest of this section briefly describes some methods of uncertainty analysis based on Monte Carlo simulation. For more information on uncertainty analysis, see the section **Uncertainty Analysis Methods** in the help file **Glossary: Uncertainty and Sensitivity Analysis**.

When the distribution of Y cannot be derived analytically, it can usually be estimated via Monte Carlo simulation. Given this simulated distribution, you can construct histograms or empirical pdf plots and empirical cdf plots, as well as compute summary statistics. You can also compute confidence bounds for specific quantities, such as percentiles (see the help file for `eqnpar`). These confidence bounds are based on the assumption that the observed values of Y are randomly selected based on simple random sampling. When Latin Hypercube sampling is used to generate input variables and hence the output variable Y, the statistical theory for confidence bounds based on simple random sampling is not truly applicable (Easterling, 1986; Iman and Helton, 1991, p.593; Stein, 1987). Most of the time, confidence bounds that assume simple random sampling but are applied to the results of Latin Hypercube sampling will probably be too wide.

Quantifying Uncertainty in the Estimated CDF by Repeating the Monte Carlo Simulation

One way around the above problem with Latin Hypercube sampling is to use ***replicated Latin Hypercube sampling***, that is, repeat the Monte Carlo simulation numerous times, say N, so that you have a collection of N empirical distributions of Y, where each empirical distribution is based on n observations of the input vector $\underline{X}$. You can then use these N replicate distributions to assess the variability of the sample mean, median, 95th percentile, empirical cdf, etc. Obviously, this is a computer intensive process.

A simpler process is to repeat the simulation just twice and compare the values of certain distribution characteristics, such as the mean, median, 5th, 10th, 90th and 95th percentiles, and also to graphically compare the two empirical cdf plots. If the values between the two simulations are within a small percentage of each other, then you can be fairly confident about characterizing the distribution of the output variable. Iman and Helton (1991) did this for a very complex risk assessment for a nuclear power plant and found a remarkable agreement in the empirical cdf's. Thompson et al. (1992) did the same thing for a risk assessment for incremental lifetime cancer risk due to ingestion or dermal contact with soil contaminated with benzene.

You may also want to compare the results of the original simulation with a simulation that uses say twice as many Monte Carlo trials (e.g., Thompson et al., 1992). Barry (1996) warns that if the moments of the simulated distribution do not appear to stabilize with an increasing number of Monte Carlo trials, this can mean that they don't exist. For most risk models, however, the true distributions of any random variables involved in a denominator in **Equation (10.1)** are bounded above 0, so the moments will exist. If a random variable involved in the denominator has a mean or median that is close to 0, it is important to use a bounded or truncated distribution to assure the random variable stays sufficiently far away from 0.

Quantifying Uncertainty Based on Mixture Distributions.

To account for the uncertainty in specifying the distribution of the input variables, some authors suggest using mixture distributions to describe the distributions of the input variables (e.g., Hoffman and Hammonds, 1994; Burmaster and Wilson, 1996). For each input variable, a distribution is specified for the parameter(s) of the input variable's distribution. Some authors call random variables with this kind of distribution ***second-order random variables*** (e.g., Burmaster and Wilson, 1996).

For example we may assume the first input variable comes from a lognormal distribution with a certain mean and coefficient of variation (CV). The mean is unknown to a certain degree, and so we may specify that the mean comes from a uniform distribution with a given set of upper and lower bounds. We can also specify a distribution for the CV.

In this case, the Monte Carlo simulation can be broken down into two stages. In the first stage, a set of parameters is generated for each input distribution. In the second stage, n realizations of the input vector $\underline{X}$ are generated based on this one set of distribution parameters. This two-stage process is repeated N times, so that you end up with N different empirical distributions of the output variable Y, and each empirical distribution is based on n observations of the input vector $\underline{X}$.

For a fixed number of Monte Carlo trials nN, the optimal combination of n and N will depend on how the distribution of the output variable Y changes relative to variability in the input distribution parameter(s) vs. variability in the input variables. For example, if the distribution of Y is very sensitive to changes in the input distribution parameters, then N should be large relative to n or nN. On the other hand, if the distribution of Y is relatively insensitive to the values of the parameters of the input distributions, but varies substantially with the values of the input variables, then N may be small relative to n.

Caveat

An important point to remember is that no matter how complex the mathematical model in **Equation (10.1)** is, or how extensive the uncertainty and sensitivity analyses are, ***there is always the question of how well the mathematical model reflects reality***. The only way to attempt to answer this question is with data collected directly on the input and output variables. Often, however, it is not possible to do this, which is why the model was constructed in the first place.

For example, in order to attempt to directly verify a model for incremental lifetime cancer risk for a particular exposed population within a particular geographical region, you have to collect data on lifetime exposure for each person and the actual proportion of people who developed that particular cancer within their lifetime, accounting for competing risks as well, and compare these data to similar data collected on a proper control population. A controlled experiment that involves exposing a random subset of a particular human population to a toxin and following the exposed and control group throughout their lifetimes for the purpose of a risk assessment is not possible to perform for several reasons, including ethical and practical ones. Rodricks (1992) is an excellent text that discusses the complexities of risk assessment based on animal bioassay and epidemiological studies.

Risk Assessment

This section discusses the concepts and practices involved in risk assessment, and shows you how to use functions in ENVIRONMENTALSTATS for S-PLUS to perform probabilistic risk assessment. This information is also contained in the help file **Glossary: Risk Assessment**.

Definitions

Before we discuss how to use ENVIRONMENTALSTATS for S-PLUS to aid in probabilistic risk assessment, it will be helpful to define common terms and concepts used in risk assessment (see the help file **Glossary: Risk Assessment -- Definitions**).

Risk

The common meaning of the term ***risk*** when used as a noun is "the chance of injury, damage, or loss." Thus, risk is a probability, since "chance" is another term for "probability."

Risk Assessment

Risk assessment is the practice of gathering and analyzing information in order to predict future risk. Risk assessment has been commonly used in the fields of insurance and engineering for quite some time. In the last couple of decades it has been increasingly applied to the problems of predicting human health and ecological effects from exposure to toxicants in the environment (e.g., Hallenbeck, 1993; Suter, 1993). In this chapter, the term risk assessment is applied in the context of human health and ecological risk assessment.

The basic model that is often used as the foundation for human health and ecological risk assessment is:

$$Risk = Dose \times \Pr(Effect \text{ per } Unit\ Dose) \tag{10.7}$$

That is, the risk of injury (the ***effect***) to an individual is equal to the amount of toxicant the individual absorbs (the ***dose***) times the probability of the effect occurring for a single unit of the toxicant. If the effect is some form of cancer, the second term on the right-hand side of **Equation (10.7)** is often called the ***cancer slope factor*** (abbreviated CSF) or the ***cancer potency factor*** (abbreviated CPF).

The first term on the right-hand side of **Equation (10.7)**, the dose, is estimated by identifying sources of the toxicant and quantifying their concentrations, identifying how these sources will expose an individual to the toxicant (via fate and transport models), quantifying the amount of exposure an individual will receive, and estimating how much toxicant the individual will absorb at various levels of exposure. Sometimes the dose represents the amount of toxicant absorbed over a lifetime, and sometimes it represents the amount absorbed over a shorter period of time.

The second term on the right-hand side of **Equation (10.7)**, the probability of an effect (CSF or CPF), is estimated from a ***dose-response curve***, a model that relates the probability of the effect to the dose received. Dose-response curves are developed from controlled laboratory experiments on animals or other organisms, and/or from epidemiological studies of human populations (Hallenbeck, 1993).

Risk assessment therefore involves three major steps (Hallenbeck, 1993, p.1; USEPA, 1995c):

- **Hazard Identification.** Describe the effects (if any) of the toxicant on laboratory animals, humans, and/or wildlife species, based on documented studies. Describe the quality and relevance of the data from these studies. Describe what is known about how the toxicant produces these effects. Describe the uncertainties and subjective choices or assumptions associated with determining the degree of hazard of the toxicant.

- **Dose-Response Assessment.** Describe what is known about the biological mechanism that causes the health or ecological effect. Describe what data, models, and extrapolations have been used to develop the dose-response curve for laboratory animals, humans, and/or wildlife species. Describe the routes and levels of exposure used in the studies to determine the dose-response curve, and compare them to the expected routes and levels of exposure in the population(s) of concern. Describe the uncertainties and subjective choices or assumptions associated with characterizing the dose-response relationship.

- **Exposure Assessment.** Identify the sources of environmental exposure to the population(s) of concern. Describe what is known about the principal paths, patterns, and magnitudes of exposure. Determine average and "high end" levels of exposure. Describe the characteristics of the population(s) that is(are) potentially exposed. Determine how many members of the population are likely to be exposed. Describe the uncertainties and subjective choices or assumptions associated with characterizing the exposure for the population of concern.

Once these steps have been completed, some form of **Equation (10.7)** is usually used to estimate the risk for a particular population of concern. Both sensitivity analysis and uncertainty analysis should be applied to the risk assessment model to quantify the uncertainty associated with the estimated risk.

Risk Characterization

USEPA (1995c) distinguishes between the process of risk assessment and risk characterization. ***Risk characterization*** is the summarizing step of risk assessment that integrates all of the information from the risk assessment, including uncertainty and sensitivity analyses and a discussion of uncertainty vs. variability, to form an overall conclusion about the risk. Risk assessment is the tool that a risk assessor uses to produce a risk characterization. A risk characterization is the product that is delivered to the risk assessor's client: the risk manager.

Risk Management

Risk management is the process of using information from risk characterizations (calculated risks), perceived risks, regulatory policies and statutes, and economic and social analyses in order to make and justify a decision (USEPA, 1995c). If the risk manager decides that the risk is not acceptable, he or she will order or recommend some sort of action to decrease the risk. If the risk manager decides that the risk poses minimal danger to the population of concern, he or she may recommend that no further action is needed at the present time.

Because the risk manager is the client of the risk assessor, the risk manager must be involved in the risk assessment process from the start, helping to determine the scope and endpoints of the risk assessment (USEPA, 1995c). Also, the risk manager must interact with the risk assessor throughout the risk assessment process, so that he or she may take responsibility for critical decisions. The risk manager, however, must be careful not to let non-scientific (e.g., political) issues influence the risk assessment. Non-scientific issues are dealt with at the risk management stage, not the risk assessment stage.

Risk Communication

USEPA (1995c) defines ***risk communication*** as exchanging information with the public. While the communication of risk from the risk assessor to the risk manager is accomplished through risk characterization, the communication of risk between the risk manager (or representatives of his or her agency) and the public is accomplished through risk communication. The risk characterization will probably include highly technical information, while risk communication should concentrate on communicating basic ideas of risk to the public.

Building a Risk Assessment Model

Building a risk assessment model involves all three steps outlined in the definition of risk assessment (hazard identification, dose-response assessment, and exposure assessment). Once these steps have been completed, some form of **Equation (10.7)** is usually used to estimate the risk for a particular population of concern.

The form of the two terms on the right-hand side of **Equation (10.7)** may be very complex. Estimation of dose involves identifying sources of exposure, postulating pathways of exposure from these sources, estimating exposure concentrations, and estimating the resulting dose for a given exposure. Estimation of dose-response involves using information from controlled laboratory experiments on animals and/or epidemiological studies. Given a set of dose-response data, there are several possible statistical models that can be used to fit these data, including tolerance distribution models, mechanistic models, linear-quadratic-exponential models, and time-to-response models (Hallenbeck, 1993, Chapter 4). Probably the biggest controversy in risk assessment involves the extrapolation of dose-response data from high-dose to low-dose and from one species to another (e.g., between mice and humans).

Many risk assessment models have the general form

$$\begin{aligned} Risk &= Dose \times \Pr(Effect \text{ per } Unit\ Dose) \\ &= h(X_1, X_2, \ldots, X_k) \times CPF \end{aligned} \qquad (10.8)$$

That is, the risk is assumed to be proportional to dose, and the dose term is a function of several input variables. For example, USEPA (1991b, p.2) uses the following general equation for intake (here intake is equivalent to exposure and dose):

$$Intake = \frac{C \times IR \times EF \times ED}{BW \times AT} \qquad (10.9)$$

where C is the chemical concentration, IR is the intake or contact rate, EF is the exposure frequency, ED is the exposure duration, BW is body weight, and AT is the averaging time (equal to exposure duration for non-carcinogens and 70 years for carcinogens).

Usually, many of the input variables and sometimes the cancer potency factor (CPF) in **Equation (10.8)** are themselves assumed to be random variables because they exhibit inherent heterogeneity (variability) within the population (e.g., body weight, fluid intake, etc.), and because there is a certain amount of uncertainty associated with their values. The choice of what distribution to use for each of the input variables is based on a combination of available data and expert judgment (see the help file **References: Risk Assessment - Choosing Input Probability Distributions**).

Example 10.3: Cancer Risk from PCE in Groundwater

McKone and Bogen (1991) present a risk assessment model for lifetime cancer risk from exposure to perchloroethylene (PCE) in groundwater in California. In order to estimate dose, they estimated the concentration of PCE in California water supplies, identified three pathways of exposure (ingestion of tap water, indoor inhalation from tap water, and dermal contact from bathing and showering), estimated concentrations and absorption rates for each pathway, and estimated actual dose based on exposure concentration for each pathway. To estimate the cancer potency factor (CPF), McKone and Bogen (1991) used eight rodent-bioassay data sets and a multistage dose-response extrapolation model. They assumed the input variables for dose and the CPF follow certain probability distributions, and they present two point estimates of risk (using the mean value of each input variable and the 95th percentile of each input variable) as well as a distribution of risk based on a Monte Carlo simulation.

Example 10.4: Incremental Lifetime Cancer Risk from Ingesting Soil Contaminated with Benzene

Thompson et al. (1992) present a simplified case study of a risk assessment for incremental lifetime cancer risk (ILCR) for children (ages 8-18) ingesting soil contaminated with benzene. They use the following model:

$$ILCR = Dose \times CPF$$

$$= \frac{Cs \cdot SIngR \cdot RBA \cdot DpW \cdot WpY \cdot YpL \cdot CF}{BW \cdot DinY \cdot YinL} \times CPF \qquad (10.10)$$

Symbol	Variable Name	Units	Distribution*	Point Estimate
Cs	Benzene concentration	mg/kg	LN(0.84, 0.77)	3.39
SIngR	Soil ingestion rate	mg/d	LN(3.44, 0.80)	50
RBA	Relative bioavailability			1
DpW	Exposure days per week	d/wk		1
WpY	Exposure weeks per year	wk/yr		20
YpL	Exposure years per life	yr/life		10
CF	Conversion factor	kg/mg		1×10^{-6}
BW	Average body weight	kg	N(47,8.3)	47
DinY	Days in a year	d/yr		364
YinL	Years in a lifetime	yr/life		70
CPF	Cancer potency factor	(kg d)/mg	LN(-4.33, 0.67)	0.029

* LN denotes lognormal distribution and N denotes normal distribution. The parameters of the normal distribution are the mean and standard deviation. The parameters of the lognormal distribution are the mean and standard deviation of the log-transformed random variable.

Table 10.2: Description of terms in **Equation (10.10)** (Thompson et al., 1992).

Table 10.2 explains what the input variables are in **Equation (10.10)** and their assumed distributions.

The point estimates for the four input variables with associated probability distributions, benzene concentration (`Cs`), soil ingestion rate (`SIngR`), average body weight (`BW`), and cancer potency factor (`CPF`) are, respectively, the 95% upper confidence limit for the mean, the 72nd percentile, the mean, and the 88th percentile. More conservative and perhaps unrealistic point estimates for `SIngR` and `CPF` are their 95th percentiles, which are, respectively,

```
> qlnorm(0.95, 3.44, 0.80)
[1] 116.2663

> qlnorm(0.95, -4.33, 0.67)
[1] 0.03963875
```

Thompson et al. (1992) used the model in **Equation (10.10)** along with a model for ILCR due to dermal contact with the soil to demonstrate how to use Monte Carlo simulation to perform risk assessment.

Point Estimates of Risk

USEPA (1989c, pp.6-4 to 6-5) states that

> Actions at Superfund sites should be based on an estimate of *the reasonable maximum exposure (RME)* expected to occur under both *current* and *future* land-use conditions. The reasonable maximum exposure is defined here as the highest exposure that is reasonably expected to occur at a site. RMEs are estimated for individual pathways. If a population is exposed via more than one pathway, the combination of exposure pathways also must represent an RME. ... The intent of the RME is to estimate a conservative exposure case (i.e., well above the average case) that is still within the range of possible exposures.

In the context of estimating the exposure term of a risk assessment model, the document further specifies that

> Each intake variable in the equation has a range of values. For Superfund exposure assessments, intake variable values for a given pathway should be selected so that the combination of all intake variables results in an estimate of the reasonable maximum exposure for that pathway. ... Under this approach, some intake variables may not be at their individual maximum values but when in combination with other variables will result in estimates of the RME.

(USEPA, 1989c, p.6-19). The document then goes on to suggest using the 95 percent upper confidence limit of the mean for exposure concentration, the 95th or 90th percentile for contact rate, the 95th percentile for exposure frequency and duration, and the average body weight over the exposure period (USEPA, 1989c, pp.6-19, 6-22). This guidance is intended to yield an estimate of risk that is both "protective and reasonable." USEPA (1991b, p.2) reiterates these suggestions.

The above type risk estimate is a "conservative" point estimate of risk. This type of estimate of risk has been frequently criticized for the following reasons:

- There is no way of assessing the degree of conservatism in this kind of estimate (Thompson et al., 1992).
- The input variables may be set to a specific combination that will rarely, if ever, occur in reality (Thompson et al., 1992).
- Traditional sensitivity analyses (one-at-a-time-deviations from the baseline case) are meaningless for this baseline case since many of the input variables are already at or close to their maximum values (Thompson et al., 1992).
- This estimate of risk is often larger by one or several orders of magnitude than the 95th percentile of the population risk (McKone and Bogen, 1991; Burmaster and Harris, 1993; Cullen, 1994).

Probabilistic Risk Assessment

Because of the shortcomings of point estimates of risk, most practitioners in the field of risk assessment now advocate presenting not only the results of points estimates but also the whole distribution of risk and the characteristics of this distribution (Eschenroeder and Faeder, 1988; McKone and Bogen, 1991; Thompson et al., 1992; Burmaster and Harris, 1993; Finley and Paustenbach, 1994; Smith, 1994; McKone, 1994). Presenting the distribution of the output variable (risk in this case) and characterizing this distribution is in fact the definition of uncertainty analysis (Iman and Helton, 1988)! Probabilistic risk assessment is simply uncertainty analysis applied to a risk assessment model that assumes some of the input variables in the model are random variables.

Sometimes the distribution of risk can be derived analytically for simple models (Springer, 1979; Slob, 1994). Usually, however, the distribution of risk must be derived by Monte Carlo simulation. The final section of this chapter shows you how to use functions in ENVIRONMENTALSTATS for S-PLUS to perform a probabilistic risk assessment.

Guidelines for Conducting and Reporting a Probabilistic Risk Assessment

Because of the complexity of probabilistic risk assessment, it is important to carefully conduct the risk assessment, document exactly how the risk assessment was performed, and effectively communicate the results. Hoaglin and Andrews (1975) suggest certain procedures to follow when reporting the results of a Monte Carlo simulation. More recent references specific to probabilistic risk assessment include Burmaster and Anderson (1994), Vose (1996, Chapter 12), and USEPA (1997). The following guidelines are based on these last three references.

Conducting a Probabilistic Risk Assessment

1. Identify the purpose and scope of the risk assessment, and define the endpoints.

2. Create a model or several models for the risk assessment based on the three steps outlined in the definition of risk assessment (hazard identification, dose-response assessment, and exposure assessment). Document the assumptions of each model.

3. Conduct preliminary sensitivity analyses to determine if structural differences (different models) have substantial effects on the output distribution (in both the region of central tendency and the tails). For a given model, determine which input variables contribute the most to the variability of the output distribution. If warranted, determine the effects of changing the shape of the input distributions and/or accounting for correlations between the input variables (Smith et al., 1992; Bukowski et al., 1995; Hammed and Bedient, 1997).

4. Simplify the Monte Carlo simulation by setting unimportant input variables (determined from the previous step) to fixed values.

5. Develop distributions for important input variables based on available data and expert judgment. Document the sources of the data sets and what assumptions were used to derive the input distributions. Perform goodness-of-fit tests and graphical assessments of goodness-of-fit.

6. Quantify and distinguish between the variability and uncertainty in the input parameters. For important input parameters, it may be necessary to gather additional data to reduce uncertainty.

7. Perform the Monte Carlo simulations, always recording the random number seeds. Perform sensitivity analyses to determine which variables or group of variables contribute substantially to the variability of the output variable (exposure or risk). Distinguish between variability and uncertainty in the input parameters (one way to do this is to model the input distributions as mixture distributions; see Hoffman and Hammonds, 1994 and Burmaster and Wilson, 1996).

8. Verify the numerical stability of the Monte Carlo simulation results by performing at least two simulations (using different starting random seeds, of course) or dividing the simulation into non-overlapping sub-sequences and comparing results.

Presenting the Results of a Probabilistic Risk Assessment

1. Taylor the document to the client. Results presented to a risk manager will include many technical details that are usually omitted for presentations to the general public. Results presented to a risk manager should be detailed enough so that the risk assessment may be independently duplicated and verified.

2. Use a tiered presentation style. Put the question and results up front, put details in appendices.

3. Use both a histogram and empirical cumulative distribution function plot to present the distribution of risk on one page, using identical horizontal scales for each plot. On each plot, identify the mean, the 95th percentile, and the point estimate of risk based on established regulatory guidance. Accompany these plots with a table of summary statistics (e.g., minimum, 5th percentile, median, mean, 95th percentile, and maximum).

4. Present and explain the model. Include all equations, describe all terms and their units of measure, and, where appropriate, include schematic diagrams documenting the structure of the model. Outline model assumptions and shortcomings. Document references for the model.

5. Document how each input distribution was selected, including sources of any relevant data sets. Include results of goodness-of-fit tests and graphical assessments of goodness-of-fit. Also discuss the relative contributions of uncertainty and variability for each input distribution. For each input distribution, present a plot of the probability density function and cumulative distribution function on one page, using identical horizontal axes for each plot. Indicate the location of the mean on each plot. Also include a table of relevant statistics for the input distribution, including the minimum, 5th percentile, median, mean, 95th percentile, and maximum.

6. Discuss the results of sensitivity and uncertainty analyses, including the effects of changing the shape of input distributions and/or including correlations among the input variables, as well as analyses that separate uncertainty from variability.

7. Document the software that was used to perform the Monte Carlo simulation, and present the name and statistical quality (i.e., the period) of the random number generator used by the software (see the help file **Glossary: Random Numbers**).

Using ENVIRONMENTALSTATS for S-PLUS for Probabilistic Risk Assessment

This section gives several examples of how to use certain functions in ENVIRONMENTALSTATS for S-PLUS to perform Monte Carlo simulations and sensitivity and uncertainty analyses for a probabilistic risk assessment. All of the examples are based on Thompson et al.'s (1992) model for incremental lifetime cancer risk (ILCR) for children ingesting soil contaminated with benzene. This model is described above in **Equation (10.10)** and **Table 10.2**. The information in this section is also contained in the help file **Glossary: Risk Assessment – Using ENVIRONMENTALSTATS for S-PLUS for Probabilistic Risk Assessment.**

Creating Functions to Describe Exposure and Risk

The following functions compute exposure and ILCR based on Thompson et al.'s (1992) model:

```
> expo.fcn <- function(Cs, S.Ing.R, BW,
+    RBA = 1, DpW = 1, WpY = 20, YpL = 10,
+    CF = 1e-006, DinY = 364, YinL = 70)
+ {
+    (Cs * S.Ing.R * RBA * DpW * WpY * YpL * CF)/
+      (BW * DinY * YinL)
+ }

> ilcr.fcn <- function(Cs, S.Ing.R, BW, CPF,
+    RBA = 1, DpW = 1, WpY = 20, YpL = 10,
+    CF = 1e-006, DinY = 364, YinL = 70)
+ {
+    ((Cs * S.Ing.R * RBA * DpW * WpY * YpL * CF)/
+      (BW * DinY * YinL)) * CPF
+ }
```

Note that in both of these functions, the input variables that have associated probability distributions appear first in the argument list and have no default values. This style of defining the functions is used so that these functions may be easily used with the `eval.expr` function (which will be explained below).

Here are the values of exposure and ILCR for the point estimates given in **Table 10.2**:

```
> expo.pt.est <- expo.fcn(Cs = 3.39, S.Ing.R = 50,
+    BW = 47)

> expo.pt.est
[1] 2.830756e-008

> ilcr.pt.est <- ilcr.fcn(Cs = 3.39, S.Ing.R = 50,
+    BW = 47, CPF = 2.9e-2)

> ilcr.pt.est
[1] 8.209192e-010
```

So the point estimate for ILCR is about 8 in 10 billion.

Recall that in **Table 10.2**, the point estimates of soil ingestion rate (`SIngR`) and cancer potency factor (`CPF`) that Thompson et al. (1992) used are the 72th and 88th percentiles, respectively, of the assumed distributions. The point estimate of exposure and ILCR when both of these variables are set to their 95th percentiles are:

```
> expo.pt.est.cons <- expo.fcn(Cs = 3.39,
+    S.Ing.R = qlnorm(0.95, 3.44, 0.8), BW = 47)

> expo.pt.est.cons
[1] 6.582428e-008

> ilcr.pt.est.cons <- ilcr.fcn(Cs = 3.39,
+    S.Ing.R = qlnorm(0.95, 3.44, 0.8), BW = 47,
+    CPF = qlnorm(0.95, -4.33, 0.67))

> ilcr.pt.est.cons
[1] 2.609192e-009
```

Thus, the point estimate of ILCR increases to about 3 in 1 billion.

Characterizing the Theoretical Input Distributions

Within the final risk assessment report, it is a good idea to include graphs and statistical descriptions of the theoretical probability distributions associated with each of the input variables.

Graphical Characterization

You can use the functions `pdfplot` and `cdfplot` to create plots of the probability density functions and cumulative distribution functions for the input variables that have associated probability distributions. To put two plots on the same page, one on top of the other, you can use the command

```
> par(mfrow = c(2, 1))
```

(see the S-PLUS help file for `par`).

Here are the commands to make the pdf plot for the distribution associated with benzene concentration in the soil (Cs), and indicate where the mean lies (see the help file for **Lognormal** for a discussion of the relationship between the mean and standard deviation of the log-transformed distribution vs. the mean and coefficient of variation of the untransformed distribution):

```
> pdfplot(dist = "lnorm",
+    param.list = list(meanlog = 0.84, sdlog = 0.77),
+    right.tail.cutoff = 0.005,
+    xlab = "Benzene Concentration (mg/kg)",
+    main = "Assumed PDF for Benzene Concentration")

> avg <- exp(0.84 + (0.77^2)/2)

> f.avg <- dlnorm(avg, 0.84, 0.77)

> segments(avg, par("usr")[3], avg, f.avg)

> points(avg, f.avg, pch = 16,
+    cex = 1.25 * par("cex"))

> mtext("Mean", side = 1, at = avg)
```

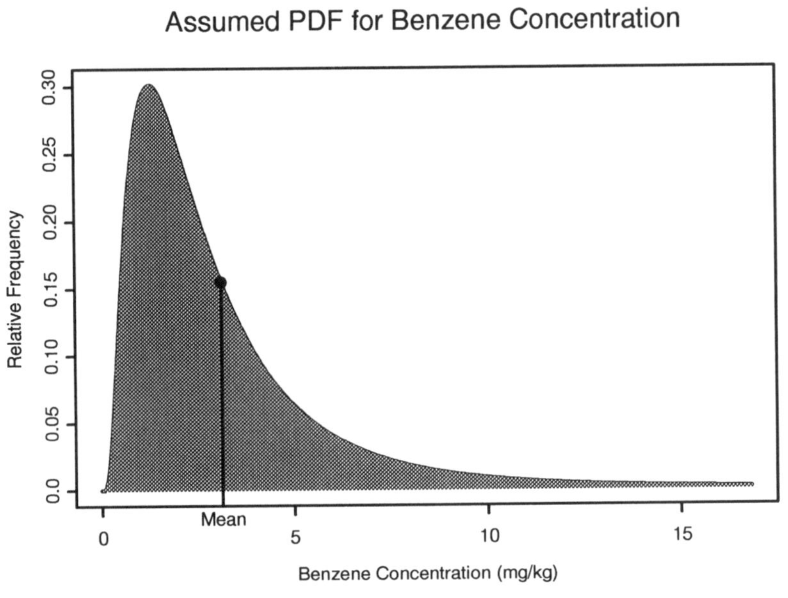

Figure 10.14: Assumed pdf for benzene concentration in soil.

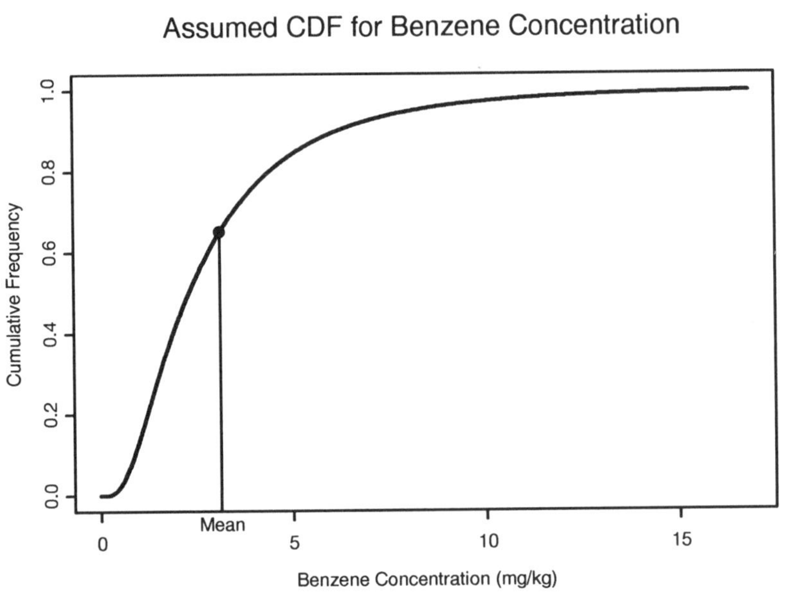

Figure 10.15: Assumed cdf for benzene concentration in soil.

Here are the commands to make the cdf plot and indicate where the mean lies:

```
> cdfplot(dist = "lnorm",
+    param.list = list(meanlog = 0.84, sdlog = 0.77),
+    right.tail.cutoff = 0.005,
+    xlab = "Benzene Concentration (mg/kg)",
+    main = "Assumed CDF for Benzene Concentration")

> F.avg <- plnorm(avg, 0.84, 0.77)

> segments(avg, par("usr")[3], avg, F.avg)

> points(avg, F.avg, pch = 16,
+    cex = 1.25 * par("cex"))

> mtext("Mean", at = avg, side = 1)
```

Figure 10.14 and **Figure 10.15** show the pdf and cdf plots created with these commands. If you prefer to show the cdf plot with the area under the curve filled in, you can use the argument `curve.fill=T` in the call to `cdfplot` above.

Statistical Characterization

To compute various quantiles for the input distributions, simply use the `q`*abb* function, where *abb* denotes the abbreviation of the probability distribution (see the help file for `.Distribution.frame` or the second column of **Table 5.1** for a list of distribution abbreviations). For example, here are the minimum, 5th percentile, median, 95th percentile, and maximum for the distribution for benzene concentration, rounded off to the 1st decimal place:

```
> round(qlnorm(c(0, 0.05, 0.5, 0.95, 1),
+    meanlog = 0.84, sdlog = 0.77), 1)
[1] 0.0 0.7 2.3 8.2 Inf
```

Monte Carlo Simulation

You can use the functions `simulate.vector` and `simulate.mv.matrix` to generate random numbers from specified probability distributions. The function `simulate.vector` generates random numbers from a specified probability distribution (theoretical or empirical) and returns them in a vector. The function `simulate.mv.matrix` generates a matrix of random numbers from several possibly different distributions, and allows you to specify rank correlations between the matrix columns. Both `simulate.vector` and `simulate.mv.matrix` allow you use either simple random sampling or Latin Hypercube sampling.

You can also use any of the functions of the form `r`*abb* (where *abb* denotes the abbreviation of the distribution) to generate random numbers from a specific distribution (see the help file **Probability Distributions and Random Numbers**).

For example, `rnorm` generates random numbers from the normal distribution, and `rmvnorm` generates random vectors from a multivariate normal distribution with a specified correlation matrix.

Simple Random Sampling

Let's generate 10,000 observations for each of the variables *Cs*, *SIngR*, *BW*, and *CPF* using the distributions specified in **Table 10.2** and assuming these variables are independent. First let's generate observations based on simple random sampling.

```
> mat.SRS <- simulate.mv.matrix(10000,
+ distributions = c(Cs      = "lnorm",
+                   S.Ing.R = "lnorm",
+                   BW      = "norm",
+                   CPF =     "lnorm"),
+ param.list = list(
+   Cs      = list(meanlog =  0.84, sdlog = 0.77),
+   S.Ing.R = list(meanlog =  3.44, sdlog = 0.80),
+   BW      = list(mean    = 47,    sd    = 8.3),
+   CPF     = list(meanlog = -4.33, sdlog = 0.67)),
+ cor.mat = diag(4), seed = 47)
```

Note that we set the argument `seed` to `47` in the call to `simulate.mv.matrix`. That way, we can reproduce the exact same matrix again if we need to.

Latin Hypercube Sampling

Now let's generate 10,000 observations for each of the variables using Latin Hypercube sampling. We must use care in generating Latin Hypercube samples from unbounded distributions. For example, the variable average body weight (*BW*) is assumed to follow a N(47, 8.3) distribution, that is, a normal distribution with a mean of 47kg and standard deviation of 8.3kg. With standard Latin Hypercube sampling, if we generate 10,000 observations, then the smallest observation will be some number between the 0th and 0.01^{th} percentile of the distribution. Similarly, the largest observation will be some number between the 99.99^{th} percentile and the 100th percentile. The 0th and 100th percentiles for any normal distribution are negative and positive infinity, respectively. Thus, we must specify some lower and upper bounds for percentiles to use when performing Latin Hypercube sampling on a normal distribution or another distribution that is unbounded at one or both ends. This problem can be avoided by using truncated distributions, in which lower and/or upper bounds are specified for the distribution (see the help files for **NormalTrunc** and **LognormalTrunc**).

For this example, we'll use the percentile bounds shown in **Table 10.3** for the four input variables that have associated probability distributions.

Symbol	Variable Name	Units	Distribution	Lower Percentile Bound	Upper Percentile Bound	Sampled Distribution Lower Bound	Sampled Distribution Upper Bound
Cs	Benzene concentration	mg/kg	LN(0.84, 0.77)	0	99.999	0	61.80
SIngR	Soil ingestion rate	mg/d	LN(3.44, 0.80)	0	99.999	0	945.69
BW	Average body weight	kg	N(47,8.3)	0.01	99.999	16.13	82.40
CPF	Cancer potency factor	(kg d)/mg	LN(-4.33, 0.67)	0	99.999	0	0.23

Table 10.3: Percentile bounds for Latin Hypercube sampling.

Now we'll generate the observations:

```
> mat.LHS <- simulate.mv.matrix(10000,
+   distributions = c(Cs      = "lnorm",
+                     S.Ing.R = "lnorm",
+                     BW      = "norm",
+                     CPF     = "lnorm"),
+   param.list = list(
+     Cs      = list(meanlog =  0.84, sdlog = 0.77),
+     S.Ing.R = list(meanlog =  3.44, sdlog = 0.80),
+     BW      = list(mean    = 47,    sd    = 8.3),
+     CPF     = list(meanlog = -4.33, sdlog = 0.67)),
+   cor.mat = diag(4), sample.method = "LHS",
+   seed = 289)
```

Characterizing the Simulated Distributions of the Input Variables

It is a good idea to look at the actual distributions of the input variables that were generated with the Monte Carlo simulation, and also at the relationships between the variables. Based on these analyses, you may decide to omit certain cases before computing the simulated distributions of exposure and risk.

Summary Statistics

Let's look at some summary statistics for the random numbers generated with simple random sampling:

```
> round(apply(mat.SRS, 2, summary), 3)
            Cs  S.Ing.R    BW   CPF
   Min.  0.085    0.926 19.07 0.001
1st Qu.  1.373   18.140 41.44 0.008
 Median  2.348   30.740 47.08 0.013
   Mean  3.137   42.440 47.07 0.016
3rd Qu.  3.959   52.610 52.66 0.020
   Max. 34.350  773.600 77.70 0.179
```

Based on these summary statistics, we may or may not want to exclude certain cases. For example, the model in **Equation (10.10)** applies to children ages 8-18. If an average body weight (*BW*) of less than 20 kg (about 44 pounds) does not make sense for this scenario, then we should exclude all cases where the value of `BW` is less than 20. (This is essentially specifying a truncated normal distribution for average body weight.)

Now let's look at some summary statistics for the Latin Hypercube sample and compare them with those for the simple random sample:

```
> round(apply(mat.LHS, 2, summary), 3)
            Cs  S.Ing.R    BW   CPF
   Min.  0.130    1.548 14.64 0.001
1st Qu.  1.378   18.180 41.40 0.008
 Median  2.316   31.190 47.00 0.013
   Mean  3.115   42.950 47.00 0.016
3rd Qu.  3.893   53.500 52.60 0.021
   Max. 46.510  761.700 79.26 0.173

> round(apply(mat.SRS, 2, summary), 3)
            Cs  S.Ing.R    BW   CPF
   Min.  0.085    0.926 19.07 0.001
1st Qu.  1.373   18.140 41.44 0.008
 Median  2.348   30.740 47.08 0.013
   Mean  3.137   42.440 47.07 0.016
3rd Qu.  3.959   52.610 52.66 0.020
   Max. 34.350  773.600 77.70 0.179
```

These summary statistics do not indicate any apparent major differences between the two samples. For small to moderate sample sizes, Latin Hypercube samples do a better job of covering the entire distribution than simple random samples (McKay et al., 1979; Iman and Conover 1980, 1982). For very large sample sizes however, there may often be little difference between the two kinds of samples.

Graphical Characterization

You can use the functions `hist` or `histogram` to create histograms of the empirical distributions, and you can use the function `ecdfplot` to create plots of the empirical cumulative distribution functions. To put two plots on the same page, one on top of the other, you can use the command

```
> par(mfrow = c(2, 1))
```

(see the S-PLUS help file for `par`).

Figure 10.16 and **Figure 10.17** display the histogram and empirical cdf of the average body weight data generated from the Monte Carlo simulation based on simple random sampling. Here are the commands to create these figures:

```
> hist(mat.SRS[, "BW"], nclass = 100, prob = T,
+    xlab = "Average Body Weight (kg)",
+    ylab = "Relative Frequency",
+    main = paste("Empirical PDF of Simulated",
+    "Average Body Weight"))

> ecdfplot(mat.SRS[, "BW"],
+    xlab = "Average Body Weight (kg)",
+    main = paste("Empirical CDF of Simulated",
+    "Average Body Weight"))
```

You can use the `pairs` or `xyplot` functions to look at scatterplot matrices to examine the relationship between the input variables.

```
> pairs(mat.SRS)
```

The result of calling the `pairs` function is shown in **Figure 10.18**. Note the relationship between soil ingestion rate (*SIngR*) and average body weight (*BW*). These two variables are assumed to be independent, so there is no monotonic relationship between these two variables in the simulated data set. In real life, there probably is a relationship between soil ingestion rate and body weight. The function `simulate.mv.matrix` allows you to specify a rank correlation between variables via the argument `cor.mat`.

> **Note:** Both Smith et al. (1992) and Bukowski et al. (1995) show that for many circumstances, the shapes of the chosen input probability distributions have much more effect on the resultant distribution of risk than does including or excluding correlations between the input variables.

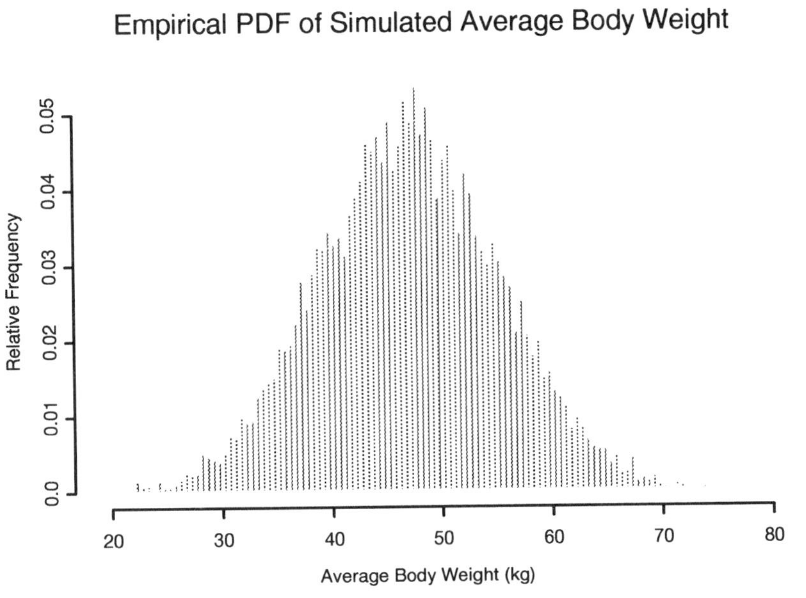

Figure 10.16: Histogram of simulated values for average body weight.

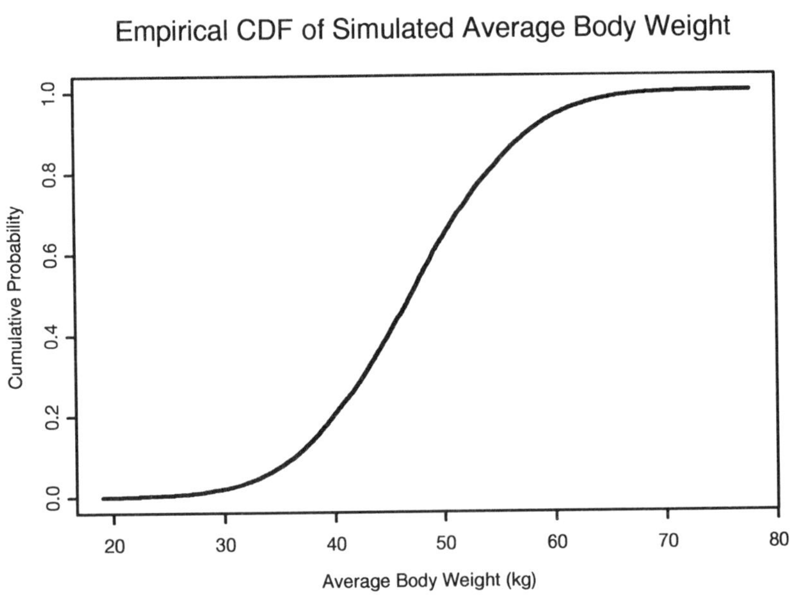

Figure 10.17: Empirical cdf of simulated values for average body weight.

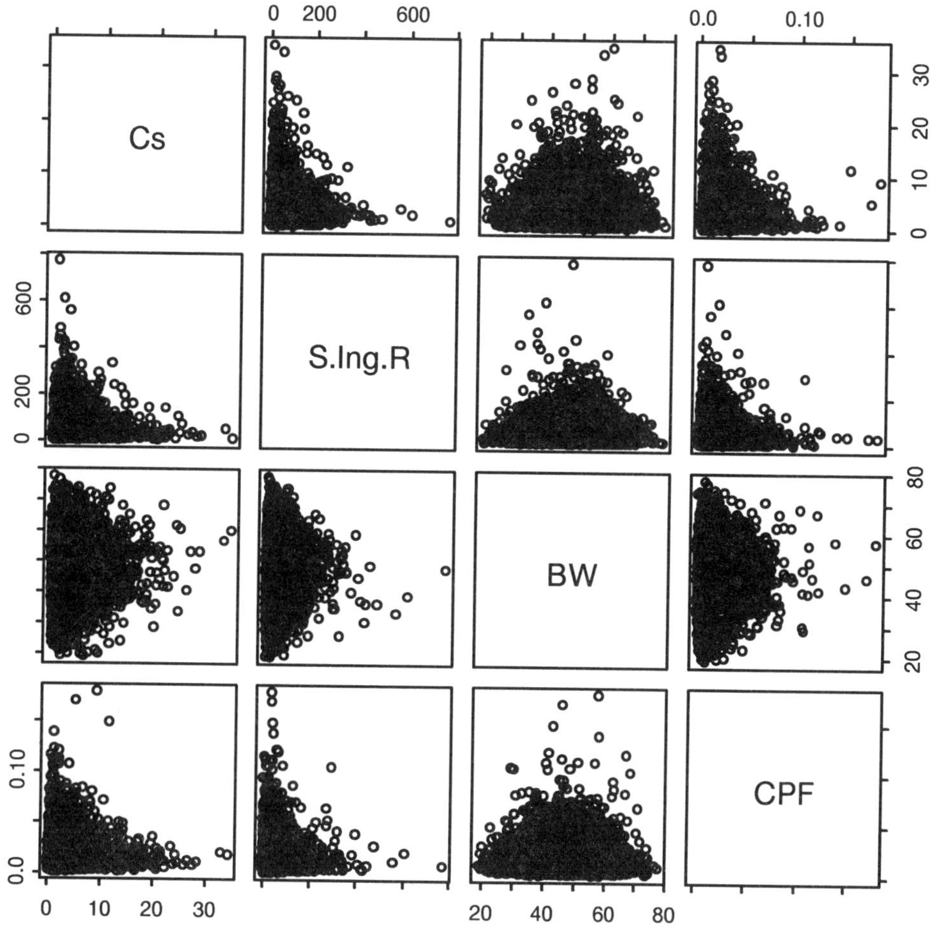

Figure 10.18: Pair-wise scatterplots for the four input variables with assumed probability distributions.

Characterizing the Simulated Distributions of Exposure and Risk

Once you are satisfied with the input distributions you've chosen and the simulated values, you can compute the exposure or risk for each outcome of the simulated vector (i.e., each row of the matrix of simulated random numbers). Then you need to characterize the distribution of exposure or risk.

Computing the Exposure and Risk Based on the Simulated Data

You can use the function `eval.expr` (evaluates an S-PLUS expression) to compute the empirical distribution of risk or exposure. The first argument to `eval.expr` is any legitimate S-PLUS expression. This expression should be the formula for exposure or risk, or an expression that calls a function that computes the exposure or risk. The second argument to `eval.expr` is a matrix, data frame, or list that contains simulated random numbers. Each column of the matrix or data frame or each component of the list should have a name corresponding to one of the variables that are used in the expression for risk or exposure.

Here are two ways to compute the empirical distribution for exposure based on the simple random sample we generated earlier and stored in `mat.SRS`:

```
> expo.vec <- eval.expr(
+     (Cs * S.Ing.R * 1 * 1 * 20 * 10 * 1e-6) /
+       (BW * 364 * 70),
+     data = mat.SRS)

> expo.vec <- eval.expr(expo.fcn(Cs, S.Ing.R, BW),
+     data = mat.SRS)
```

and here are two ways to compute the empirical distribution for ILCR:

```
> ilcr.vec <- eval.expr(
+     ((Cs * S.Ing.R * 1 * 1 * 20 * 10 * 1e-6) /
+       (BW * 364 * 70)) * CPF,
+     data = mat.SRS)

> ilcr.vec <- eval.expr(ilcr.fcn(Cs, S.Ing.R, BW, CPF),
+     data = mat.SRS)
```

Histograms, Empirical CDFs, and Summary Statistics of Exposure and Risk

Now let's plot the histogram and empirical cdf of exposure, and on each plot identify the mean and 95th percentile of the distribution. (You can use either the `quantile` or `qemp` function to compute the empirical quantile of a distribution based on a sample of observations.)

First let's compute the average and 95th percentile of exposure:

```
> expo.avg <- mean(expo.vec)

> expo.95.pct <- qemp(0.95, expo.vec)
```

Now let's plot the histogram and empirical cdf. First, set up the graphics device to put two plots on one page:

```
> par(mfrow = c(2, 1))
```

Here are the commands to produce the histogram:

```
> hist(expo.vec, nclass = 100, prob = T,
+    xlab = "Exposure [ mg/(kg-day) ]",
+    ylab = "Relative Frequency",
+    main = "Empirical PDF of Simulated Exposure")

> abline(v = expo.avg, col = 3)

> abline(v = expo.95.pct, col = 3)

> box()

> mtext("Mean", at = expo.avg)

> mtext("95%", at = expo.95.pct)
```

Here are the commands to produce the empirical cdf plot:

```
> ecdfplot(expo.vec,
+    xlab = "Exposure [ mg/(kg-day) ]",
+    main = "Empirical CDF of Simulated Exposure")

> abline(v = expo.avg, col = 3)

> abline(v = expo.95.pct, col = 3)

> mtext("Mean", at = expo.avg)

> mtext("95%", at = expo.95.pct)
```

Figure 10.19 and **Figure 10.20** display the empirical pdf and cdf for exposure. The *x*-axis extends to over 5×10^{-7} because there are observations larger than this value. You can control the limits of the *x*-axis by using the `xlim` argument in the call to `hist` or `ecdfplot`.

Here are some summary statistics for the simulated exposure distribution, where we've multiplied by 1×10^8 to make it easier to read:

```
> round(summary(1e8 * expo.vec), 2)
 Min. 1st Qu. Median Mean 3rd Qu.  Max.
 0.02    0.57   1.22 2.28     2.6 56.93

> round(quantile(1e8 * expo.vec ,
+ c(0, 0.05, 0.5, 0.95, 1)), 2)
   0%   5%  50%  95%  100%
 0.02 0.19 1.22 7.69 56.93
```

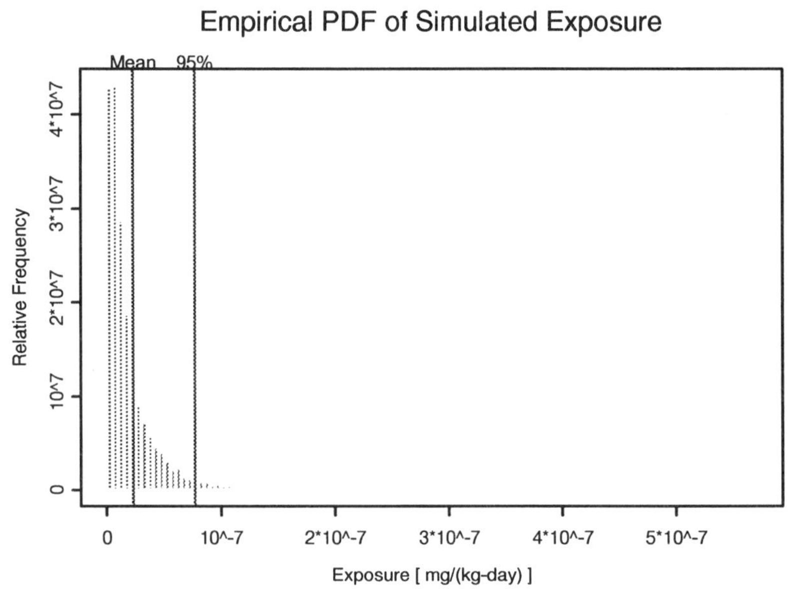

Figure 10.19: Empirical probability density of exposure.

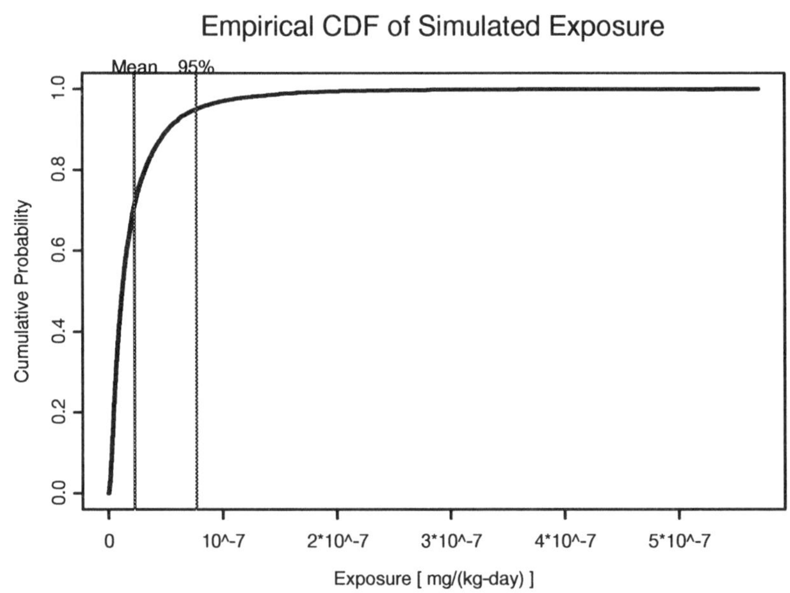

Figure 10.20: Empirical cdf of exposure.

Now let's plot the histogram and empirical cdf of ILCR, and on each plot identify the mean and 95th percentile of the distribution, along with the point estimate of ILCR used by Thompson et al. (1992).

First let's compute the average and 95th percentile and compare them to the point estimate:

```
> ilcr.avg <- mean(ilcr.vec)

> ilcr.95.pct <- qemp(0.95, ilcr.vec)

> c(Avg = ilcr.avg, Pt.Est = ilcr.pt.est,
+    "95%" = ilcr.95.pct)
           Avg          Pt.Est             95%
 3.697621e-010 8.209192e-010 1.365896e-009
```

So the average ILCR is about 4 in 10 billion, the point estimate is about 8 in 10 billion, and the 95th percentile is about 1 in one billion.

Now let's plot the histogram and empirical cdf. First, set up the graphics device to put two plots on one page:

```
> par(mfrow = c(2, 1))
```

Here are the commands to create the histogram (see **Figure 10.21**).

```
> hist(ilcr.vec, nclass = 500, prob = T,
+    xlim = c(0, 3e-9),
+    xlab = "Incremental Lifetime Cancer Risk (ILCR)",
+    ylab = "Relative Frequency",
+    main = "Empirical PDF of Simulated ILCR")

> abline(v = ilcr.avg, col = 3)

> abline(v = ilcr.95.pct, col = 3)

> abline(v = ilcr.pt.est, col = 3)

> box()

> mtext("Mean", at = ilcr.avg)

> mtext("95%", at = ilcr.95.pct)

> mtext("Pt Est", at = ilcr.pt.est)
```

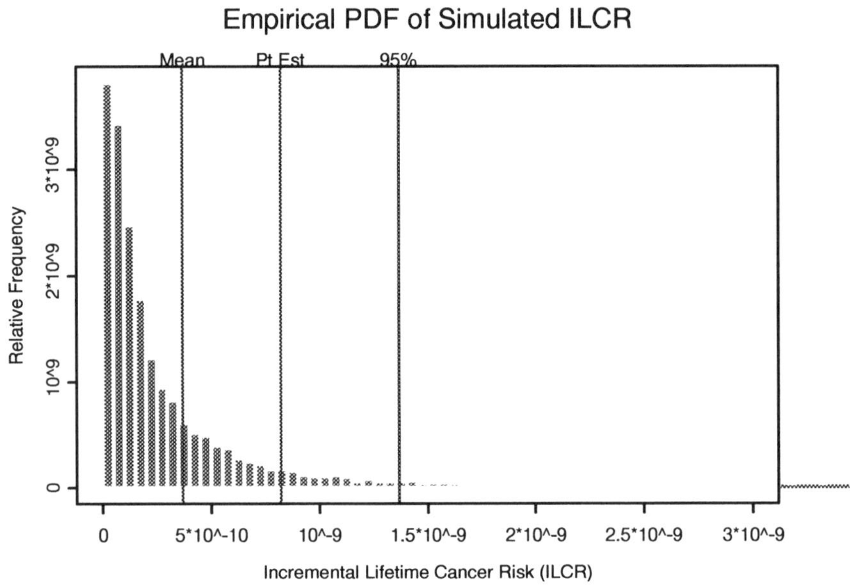

Figure 10.21: Empirical probability density of ILCR.

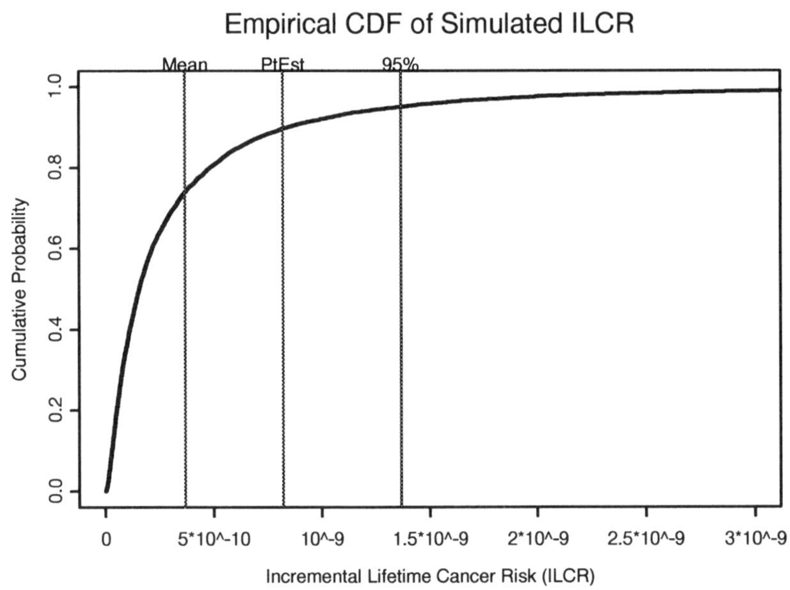

Figure 10.22: Empirical cdf of ILCR.

Here are the commands to create the empirical cdf plot (see **Figure 10.22**).

```
> ecdfplot(ilcr.vec,
+    xlim = c(0, 3e-9),
+    xlab = "Incremental Lifetime Cancer Risk (ILCR)",
+    main = "Empirical CDF of Simulated ILCR")

> abline(v = ilcr.avg, col = 3)

> abline(v = ilcr.95.pct, col = 3)

> abline(v = ilcr.pt.est, col = 3)

> mtext("Mean", at = ilcr.avg)

> mtext("95%", at = ilcr.95.pct)

> mtext("PtEst", at = ilcr.pt.est)
```

Here are some summary statistics for the simulated ILCR distribution, where we've multiplied by 1×10^9 to make it easier to read:

```
> round(summary(1e9 * ilcr.vec), 3)
  Min. 1st Qu. Median Mean 3rd Qu.   Max.
 0.001   0.066  0.157 0.37   0.386 15.93

> round(quantile(1e9 * ilcr.vec,
+    c(0, 0.05, 0.5, 0.95, 1)), 3)
    0%    5%   50%    95%   100%
 0.001 0.018 0.157 1.364 15.934
```

The point estimate of risk based on the point estimates used by Thompson et al. (1992) (see **Table 10.2** above) falls at the 90th percentile of the simulated ILCR distribution:

```
> round(100 * pemp(ilcr.pt.est, obs = ilcr.vec), 0)
[1] 90
```

The more conservative point estimate of risk based on setting the soil ingestion rate (*SIngR*) and cancer potency factor (*CPF*) to their respective 95th percentiles falls at the 98th percentile of the simulated ILCR distribution:

```
> round(100 * pemp(ilcr.pt.est.cons,
+    obs = ilcr.vec), 0)
[1] 98
```

Sensitivity Analysis

As discussed earlier in this chapter, there are several methods of performing sensitivity analysis for risk assessment (also see the section **Sensitivity Analysis Methods** in the help file for **Glossary: Uncertainty and Sensitivity Analysis**). For this example here, we'll simply look at the change in distribution of ILCR as we let one input variable vary and keep the other variables fixed at their point estimates (shown in **Table 10.2**).

First we'll re-plot the histogram for ILCR, then we'll compare it to the other histograms when we vary one input variable at a time (we can do the same thing for the empirical cdf plots as well, but won't do it here in this example). We'll vary the input variables in the following order: average body weight (*BW*), benzene concentration in soil (*Cs*), soil ingestion rate (*SIngR*), and cancer potency factor (*CPF*). Similar figures appear in Thompson et al. (1992).

First plot the histogram with all variables varying.

```
> hist(ilcr.vec, nclass = 500, prob = T,
+    xlim = c(0, 3e-9),
+    xlab = "Incremental Lifetime Cancer Risk (ILCR)",
+    ylab = "Relative Frequency",
+    main = "Empirical PDF of Simulated ILCR")

> abline(v = ilcr.pt.est, col = 3)

> box()

> mtext("Pt Est", at = ilcr.pt.est)
```

Now plot the histogram resulting from varying only *BW*.

```
> hist(ilcr.fcn(Cs = 3.39, S.Ing.R = 50,
+    BW = mat.SRS[, "BW"], CPF = 0.029),
+    nclass = 100, prob = T, xlim = c(0, 3e-9),
+    xlab = "Incremental Lifetime Cancer Risk (ILCR)",
+    ylab = "Relative Frequency",
+    main = "Simulated ILCR Varying BW")

> abline(v = ilcr.pt.est, col = 3)

> box()

> mtext("Pt Est", at = ilcr.pt.est)
```

Now plot the histogram resulting from varying only `Cs`.

```
> hist(ilcr.fcn(Cs = mat.SRS[, "Cs"], S.Ing.R = 50,
+    BW = 47, CPF = 0.029),
+    nclass = 500, prob = T, xlim = c(0, 3e-9),
+    xlab = "Incremental Lifetime Cancer Risk (ILCR)",
+    ylab = "Relative Frequency",
+    main = "Simulated ILCR Varying Cs")

> abline(v = ilcr.pt.est, col = 3)

> box()

> mtext("Pt Est", at = ilcr.pt.est)
```

Now plot the histogram resulting from varying only `SIngR`.

```
> hist(ilcr.fcn(Cs = 3.39,
+    S.Ing.R = mat.SRS[, "S.Ing.R"],
+    BW = 47, CPF = 0.029),
+    nclass = 500, prob = T, xlim = c(0, 3e-9),
+    xlab = "Incremental Lifetime Cancer Risk (ILCR)",
+    ylab = "Relative Frequency",
+    main = "Simulated ILCR Varying S.Ing.R")

> abline(v = ilcr.pt.est, col = 3)

> box()

> mtext("Pt Est", at = ilcr.pt.est)
```

Now plot the histogram resulting from varying only `CPF`.

```
> hist(ilcr.fcn(Cs = 3.39, S.Ing.R = 50,
+    BW = 47, CPF = mat.SRS[, "CPF"]),
+    nclass = 200, prob = T, xlim = c(0, 3e-9),
+    xlab = "Incremental Lifetime Cancer Risk (ILCR)",
+    ylab = "Relative Frequency",
+    main = "Simulated ILCR Varying CPF")

> abline(v = ilcr.pt.est, col = 3)

> box()

> mtext("Pt Est", at = ilcr.pt.est)
```

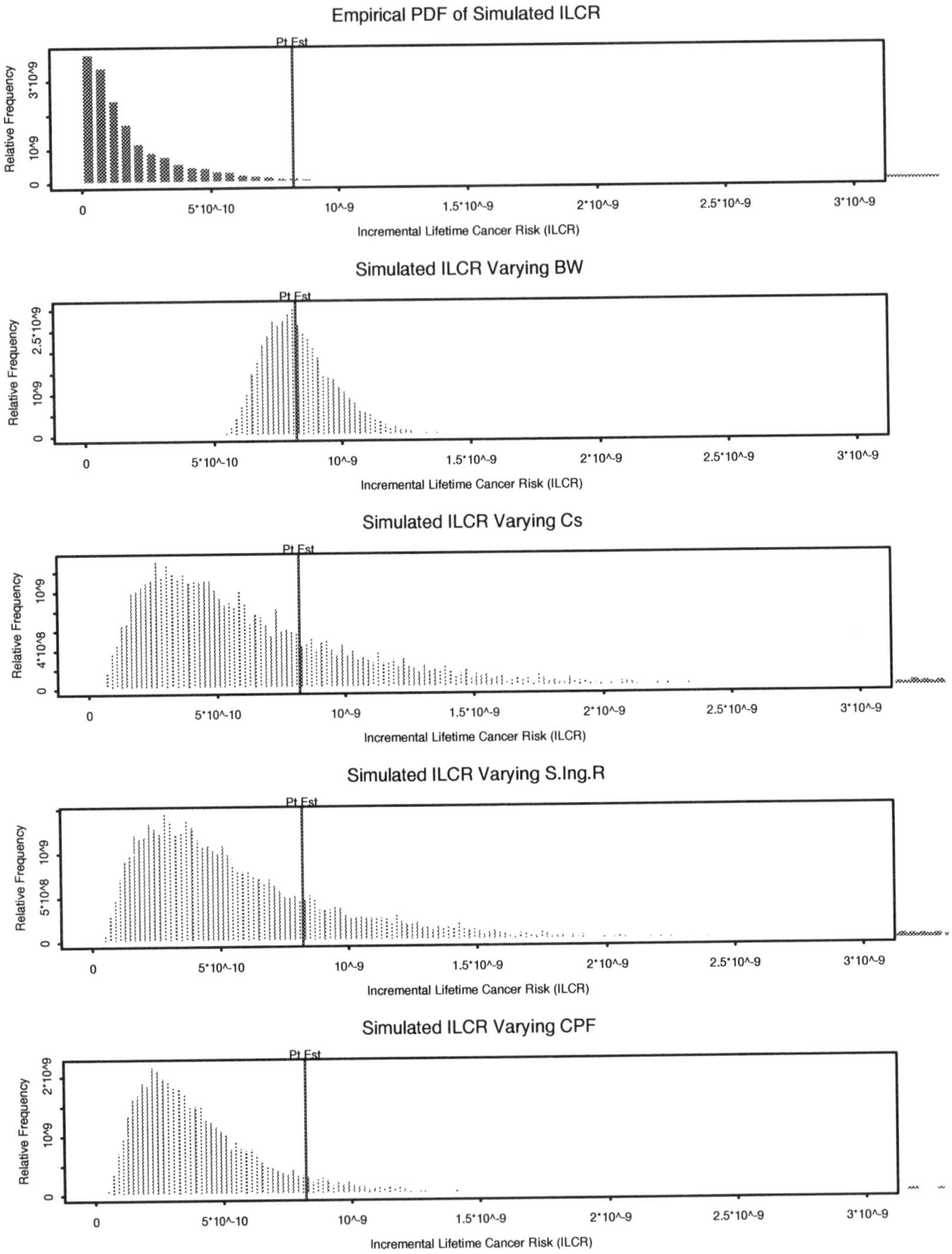

Figure 10.23: Sensitivity analysis for ILCR.

The five plots are shown in **Figure 10.23**. Based on these figures, you can see that average body weight (*BW*) has the smallest effect on the variability of ILCR compared to the other three input variables. Also, varying the other three variables brings the bulk of the distribution down below the point estimate of risk, with the cancer potency factor (*CPF*) having the greatest influence.

Uncertainty Analysis

As discussed earlier in this chapter, there are several methods of performing uncertainty analysis for risk assessment (also see the section **Uncertainty Analysis Methods** in the help file for **Glossary: Uncertainty and Sensitivity Analysis**). For this example here, we'll discuss three methods: graphical and statistical summaries, comparing statistics between replicate simulations, and running a simulation using mixture distributions for the input variables (i.e., second-order random variables).

Graphical and Statistical Summaries

We've already graphically displayed the simulated distributions of exposure and risk and computed summary statistics for these distributions in the sub-section **Characterizing the Simulated Distributions of Exposure and Risk** above. We can also, if we choose, compute confidence limits for certain quantities, such as the mean, median, or 95th percentile. Here is the nonparametric two-sided 95% confidence interval for the median exposure, based on the simulated distribution of exposure (see the help file for eqnpar):

```
> eqnpar(expo.vec, ci = T)

Results of Distribution Parameter Estimation
--------------------------------------------

Assumed Distribution:          None

Estimated Quantile(s):         Median = 1.217818e-8

Quantile Estimation Method:    Nonparametric

Data:                          expo.vec

Sample Size:                   10000

Confidence Interval for:       50'th %ile

Confidence Interval Method:    exact

Confidence Interval Type:      two-sided

Confidence Level:              95%

Confidence Limit Rank(s):      4903 5099
```

```
Confidence Interval:          LCL = 1.17857e-8
                              UCL = 1.253146e-8
```

and here is the one-sided upper 90% confidence interval for the 95th percentile of risk, based on the simulated distribution of risk:

```
> eqnpar(ilcr.vec, p = 0.95, ci = T,
+   ci.type = "upper", approx.conf.level = 0.9)

Results of Distribution Parameter Estimation
--------------------------------------------

Assumed Distribution:            None

Estimated Quantile(s):           95'th %ile = 1.363588e-9

Quantile Estimation Method:      Nonparametric

Data:                            ilcr.vec

Sample Size:                     10000

Confidence Interval for:         95'th %ile

Confidence Interval Method:      exact

Confidence Interval Type:        upper

Confidence Level:                90%

Confidence Limit Rank(s):        9528

Confidence Interval:             LCL = -Inf
                                 UCL = 1.415459e-9
```

So the upper 90% confidence limit for the 95th percentile of ILCR is about 1.4 in one billion.

Running Another Simulation With the Same Number of Monte Carlo Trials and Comparing Results

Here we'll run another Monte Carlo simulation, starting with a different random seed, then compare the results between these two simulations. Probably the best way to do this is to run the original simulation with the specific seed (47 in this case), then run the new simulation. This way, we know that the random numbers being generated are sequential in the random number generating algorithm.

```
> mat.SRS <- simulate.mv.matrix(10000,
+   distributions = c(Cs       = "lnorm",
+                     S.Ing.R = "lnorm",
+                     BW      = "norm",
+                     CPF     =     "lnorm"),
+   param.list = list(
+     Cs      = list(meanlog =  0.84, sdlog = 0.77),
+     S.Ing.R = list(meanlog =  3.44, sdlog = 0.80),
+     BW      = list(mean    = 47,    sd    = 8.3),
+     CPF     = list(meanlog = -4.33, sdlog = 0.67)),
+   cor.mat = diag(4), seed = 47)

> mat.SRS.2 <- simulate.mv.matrix(10000,
+   distributions = c(Cs       = "lnorm",
+                     S.Ing.R = "lnorm",
+                     BW      = "norm",
+                     CPF     = "lnorm"),
+   param.list = list(
+     Cs      = list(meanlog =  0.84, sdlog = 0.77),
+     S.Ing.R = list(meanlog =  3.44, sdlog = 0.80),
+     BW      = list(mean    = 47,    sd    = 8.3),
+     CPF     = list(meanlog = -4.33, sdlog = 0.67)),
+   cor.mat = diag(4))
```

Note that in the second call to `simulate.mv.matrix` we did not set a value for the seed argument. This way, the random number generator uses the current value of the seed, which was set after generating all the random numbers for `mat.SRS`.

Now we must compute the distribution of ILCR for the second simulation:

```
> ilcr.vec.2 <- eval.expr(ilcr.fcn(Cs, S.Ing.R, BW, CPF),
+   data = mat.SRS.2)
```

Now let's compare the means, medians, standard deviations, and the 5th, 10th, 90th, and 95th percentiles from these two simulations.

```
> stat.vec.1 <- c(mean(ilcr.vec), median(ilcr.vec),
+   sd = sd(ilcr.vec),
+   qemp(c(0.05, 0.1, 0.9, 0.95), ilcr.vec))

> names(stat.vec.1) <- c("mean", "median", "sd",
+   "5.Pct", "10.Pct", "90.Pct", "95.Pct")

> stat.vec.2 <- c(mean(ilcr.vec.2), median(ilcr.vec.2),
+   sd(ilcr.vec.2),
+   qemp(c(0.05, 0.1, 0.9, 0.95), ilcr.vec.2))

> names(stat.vec.2) <- names(stat.vec.1)
```

```
> round(100 * abs(stat.vec.2 - stat.vec.1) / stat.vec.1,
+    digits = 1)
 mean median  sd 5.Pct 10.Pct 90.Pct 95.Pct
  0.7    2.7 4.7   0.3    3.3    0.4    0.1
```

We see that all of these summary statistics agree within 5%. If this amount of discrepancy is not satisfactory, then you'll need to run a simulation with a larger number of trials. (**Note:** Thompson et al. (1992) found agreement to within 1% when the cancer potency factor (*CPF*) was fixed at its point estimate.)

We can also graphically compare the two empirical cdf's (see the help file for `cdf.compare`):

```
> cdf.compare(ilcr.vec, ilcr.vec.2,
+    xlim = c(0, 3e-9),
+    xlab = "Order Statistics for ILCR",
+    main = paste("Empirical CDF of ILCR for\n",
+    "Simulation 1 (Solid Line) and ",
+    "Simulation 2 (Dashed Line)", sep = ""))
```

Here the lines lie on top of each other in **Figure 10.24**. You could also plot the differences between the empirical cdf's.

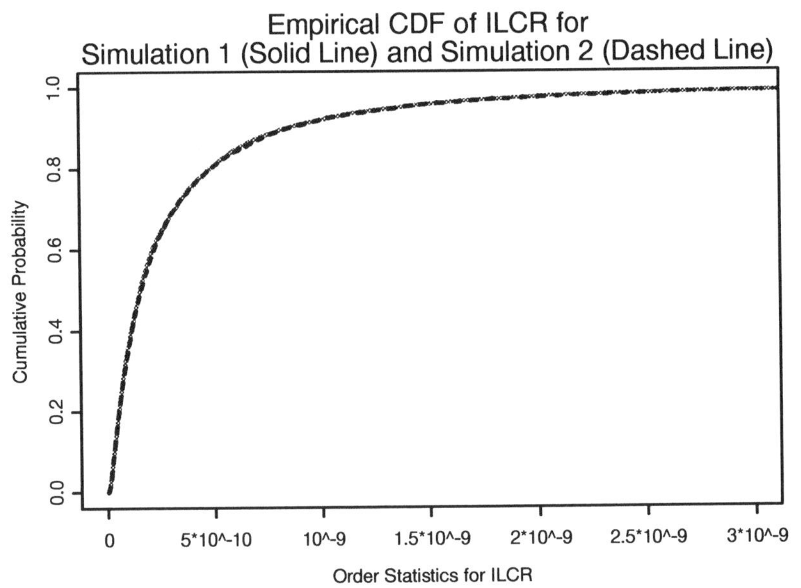

Figure 10.24: Comparison of the empirical cdf of ILCR for two simulations with 10,000 Monte Carlo trials.

Running Another Simulation With An Increased Number of Monte Carlo Trials and Comparing Results

Now let's run another simulation in which we double the number of trials from 10,000 to 20,000.

```
> mat.SRS.3 <- simulate.mv.matrix(20000,
+    distributions = c(Cs       = "lnorm",
+                      S.Ing.R = "lnorm",
+                      BW       = "norm",
+                      CPF      = "lnorm"),
+    param.list = list(
+      Cs      = list(meanlog =  0.84, sdlog = 0.77),
+      S.Ing.R = list(meanlog =  3.44, sdlog = 0.80),
+      BW      = list(mean    = 47,    sd    = 8.3),
+      CPF =     list(meanlog = -4.33, sdlog = 0.67)),
+    cor.mat = diag(4))

> ilcr.vec.3 <- eval.expr(ilcr.fcn(Cs, S.Ing.R, BW, CPF),
+    data = mat.SRS.3)

> stat.vec.3 <- c(mean(ilcr.vec.3), median(ilcr.vec.3),
+    sd(ilcr.vec.3),
+    qemp(c(0.05, 0.1, 0.9, 0.95), ilcr.vec.3))

> names(stat.vec.3) <- names(stat.vec.1)

> round(100 * abs(stat.vec.3 - stat.vec.1) /
+    stat.vec.1, 1)
 mean median  sd 5.Pct 10.Pct 90.Pct 95.Pct
  3.7    4.6 5.5   2.1    1.6    2.9    3.8
```

So the summary statistics for the simulation based on 10,000 trials vs. the summary statistics for the simulation based on 20,000 trials agree to within 5%, except for the standard deviation, which agrees within 5.5%. Again, if this amount of discrepancy is not satisfactory, then you'll need to run a simulation with a larger number of trials than the original simulation.

Barry (1996) warns that if the moments of the simulated distribution do not appear to stabilize with an increasing number of Monte Carlo trials, this can mean that they don't exist. In most cases, however, the true distributions of any random variables involved in a denominator in **Equation (10.1)** are bounded above 0, so the moments will exist. If a random variable involved in the denominator has a mean or median that is close to 0, it is important to use a bounded or truncated distribution to assure the random variable stays sufficiently far away from 0.

Here is the graphical comparison between the empirical cdf based on 10,000 trials and the empirical cdf based on 20,000 trials:

```
> cdf.compare(ilcr.vec, ilcr.vec.3,
+    xlim = c(0, 3e-9),
+    xlab = "Order Statistics for ILCR",
+    main = paste("Empirical CDF of ILCR Based on\n",
+    "10,000 (Solid Line) and ",
+    "20,000 (Dashed Line) Trials", sep = ""))
```

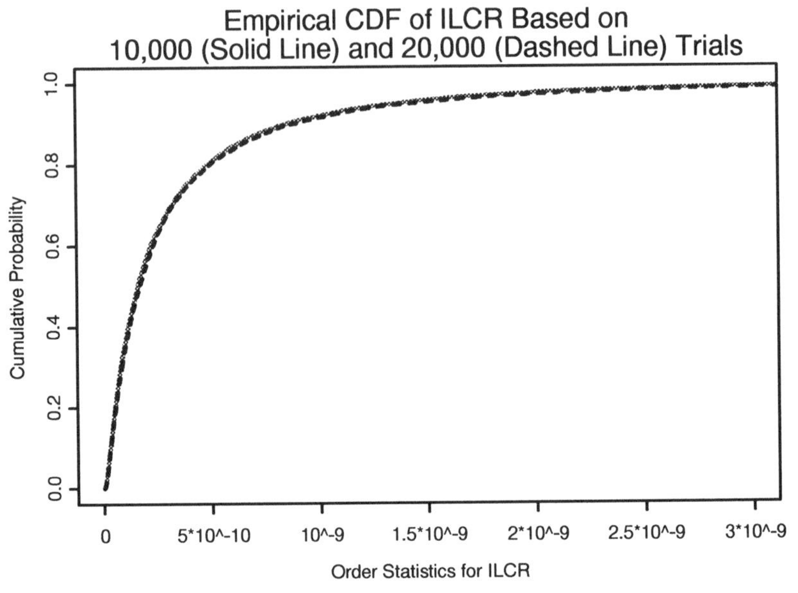

Figure 10.25: Empirical cdf of ILCR based on 10,000 and 20,000 Monte Carlo trials.

As in **Figure 10.24**, the two lines in **Figure 10.25** lie on top of each other. Again, you could also look at the differences between the two cdf's.

Using Mixture Distributions for the Input Variables

In all of the Monte Carlo simulations we've performed so far, we've assumed that we know the distribution of the four input variables that have associated probability distributions. One way to account for the uncertainty in how well we have adequately characterized these probability distributions is to use a mixture distribution for the probability distributions (Hoffman and Hammonds, 1994; Burmaster and Wilson, 1996). To do this, for each input variable, a distribution is specified for the parameter(s) of the input variable's distribution. Some authors call random variables with this kind of distribution second-order random variables (e.g., Burmaster and Wilson, 1996).

For example, average body weight (*BW*) is assumed to follow a normal distribution with mean 47kg and standard deviation 8.3kg. The true value of the mean may be unknown to a certain degree, so we may specify that the mean comes from some distribution (e.g., a uniform distribution with a given set of upper and lower bounds).

In this case, we can think of the Monte Carlo simulation as being broken down into two stages or two loops. In the first stage or outer loop, a set of parameters is generated for each input distribution. In the second stage or inner loop, *n* realizations of the input vector $\underline{X}$ are generated based on this one set of distribution parameters. This two-stage process is repeated *N* times, so that you end up with *N* different empirical distributions of risk, and each empirical distribution is based on *n* observations of the input vector $\underline{X}$.

For a fixed number of Monte Carlo trials *nN*, the optimal combination of *n* and *N* will depend on how the empirical distribution of risk changes relative to variability in the input distribution parameter(s) vs. variability in the input variables themselves. For example, if the distribution of risk is very sensitive to changes in the input distribution parameters, then *N* should be large relative to *n* or *nN*. On the other hand, if the distribution of risk is relatively insensitive to the values of the parameters of the input distributions, but varies substantially with the values of the input variables, then *N* may be small relative to *n*.

Using Mixture Distributions: Simple Example

As a simple example, we'll allow the true mean of average body weight (*BW*) to follow a uniform distribution with a minimum of 40 and a maximum of 55. We'll perform a Monte Carlo simulation with 10,000 trials, where in each trial we randomly sample from a U[40, 55] distribution to determine the mean μ of the distribution of *BW*, then take one sample from a N(μ, 8.3) distribution to simulate *BW*. We'll sample from the same distributions for the other three input variable that we used before. Thus, in this example, *N*=10,000 and *n*=1.

Because S-PLUS performs vector arithmetic, in this case we can avoid explicitly setting up a nested loop. First we'll generate the 10,000 means for the distribution for average body weight from the U[40, 55] distribution. We can do this two ways:

```
> BW.means <- simulate.vector(10000, "unif",
+ list(min = 40, max = 55), seed = 82)
```

or

```
> set.seed(82)

> BW.means <- runif(10000, 40, 55)
```

Now we'll generate the 10,000 random vectors of (*Cs*, *SIngR*, *BW*, *CPF*) and then compute the ILCR based on these random vectors:

```
> mat.SRS.4 <- simulate.mv.matrix(10000,
+   distributions = c(Cs      = "lnorm",
+                     S.Ing.R = "lnorm",
+                     BW      = "norm",
+                     CPF     =     "lnorm"),
+   param.list = list(
+     Cs       = list(meanlog =  0.84, sdlog = 0.77),
+     S.Ing.R = list(meanlog =  3.44, sdlog = 0.80),
+     BW       = list(mean = BW.means, sd = 8.3),
+     CPF =     list(meanlog = -4.33, sdlog = 0.67)),
+   cor.mat = diag(4), seed = 47)

> ilcr.vec.4 <- eval.expr(ilcr.fcn(Cs, S.Ing.R, BW, CPF),
+   data = mat.SRS.4)
```

Note that in the call to `simulate.mv.matrix`, when we defined the value of the argument `param.list`, the component named `BW` is a list defining the mean and standard deviation for average body weight. The value of the `mean` component in this list is `BW.means`, the vector of 10,000 means we generated above.

Now let's compare summary statistics of ILCR for the original simulation and this new simulation:

```
> round(summary(1e8 * ilcr.vec), 3)
 Min. 1st Qu. Median  Mean 3rd Qu.  Max.
    0   0.007  0.016 0.037   0.039 1.593

> round(summary(1e8 * ilcr.vec.4), 3)
 Min. 1st Qu. Median  Mean 3rd Qu.  Max.
    0   0.006  0.016 0.037   0.039 1.605
```

It looks like the distribution of ILCR is very insensitive to the uncertainty in the true mean of average body weight. This is not surprising since the sensitivity analysis we performed previously showed that the variability of the distribution of ILCR depended more on the other three input variables. We may get a very different behavior if we run a simulation using a mixture distribution for one of the other input variables.

Now we'll plot the histogram for the original ILCR and the histogram based on this new way of computing ILCR. First set up the graphics device to put two plots on one page:

```
> par(mfrow = c(2, 1))
```

The commands to produce the histogram for the original simulation of ILCR are shown on page 319 and the histogram is shown in **Figure 10.21** on page 320. We'll reproduce this histogram in **Figure 10.26** to make it easier to compare with the histogram for the simulation based on the mixture distribution.

Here are the commands to produce the histogram of ILCR based on the mixture distribution (see **Figure 10.27**).

```
> ilcr.avg.4 <- mean(ilcr.vec.4)

> ilcr.95.pct.4 <- qemp(0.95, ilcr.vec.4)

> hist(ilcr.vec.4, nclass = 500, prob = T,
+    xlim = c(0, 3e-9),
+    xlab = "Incremental Lifetime Cancer Risk (ILCR)",
+    ylab = "Relative Frequency", main = "")

> mtext(paste("Empirical PDF of Simulated ILCR\n",
+    "Based on Mixture Distribution for BW",
+    sep = ""), line = 3, cex = 1.25 * par("cex"))

> abline(v = ilcr.avg.4, col = 3)

> abline(v = ilcr.95.pct.4, col = 3)

> abline(v = ilcr.pt.est, col = 3)

> box()

> mtext("Mean", at = ilcr.avg.4)

> mtext("95%", at = ilcr.95.pct.4)

> mtext("Pt Est", at = ilcr.pt.est)
```

Just like the summary statistics, these histograms show there is not much change in the simulated distribution of ILCR based on the original simulation and this new one based on a mixture distribution for average body weight. We could also compare the empirical cdf's as well.

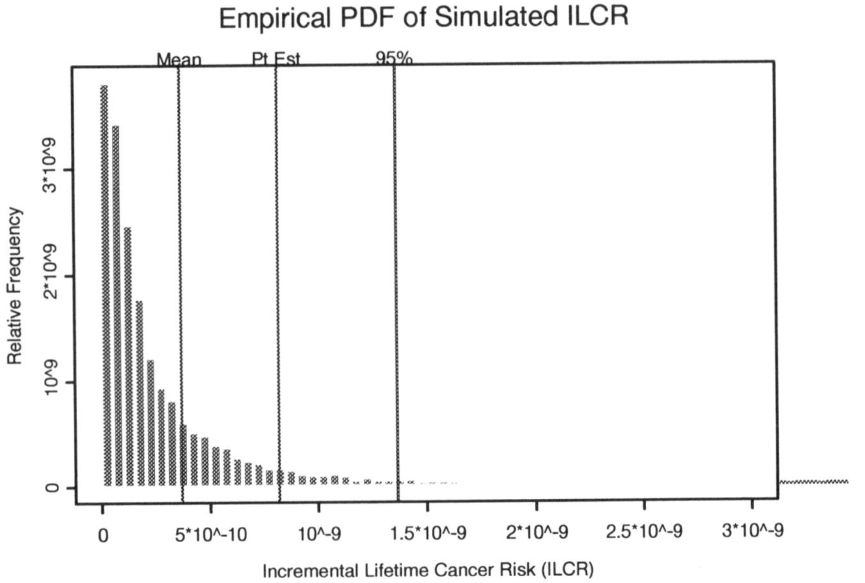

Figure 10.26: Empirical probability density function of original simulated ILCR.

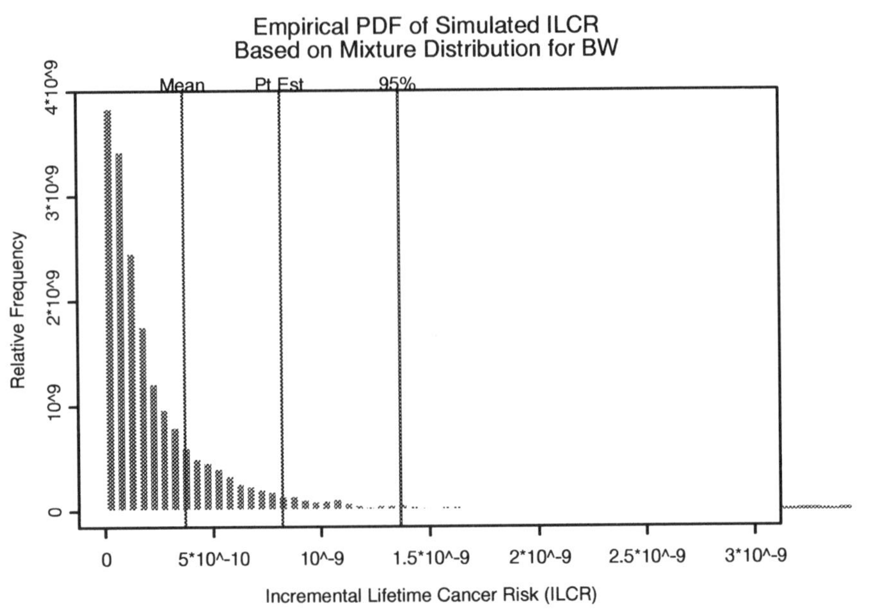

Figure 10.27: Empirical probability density function of simulated ILCR based on a mixture distribution for average body weight.

Using Mixture Distributions: Complex Example

As a final and more complicated example, we'll allow the true mean of average body weight (*BW*) to follow a uniform distribution with a minimum of 40 and a maximum of 55 as before, but we'll also allow the true mean of the logarithm of soil ingestion rate (*SIngR*) to follow a triangular distribution with a minimum of 2.64, a maximum of 4.24, and a mode of 3.44. Also, we'll assume a rank correlation between average body weight and soil ingestion rate of 0.8.

We'll perform a Monte Carlo simulation with *N*=10 and *n*=1,000, for a total of 10,000 trials. For each of the *N*=10 trials, we'll sample from the uniform and triangular distributions (using Latin Hypercube sampling) to generate values for the mean of the distribution of *BW* and the mean of the distribution of log(*SIngR*). Then, for a given set of parameter values for the distributions of *BW* and *SIngR*, we'll generate 1,000 random vectors (*Cs*, *SIngR*, *BW*, *CPF*) based on the assumed distributions, assuming a rank correlation of 0.8 between *BW* and *SIngR*.

Unlike the previous example, here we cannot avoid explicitly setting up a loop. First we'll generate the 10 means for the distribution of average body weight (from the uniform distribution), and the 10 means for the distribution of the logarithm of soil ingestion rate (from the triangular distribution). There are two ways to do this:

```
> set.seed(989)

> BW.means <- simulate.vector(10, "unif",
+    param.list = list(min = 40, max = 55),
+    sample.method = "LHS", sorted = F)

> S.Ing.R.meanlogs <- simulate.vector(10, "tri",
+    param.list = list(min = 2.64, max = 4.24,
+    mode = 3.44), sample.method = "LHS", sorted = F)
```

or

```
> param.mat <- simulate.mv.matrix(10,
+    distributions = c(BW = "unif", S.Ing.R = "tri"),
+    param.list = list(BW = list(min = 40, max = 55),
+      S.Ing.R = list(min = 2.64, max = 4.24,
+                     mode = 3.44)),
+    sample.method = "LHS", seed = 989)

> BW.means <- param.mat[, "BW"]

> S.Ing.R.meanlogs <- param.mat[, "S.Ing.R"]
```

For this example we'll use the second way.

Now let's generate the rank correlation matrix for the four input variables with associated distributions:

```
> cor.mat <- matrix(c(1, 0, 0, 0, 0, 1, 0.8, 0,
+         0, 0.8, 1, 0, 0, 0, 0, 1),
+     nrow = 4, ncol = 4,
+     dimnames = list(
+       c("Cs", "S.Ing.R", "BW", "CPF"),
+       c("Cs", "S.Ing.R", "BW", "CPF")))

> cor.mat
          BW S.Ing.R  BW CPF
      BW  1      0.0 0.0   0
 S.Ing.R  0      1.0 0.8   0
      BW  0      0.8 1.0   0
     CPF  0      0.0 0.0   1
```

Now for each of the N=`10` sets of distribution parameters, we'll generate the n=`1,000` random vectors of (*Cs*, *SIngR*, *BW*, *CPF*) and then compute the ILCR based on these random vectors. There are two ways to do this, by either using the `for` function or the `sapply` function. The `sapply` function is usually a bit more computationally efficient than the `for` function, but here there isn't much difference between the two methods. Here is the first way using the `for` function:

```
> set.seed(87)

> ilcr.mat.5 <- matrix(0, nrow = 1000, ncol = 10)

> for (i in 1:10) {
+  mat <- simulate.mv.matrix(1000,
+    distributions = c(Cs        = "lnorm",
+                      S.Ing.R = "lnorm",
+                      BW      = "norm",
+                      CPF =     "lnorm"),
+    param.list = list(
+      Cs      = list(meanlog = 0.84, sdlog = 0.77),
+      S.Ing.R = list(meanlog = S.Ing.R.meanlogs[i],
+                     sdlog = 0.80),
+      BW      = list(mean = BW.means[i], sd = 8.3),
+      CPF     = list(meanlog = -4.33, sdlog = 0.67)),
+    cor.mat = cor.mat)
+  ilcr.mat.5[, i] <- eval.expr(
+    ilcr.fcn(Cs, S.Ing.R, BW, CPF), data = mat)
+ }
```

and here is the second way using the `sapply` function:

```
> set.seed(87)

> ilcr.mat.5 <- sapply(1:10,
+   function(i, BW.m, S.Ing.R.ml, cor.mat) {
+     mat <- simulate.mv.matrix(1000,
+       distributions = c(Cs       = "lnorm",
+                         S.Ing.R = "lnorm",
+                         BW       = "norm",
+                         CPF      = "lnorm"),
+       param.list = list(
+         Cs = list(meanlog = 0.84, sdlog = 0.77),
+         S.Ing.R = list(meanlog = S.Ing.R.ml[i],
+                        sdlog = 0.80),
+         BW = list(mean = BW.m[i], sd = 8.3),
+         CPF = list(meanlog = -4.33, sdlog = 0.67)),
+       cor.mat = cor.mat)
+     eval.expr(ilcr.fcn(Cs, S.Ing.R, BW, CPF),
+       data = mat)
+   },
+   BW.m = BW.means, S.Ing.R.ml = S.Ing.R.meanlogs,
+   cor.mat = cor.mat)
```

The object `ilcr.mat.5` is a matrix with 1,000 rows and 10 columns. Each column is a vector of 1,000 observations of ILCR based on a different set of distribution parameters. Here are commands to produce a plot showing the 10 empirical cumulative distribution functions:

```
> ecdfplot(ilcr.mat.5[, 1], xlim = c(0, 3e-9),
+   xlab = "ILCR", ecdf.lwd = 1,
+   main = "Empirical CDFs of ILCR")

> for (i in 2:10)
+   ecdfplot(ilcr.mat.5[, i], add = T,
+   ecdf.lwd = 1, ecdf.lty = i)
```

Figure 10.28 shows quite a bit of variability in the empirical cdf around the 80th percentile, but there is not as much variability at the 95th percentile.

```
> vec.95.pct.5 <- apply(ilcr.mat.5, 2,
+   function(vec) qemp(0.95, vec))

> print(vec.95.pct.5, 2)
 [1] 1.3e-009 1.2e-009 9.6e-010 1.3e-009 8.6e-010
 [6] 5.9e-010 1.6e-009 1.5e-009 2.1e-009 1.1e-009

> round(1e10 * vec.95.pct.5, 0)
 [1] 13 12 10 13  9  6 16 15 21 11
```

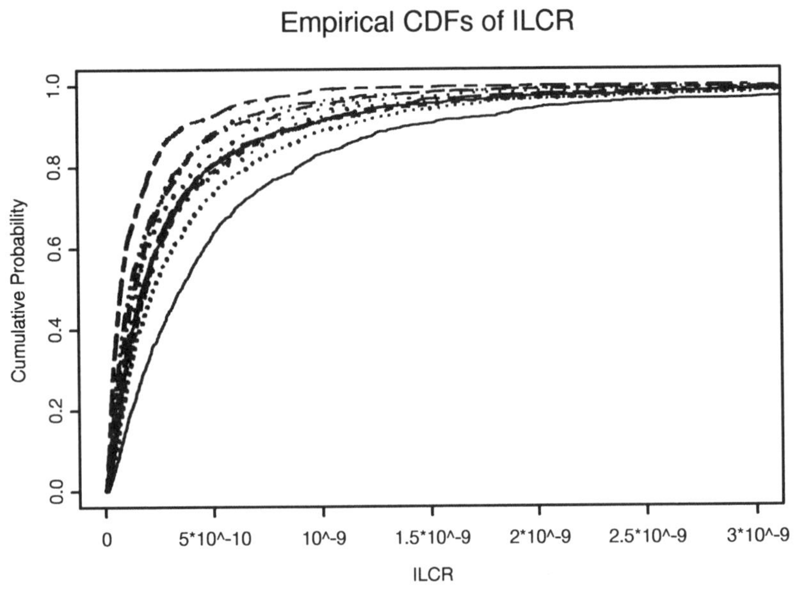

Figure 10.28: Ten different empirical cdf's of ILCR.

Finally, let's compare the histogram for the original simulation of ILCR and the histogram based on this new way of simulating ILCR. First set up the graphics device to put two plots on a single page:

```
> par(mfrow = c(2, 1))
```

The commands to produce the histogram for the original simulation of ILCR are shown on page 319 and the histogram is shown in **Figure 10.21** on page 320. We'll reproduce this histogram in **Figure 10.29** to make it easier to compare with the histogram for the simulation based on using the new mixture distribution and correlation between the variables.

Here are the commands to produce the histogram for the new simulation of ILCR (see **Figure 10.30**).

```
> ilcr.vec.5 <- as.vector(ilcr.mat.5)

> ilcr.avg.5 <- mean(ilcr.vec.5)

> ilcr.95.pct.5 <- qemp(0.95, ilcr.vec.5)
```

```
> hist(ilcr.vec.5, nclass = 500, prob = T,
+   xlim = c(0, 3e-9),
+   xlab = "Incremental Lifetime Cancer Risk (ILCR)",
+   ylab = "Relative Frequency",
+   main = "")

> mtext(paste(
+   "Empirical PDF of Simulated ILCR Based on",
+   "Mixture Distributions for BW and S.Ing.R",
+   sep = "\n"), line = 3, cex = 1.25 * par("cex"))

> abline(v = ilcr.avg.5, col = 3)

> abline(v = ilcr.95.pct.5, col = 3)

> abline(v = ilcr.pt.est, col = 3)

> box()

> mtext("Mean", at = ilcr.avg.5)

> mtext("95%", at = ilcr.95.pct.5)

> mtext("Pt Est", at = ilcr.pt.est)
```

The difference between the two distributions is shown in **Figure 10.29** and **Figure 10.30** is negligible. Here is a comparison of the two averages:

```
> ilcr.avg
[1] 3.697621e-010

> ilcr.avg.5
[1] 3.669952e-010
```

and here is a comparison of the two 95th percentiles:

```
> ilcr.95.pct
[1] 1.365896e-009

> ilcr.95.pct.5
[1] 1.266494e-009
```

We could also compare the empirical cdf's using `cdf.compare`, and look at the differences between the empirical cdf's as well.

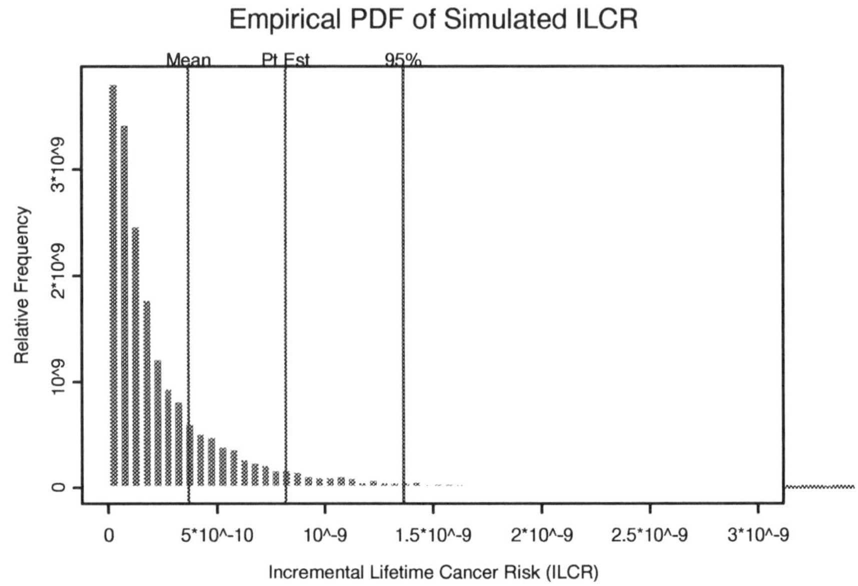

Figure 10.29: Empirical probability density function of original simulated ILCR.

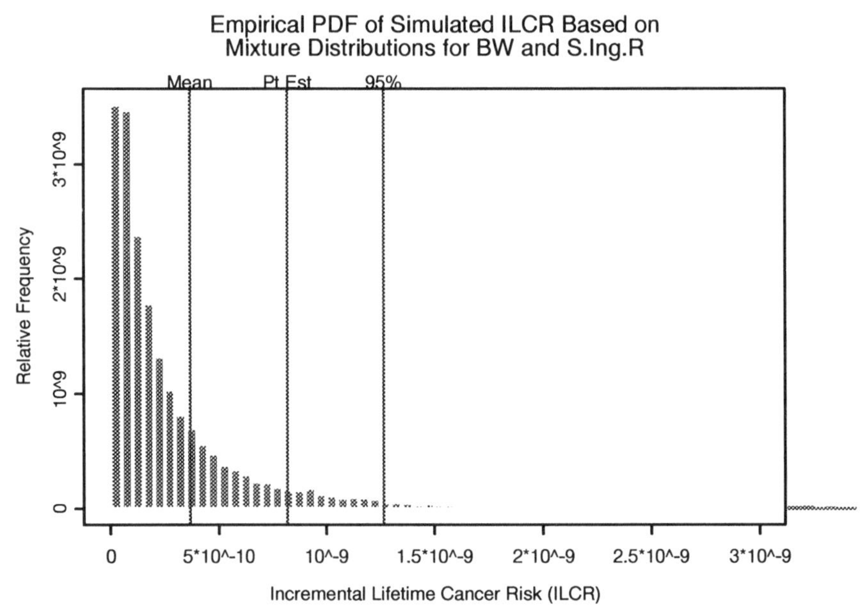

Figure 10.30: Empirical probability density function of simulated ILCR based on a mixture distribution for average body weight and a mixture distribution for soil ingestion rate.

Summary

- Human and ecological risk assessment involves characterizing the exposure to a toxicant for one or several populations, quantifying the relationship between exposure (dose) and health or ecological effects (response), determining the risk (probability) of a health or ecological effect given the observed level(s) of exposure, and characterizing the uncertainty associated with the estimated risk.

- In the past, estimates of risk were often based solely on setting values of the input variables (e.g., body weight, dose, etc.) to particular point estimates and producing a single point estimate of risk, with little, if any, quantification of the uncertainty associated with the estimated risk. More recently, several practitioners have advocated "probabilistic" risk assessment, in which the input variables are considered random variables, so the result of the risk assessment is a probability distribution for predicted risk or exposure.

- **Table 10.1** lists several functions available in S-PLUS and ENVIRONMENTALSTATS for S-PLUS that are useful for performing Monte Carlo simulation and probabilistic risk assessment. These functions include methods to generate random numbers with simple random sampling or Latin Hypercube sampling, generate random vectors from arbitrary distributions with a specified rank correlation matrix, and methods to evaluate an S-PLUS expression for risk or exposure based on simulated data.

- Most risk assessment models follow the form of **Equation (10.1)** and **(10.8)**, where the output variable (risk or cxposure) is a function of several input variables, and some or all of the input variables are considered to be random variables.

- ***Uncertainty analysis*** involves describing the variability or distribution of values of the output variable that is due to the collective variation in the input variables. This description usually involves graphical displays such as histograms, empirical pdf plots, and empirical cdf plots, as well as summary statistics such as the mean, median, standard deviation, coefficient of variation, 95th percentile, etc.

- ***Sensitivity analysis*** involves determining how the distribution of the output variable changes with changes in the individual input variables. It is used to identify which input variables contribute the most to the variation or uncertainty in the output variable. Sensitivity analysis is also used in a broader sense to determine how changing the distributions of the input variables and/or their assumed correlations or even changing the form of the model affects the output.

- The term ***variability*** refers to the inherent heterogeneity of a particular variable (parameter). For example, there is natural variation in body weight and height between individuals in a given population. The term ***uncertainty*** refers to a lack of knowledge about specific parameters, models, or factors. We can usually reduce uncertainty through further measurement or study. We cannot reduce variability, since it is inherent in the variable.

- Sensitivity analysis methods include one-at-a-time deviations from a baseline case, factorial design and response surface modeling, and Monte Carlo simulation.

- Uncertainty analysis methods include describing the empirical distribution of risk, repeating the simulation using a different set of random numbers, and using mixture distributions.

- An important point to remember is that no matter how complex the mathematical model in **Equation (10.1)** or **(10.8)** is, or how extensive the uncertainty and sensitivity analyses are, ***there is always the question of how well the mathematical model reflects reality***. The only way to attempt to answer this question is with data collected directly on the input and output variables.

- The section **Using EnvironmentalStats for S-Plus for Probabilistic Risk Assessment** illustrates several ways of using functions in S-Plus and EnvironmentalStats for S-Plus to perform Monte Carlo simulation and probabilistic risk assessment.

- More information is available in the help files **Glossary: Risk Assessment**, **Glossary: Random Numbers**, and **Glossary: Monte Carlo Simulation and Bootstrapping**.

APPENDIX A:

References

The following is an alphabetical list of references for environmental statistics. The help system for ENVIRONMENTALSTATS for S-PLUS lists references both alphabetically (**References Listed Alphabetically**) and by category (**References Listed by Category**).

Aitchison, J. (1955). On the Distribution of a Positive Random Variable Having a Discrete Probability Mass at the Origin. *Journal of the American Statistical Association* **50**, 901-908.

Aitchison, J., and J.A.C. Brown (1957). **The Lognormal Distribution (with special references to its uses in economics)**. Cambridge University Press, London, 176pp.

Akritas, M., T. Ruscitti, and G.P. Patil. (1994). Statistical Analysis of Censored Environmental Data. In Patil, G.P., and C.R. Rao, eds., **Handbook of Statistics, Vol. 12: Environmental Statistics**. North-Holland, Amsterdam, a division of Elsevier, New York, NY, Chapter 7, 221-242.

Amemiya, T. (1985). **Advanced Econometrics**. Harvard University Press, Cambridge, MA, 521pp.

Arlinghaus, S.L., ed. (1996). **Practical Handbook of Spatial Statistics**. CRC Press, Boca Raton, Florida, 307pp.

Bain, L.J., and M. Engelhardt. (1991). **Statistical Analysis of Reliability and Life-Testing Models**. Marcel Dekker, New York, 496pp.

Balakrishnan, N., and A.C. Cohen. (1991). **Order Statistics and Inference: Estimation Methods**. Academic Press, San Diego, CA, 375pp.

Barnett, V., and K.F. Turkman, eds. (1997). **Statistics for the Environment 3: Pollution Assessment and Control**. John Wiley and Sons, New York, 345pp.

Barr, D.R., and T. Davidson. (1973). A Kolmogorov-Smirnov Test for Censored Samples. *Technometrics* **15**(4), 739-757.

Barry, T.M. (1996). Recommendations on the Testing and Use of Pseudo-Random Number Generators Used in Monte Carlo Analysis for Risk Assessment. *Risk Analysis* **16**(1), 93-105.

Berthouex, P.M., and I. Hau. (1991). A Simple Rule for Judging Compliance Using Highly Censored Samples. *Research Journal of the Water Pollution Control Federation* **63**(6), 880-886.

Berthouex, P.M., and I. Hau. (1991). Difficulties Related to Using Extreme Percentiles for Water Quality Regulations. *Research Journal of the Water Pollution Control Federation* **63**(6), 873-879.

Berthouex, P.M., and L.C. Brown. (1994). **Statistics for Environmental Engineers**. Lewis Publishers, Boca Raton, FL, 335pp.

Birnbaum, Z. W., and F.H. Tingey. (1951). One-Sided Confidence Contours for Probability Distribution Functions. *Annals of Mathematical Statistics* **22**, 592-596.

Blom, G. (1958). **Statistical Estimates and Transformed Beta Variables**. John Wiley and Sons, New York.

Bogen, K.T. (1995). Methods to Approximate Joint Uncertainty and Variability in Risk. *Risk Analysis* **15**(3), 411-419.

Box, G.E.P., W.G. Hunter, and J.S. Hunter. (1978). **Statistics for Experimenters: An Introduction to Design, Data Analysis, and Model Building**. John Wiley and Sons, New York, 653pp.

Bradu, D., and Y. Mundlak. (1970). Estimation in Lognormal Linear Models. *Journal of the American Statistical Association* **65**, 198-211.

Brainard, J., and D.E. Burmaster. (1992). Bivariate Distributions for Height and Weight of Men and Women in the United States. *Risk Analysis* **12**(2), 267-275.

Bukowski, J., L. Korn, and D. Wartenberg. (1995). Correlated Inputs in Quantitative Risk Assessment: The Effects of Distributional Shape. *Risk Analysis* **15**(2), 215-219.

Burmaster, D.E., and P.D. Anderson. (1994). Principles of Good Practice for the Use of Monte Carlo Techniques in Human Health and Ecological Risk Assessments. *Risk Analysis* **14**(4), 477-481.

Burmaster, D.E., and R.H. Harris. (1993). The Magnitude of Compounding Conservatisms in Superfund Risk Assessments. *Risk Analysis* **13**(2), 131-134.

Burmaster, D.E., and A.M. Wilson. (1996). An Introduction to Second-Order Random Variables in Human Health Risk Assessments. *Human and Ecological Risk Assessment* **2**(4), 892-919.

Callahan, B.G., ed. (1996). Commemoration of the 50th Anniversary of Monte Carlo. *Human and Ecological Risk Assessment* **4**(2).

Calitz, F. (1973). Maximum Likelihood Estimation of the Parameters of the Three-Parameter Lognormal Distribution--a Reconsideration. *Australian Journal of Statistics* **15**(3), 185-190.

Castillo, E. (1988). **Extreme Value Theory in Engineering**. Academic Press, New York, 389pp.

Castillo, E., and A. Hadi. (1994). Parameter and Quantile Estimation for the Generalized Extreme-Value Distribution. *Environmetrics* **5**, 417-432.

Chambers, J.M., W.S. Cleveland, B. Kleiner, and P.A. Tukey. (1983). **Graphical Methods for Data Analysis**. Duxbury Press, Boston, MA, 395pp.

Chen, J.J., D.W. Gaylor, and R.L. Kodell. (1990). Estimation of the Joint Risk from Multiple-Compound Exposure Based on Single-Compound Experiments. *Risk Analysis* **10**(2), 285-290.

Cheng, R.C.H., and N.A.K. Amin. (1982). Maximum Product-of-Spacings Estimation with Application to the Lognormal Distribution. Journal of the Royal Statistical Society, Series B, 44, 394-403.

Chew, V. (1968). Simultaneous Prediction Intervals. *Technometrics* **10**(2), 323-331.

Chou, Y.M., and D.B. Owen. (1986). One-sided Distribution-Free Simultaneous Prediction Limits for p Future Samples. *Journal of Quality Technology* **18**, 96-98.

Chowdhury, J.U., J.R. Stedinger, and L. H. Lu. (1991). Goodness-of-Fit Tests for Regional Generalized Extreme Value Flood Distributions. *Water Resources Research* **27**(7), 1765-1776.

Clark, M.J.R., and P.H. Whitfield. (1994). Conflicting Perspectives About Detection Limits and About the Censoring of Environmental Data. *Water Resources Bulletin* **30**(6), 1063-1079.

Clayton, C.A., J.W. Hines, and P.D. Elkins. (1987). Detection Limits with Specified Assurance Probabilities. *Analytical Chemistry* **59**, 2506-2514.

Cleveland, W.S. (1993). **Visualizing Data**. Hobart Press, Summit, New Jersey, 360pp.

Cleveland, W.S. (1994). **The Elements of Graphing Data**. Revised Edition. Hobart Press, Summit, New Jersey, 297pp.

Cochran, W.G. (1977). **Sampling Techniques**. Third Edition. John Wiley and Sons, New York, 428pp.

Code of Federal Regulations. (1996). Definition and Procedure for the Determination of the Method Detection Limit--Revision 1.11. **Title 40, Part 136, Appendix B**, 7-1-96 Edition, pp.265-267.

Cogliano, V.J. (1997). Plausible Upper Bounds: Are Their Sums Plausible? *Risk Analysis* **17**(1), 77-84.

Cohen, A.C. (1951). Estimating Parameters of Logarithmic-Normal Distributions by Maximum Likelihood. *Journal of the American Statistical Association* **46**, 206-212.

Cohen, A.C. (1959). Simplified Estimators for the Normal Distribution When Samples are Singly Censored or Truncated. *Technometrics* **1**(3), 217-237.

Cohen, A.C. (1963). Progressively Censored Samples in Life Testing. *Technometrics* **5**, 327-339.

Cohen, A.C. (1988). Three-Parameter Estimation. In Crow, E.L., and K. Shimizu, eds. **Lognormal Distributions: Theory and Applications**. Marcel Dekker, New York, Chapter 4.

Cohen, A.C. (1991). **Truncated and Censored Samples**. Marcel Dekker, New York, New York, 312pp.

Cohen, A.C., and B.J. Whitten. (1980). Estimation in the Three-Parameter Lognormal Distribution. *Journal of the American Statistical Association* **75**, 399-404.

Cohen, A.C., B.J. Whitten, and Y. Ding. (1985). Modified Moment Estimation for the Three-Parameter Lognormal Distribution. *Journal of Quality Technology* **17**, 92-99.

Cohen, M.A., and P.B. Ryan. (1989). Observations Less than the Analytical Limit of Detection: A New Approach. *JAPCA* **39**(3), 328-329.

Cohn, T.A. (1988). Adjusted Maximum Likelihood Estimation of the Moments of Lognormal Populations form Type I Censored Samples. **U.S. Geological Survey Open-File Report 88-350**, 34pp.

Cohn, T.A., E.J. Gilroy, and W.G. Baier. (1992). Estimating Fluvial Transport of Trace Constituents Using a Regression Model with Data Subject to Censoring. *Proceedings of the American Statistical Association, Section on Statistics and the Environment.*

Cohn, T.A., L.L. DeLong, E.J. Gilroy, R.M. Hirsch, and D.K. Wells. (1989). Estimating Constituent Loads. *Water Resources Research* **25**(5), 937-942.

Conover, W.J. (1980). **Practical Nonparametric Statistics**. Second Edition. John Wiley and Sons, New York, 493 pp.

Cothern, C.R. (1994). How Does Scientific Information in General and Statistical Information in Particular Input to the Environmental Regulatory Process? In Patil, G.P., and C.R. Rao, eds., **Handbook of Statistics, Vol. 12: Environmental Statistics**. North-Holland, Amsterdam, a division of Elsevier, New York, NY, Chapter 25, 791-816.

Cothern, C.R., and N.P. Ross. (1994). **Environmental Statistics, Assessment, and Forecasting**. Lewis Publishers, Boca Raton, FL, 418pp.

Cox, D.R., and D.V. Hinkley. (1974). **Theoretical Statistics**. Chapman and Hall, New York, 511pp.

Cox, L.H., and W.W. Piegorsch. (1996). Combining Environmental Information, I: Environmental Monitoring, Measurement and Assessment. *Environmetrics* **7**, 299-308.

Cressie, N.A.C. (1993). **Statistics for Spatial Data, Revised Edition**. John Wiley and Sons, New York, 900 pp.

Crow, E.L., and K. Shimizu, eds. (1988). **Lognormal Distributions: Theory and Applications**. Marcel Dekker, New York, 387pp.

Cullen, A.C. (1994). Measures of Compounding Conservatism in Probabilistic Risk Assessment. *Risk Analysis* **14**(4), 389-393.

Cunnane, C. (1978). Unbiased Plotting Positions -- A Review. *Journal of Hydrology* **37**(3/4), 205-222.

Currie, L.A. (1968). Limits for Qualitative Detection and Quantitative Determination: Application to Radiochemistry. *Annals of Chemistry* **40**, 586-593.

Currie, L.A. (1988). **Detection in Analytical Chemistry: Importance, Theory, and Practice**. American Chemical Society, Washington, D.C.

Currie, L.A. (1995). Nomenclature in Evaluation of Analytical Methods Including Detection and Quantification Capabilities. *Pure & Applied Chemistry* **67**(10), 1699-1723.

D'Agostino, R.B. (1970). Transformation to Normality of the Null Distribution of $g1$. *Biometrika* **57**, 679-681.

D'Agostino, R.B. (1971). An Omnibus Test of Normality for Moderate and Large Size Samples. *Biometrika* **58**, 341-348.

D'Agostino, R.B. (1986a). Graphical Analysis. In: D'Agostino, R.B., and M.A. Stephens, eds. **Goodness-of Fit Techniques**. Marcel Dekker, New York, Chapter 2, pp.7-62.

D'Agostino, R.B. (1986b). Tests for the Normal Distribution. In: D'Agostino, R.B., and M.A. Stephens, eds. **Goodness-of Fit Techniques**. Marcel Dekker, New York, Chapter 9, pp.367-419.

D'Agostino, R.B., and E.S. Pearson (1973). Tests for Departures from Normality. Empirical Results for the Distributions of $b2$ and $\sqrt{b1}$. *Biometrika* **60**(3), 613-622.

D'Agostino, R.B., and G.L. Tietjen (1973). Approaches to the Null Distribution of $\sqrt{b1}$. *Biometrika* **60**(1), 169-173.

D'Agostino, R.B., and M.A. Stephens, eds. (1986). **Goodness-of Fit Techniques**. Marcel Dekker, New York, 560pp.

Dallal, G. E., and L. Wilkinson. (1986). An Analytic Approximation to the Distribution of Lilliefor's Test for Normality. *The American Statistician* **40**, 294-296.

Danziger, L., and S. Davis. (1964). Tables of Distribution-Free Tolerance Limits. *Annals of Mathematical Statistics* **35**(5), 1361-1365.

David, M. (1977). **Geostatistical Ore Reserve Estimation**. Elsevier, New York, 364pp.

Davidson, J.R. (1994). **Elipgrid-PC: A PC Program for Calculating Hot Spot Probabilities**. ORNL/TM-12774. Oak Ridge National Laboratory.

Davis, C.B. (1994). Environmental Regulatory Statistics. In Patil, G.P., and C.R. Rao, eds., **Handbook of Statistics, Vol. 12: Environmental Statistics**. North-Holland, Amsterdam, a division of Elsevier, New York, NY, Chapter 26, 817-865.

Davis, C.B., and R.J. McNichols. (1987). One-sided Intervals for at Least *p* of *m* Observations from a Normal Population on Each of *r* Future Occasions. *Technometrics* **29**, 359-370.

Davis, C.B., and R.J. McNichols. (1994a). Ground Water Monitoring Statistics Update: Part I: Progress Since 1988. *Ground Water Monitoring and Remediation* **14**(4), 148-158.

Davis, C.B., and R.J. McNichols. (1994b). Ground Water Monitoring Statistics Update: Part II: Nonparametric Prediction Limits. *Ground Water Monitoring and Remediation* **14**(4), 159-175.

Davis, C.S., and M.A. Stephens. (1978). Algorithm AS 128. Approximating the Covariance Matrix of Normal Order Statistics. *Applied Statistics* **27**, 206-212.

Diggle, P.J. (1983). **Statistical Analysis of Spatial Point Patterns**. Academic Press, London.

Downing, D.J., R.H. Gardner, and F.O. Hoffman. (1985). An Examination of Response-Surface Methodologies for Uncertainty Analysis in Assessment Models. *Technometrics* **27**(2), 151-163. See also Easterling, R.G. (1986). Discussion of Downing, Gardner, and Hoffman (1985). *Technometrics* **28**(1), 91-93.

Downton, F. (1966). Linear Estimates of Parameters in the Extreme Value Distribution. *Technometrics* **8**(1), 3-17.

Draper, N., and H. Smith. (1981). **Applied Regression Analysis**. Second Edition. John Wiley and Sons, New York, 709pp.

Dunnett, C.W. (1955). A Multiple Comparisons Procedure for Comparing Several Treatments with a Control. *Journal of the American Statistical Association* **50**, 1096-1121.

Dunnett, C.W. (1964). New Tables for Multiple Comparisons with a Control. *Biometrics* **20**, 482-491.

Dunnett, C.W., and M. Sobel. (1955). Approximations to the Probability Integral and Certain Percentage Points of a Multivariate Analogue of Student's t-Distribution. *Biometrika* **42**, 258-260.

Edland, S.D., and G. Van Belle. (1994). Decreased Sampling Costs and Improved Accuracy with Composite Sampling. In Cothern, C.R., and N.P. Ross, eds., **Environmental Statistics, Assessment, and Forecasting**. Lewis Publishers, Boca Raton, FL, Chapter 2, pp.29-55.

Efron, B., and R.J. Tibshirani. (1993). **An Introduction to the Bootstrap**. Chapman and Hall, New York, 436pp.

Ellison, B.E. (1964). On Two-Sided Tolerance Intervals for a Normal Distribution. *Annals of Mathematical Statistics* **35**, 762-772.

El-Shaarawi, A.H. (1989). Inferences About the Mean from Censored Water Quality Data. *Water Resources Research* **25**(4) 685-690.

El-Shaarawi, A.H. (1989). Inferences About the Mean from Censored Water Quality Data. *Water Resources Research* **25**(4) 685-690.

El-Shaarawi, A.H. (1993). Environmental Monitoring, Assessment, and Prediction of Change. *Environmetrics* **4**(4), 381-398.

El-Shaarawi, A.H., and A. Naderi. (1991). Statistical Inference from Multiply Censored Environmental Data. *Environmental Monitoring and Assessment* **17**, 339-347.

El-Shaarawi, A.H., and D.M. Dolan. (1989). Maximum Likelihood Estimation of Water Quality Concentrations from Censored Data. *Canadian Journal of Fisheries and Aquatic Sciences* **46**, 1033-1039.

El-Shaarawi, A.H., and S.R. Esterby. (1992). Replacement of Censored Observations by a Constant: An Evaluation. *Water Research* **26**(6), 835-844.

Eschenroeder, A.Q., and E.J. Faeder. (1988). A Monte Carlo Analysis of Health Risks From PCB-Contaminated Mineral Oil Transformer Fires. *Risk Analysis* **8**(2), 291-297.

Esterby, S.R. (1993). Trend Analysis Methods for Environmental Data. *Environmetrics* **4**(4), 459-481.

Evans, M., N. Hastings, and B. Peacock. (1993). **Statistical Distributions.** Second Edition. John Wiley and Sons, New York, 170pp.

Fill, H.D., and J.R. Stedinger. (1995). *L* Moment and Probability Plot Correlation Coefficient Goodness-of-Fit Tests for the Gumbel Distribution and Impact of Autocorrelation. *Water Resources Research* **31(1)**, 225-229.

Filliben, J.J. (1975). The Probability Plot Correlation Coefficient Test for Normality. *Technometrics* **17**(1), 111-117.

Finley, B., and D. Paustenbach. (1994). The Benefits of Probabilistic Exposure Assessment: Three Case Studies Involving Contaminated Air, Water, and Soil. *Risk Analysis* **14**(1), 53-73.

Finley, B., D. Proctor, P. Scott, N. Harrington, D. Paustenback, and P. Price. (1994). Recommended Distributions for Exposure Factors Frequently Used in Health Risk Assessment. *Risk Analysis* **14**(4), 533-553.

Finney, D.J. (1941). On the Distribution of a Variate Whose Logarithm is Normally Distributed. *Supplement to the Journal of the Royal Statistical Society* **7**, 155-161.

Fleiss, J. L. (1981). **Statistical Methods for Rates and Proportions**. Second Edition. John Wiley and Sons, New York, 321pp.

Fleming, T.R., and D.P. Harrington. (1981). A Class of Hypothesis Tests for One and Two Sample Censored Survival Data. *Communications in Statistics -- Theory and Methods* **A10**(8), 763-794.

Fleming, T.R., and D.P. Harrington. (1991). **Counting Processes & Survival Analysis**. John Wiley and Sons, New York, 429pp.

Fletcher, D.J., and B.F.J. Manly, eds. (1994). **Statistics in Ecology and Environmental Monitoring, Otago Conference Series No. 2**. University of Otago Press, Dunedin, New Zealand, 269pp.

Flora, J.D. (1991). Statistics and Environmental Regulations. *Environmetrics* **2**(2), 129-137.

Freudenburg, W.R., and J.A. Rursch. (1994). The Risks of "Putting the Numbers in Context": A Cautionary Tale. *Risk Analysis* **14**(6), 949-958.

Gaylor, D.W., and J.J. Chen. (1996). A Simple Upper Limit for the Sum of the Risks of the Components in a Mixture. *Risk Analysis* **16**(3), 395-398.

Gentle, J.E. (1985). Monte Carlo Methods. In Kotz, S., and N.L. Johnson, eds. **Encyclopedia of Statistics, Volume 5**. John Wiley and Sons, New York, 612-617.

Gibbons, R.D. (1987a). Statistical Prediction Intervals for the Evaluation of Ground-Water Quality. *Ground Water* **25**, 455-465.

Gibbons, R.D. (1987b). Statistical Models for the Analysis of Volatile Organic Compounds in Waste Disposal Sites. *Ground Water* **25**, 572-580.

Gibbons, R.D. (1990). A General Statistical Procedure for Ground-Water Detection Monitoring at Waste Disposal Facilities. *Ground Water* **28**, 235-243.

Gibbons, R.D. (1991a). Some Additional Nonparametric Prediction Limits for Ground-Water Detection Monitoring at Waste Disposal Facilities. *Ground Water* **29**, 729-736.

Gibbons, R.D. (1991b). Statistical Tolerance Limits for Ground-Water Monitoring. *Ground Water* **29**, 563-570.

Gibbons, R.D. (1994). **Statistical Methods for Groundwater Monitoring**. John Wiley and Sons, New York, NY, 286pp.

Gibbons, R.D. (1994). The Folly of Subtitle D Statistics: When Green Field Sites Fail. **Proceedings of Waste Technology '94**. National Solid Waste Management Association, Washington, D.C., January 13-15.

Gibbons, R.D. (1995). Some Statistical and Conceptual Issues in the Detection of Low-Level Environmental Pollutants (with Discussion). *Environmetrics* **2**, 125-167.

Gibbons, R.D. (1996). Some Conceptual and Statistical Issues in Analysis of Groundwater Monitoring Data. *Environmetrics* **7**, 185-199.

Gibbons, R.D., and J. Baker. (1991). The Properties of Various Statistical Prediction Intervals for Ground-Water Detection Monitoring. *Journal of Environmental Science and Health* **A26**(4), 535-553.

Gibbons, R.D., F.H. Jarke, and K.P. Stoub. (1991). Detection Limits: For Linear Calibration Curves with Increasing Variance and Multiple Future Detection Decisions. In Tatsch, D.E., editor. **Waste Testing and Quality Assurance: Volume 3**. American Society for Testing and Materials, Philadelphi, PA.

Gibbons, R.D., N.E. Grams, F.H. Jarke, and K.P. Stoub. (1992). *Practical Quantitation Limits. Chemometrics Intelligent Laboratory Systems* **12**, 225-235.

Gibrat, R. (1930). Une loi des repartition economiques: l'effect proportionnel. *Bull. Statist. Gen. Fr.* **19**, 469ff.

Gilbert, R.O. (1987). **Statistical Methods for Environmental Pollution Monitoring**. Van Nostrand Reinhold, New York, NY, 320pp.

Gilbert, R.O. (1994). An Overview of Statistical Issues Related to Environmental Cleanup. In Patil, G.P., and C.R. Rao, eds., **Handbook of Statistics, Vol. 12: Environmental Statistics**. North-Holland, Amsterdam, a division of Elsevier, New York, NY, Chapter 27, 867-880.

Gilliom, R.J., and D.R. Helsel. (1986). Estimation of Distributional Parameters for Censored Trace Level Water Quality Data: 1. Estimation Techniques. *Water Resources Research* **22**, 135-146.

Gilliom, R.J., R.M. Hirsch, and E.J. Gilroy. (1984). Effect of Censoring Trace-Level Water-Quality Data on Trend-Detection Capability. *Environmental Science and Technology* **18**, 530-535.

Gilroy, E.J., R.M. Hirsch, and T.A. Cohn. (1990). Mean Square Error of Regression-Based Constituent Transport Estimates. *Water Resources Research* **26**(9), 2069-2077.

Glasser, J.A., D.L. Foerst, G.D. McKee, S.A. Quane, and W.L. Budde. (1981). Trace Analyses for Wastewaters. *Environmental Science and Technology* **15**, 1426-1435.

Gleit, A. (1985). Estimation for Small Normal Data Sets with Detection Limits. *Environmental Science and Technology* **19**, 1201-1206.

Greenwood, J.A., J.M. Landwehr, N.C. Matalas, and J.R. Wallis. (1979). Probability Weighted Moments: Definition and Relation to Parameters of Several Distributions Expressible in Inverse Form. *Water Resources Research* **15**(5), 1049-1054.

Griffiths, D.A. (1980). Interval Estimation for the Three-Parameter Lognormal Distribution via the Likelihood Function. *Applied Statistics* **29**, 58-68.

Gringorten, I.I. (1963). A Plotting Rule for Extreme Probability Paper. *Journal of Geophysical Research* **68**(3), 813-814.

Gupta, A.K. (1952). Estimation of the Mean and Standard Deviation of a Normal Population from a Censored Sample. *Biometrika* **39**, 260-273.

Guttman, I. (1970). **Statistical Tolerance Regions: Classical and Bayesian**. Hafner Publishing Co., Darien, CT, 150pp.

Haas, C.N. (1997). Importance of Distributional Form in Characterizing Inputs to Monte Carlo Risk Assessments. *Risk Analysis* **17**(1), 107-113.

Haas, C.N., and P.A. Scheff. (1990). Estimation of Averages in Truncated Samples. *Environmental Science and Technology* **24**(6), 912-919.

Hahn, G.J. (1969). Factors for Calculating Two-Sided Prediction Intervals for Samples from a Normal Distribution. *Journal of the American Statistical Association* **64**(327), 878-898.

Hahn, G.J. (1970a). Additional Factors for Calculating Prediction Intervals for Samples from a Normal Distribution. *Journal of the American Statistical Association* **65**(332), 1668-1676.

Hahn, G.J. (1970b). Statistical Intervals for a Normal Population, Part I: Tables, Examples and Applications. *Journal of Quality Technology* **2**(3), 115-125.

Hahn, G.J. (1970c). Statistical Intervals for a Normal Population, Part II: Formulas, Assumptions, Some Derivations. *Journal of Quality Technology* **2**(4), 195-206.

Hahn, G., and W. Nelson. (1973). A Survey of Prediction Intervals and Their Applications. *Journal of Quality Technology* **5**, 178-188.

Hahn, G.J., and W.Q. Meeker. (1991). **Statistical Intervals: A Guide for Practitioners**. John Wiley and Sons, New York, 392pp.

Haimes, Y.Y., T. Barry, and J.H. Lambert, eds. (1994). When and How Can You Specify a Probability Distribution When You Don't Know Much? *Risk Analysis*, **14**(5), 661-706.

Hall, I.J., R.R. Prairie, and C.K. Motlagh. (1975). Non-Parametric Prediction Intervals. *Journal of Quality Technology* **7**(3), 109-114.

Hallenbeck, W.H. (1993). **Quantitative Risk Assessment for Environmental and Occupational Health**. Second Edition. Lewis Publishers, Boca Raton, FL, 224pp.

Hamby, D.M. (1994). A Review of Techniques for Parameter Sensitivity Analysis of Environmental Models. *Environmental Monitoring and Assessment* **32**, 135-154.

Hamed, M., and P.B. Bedient. (1997). On the Effect of Probability Distributions of Input Variables in Public Health Risk Assessment. *Risk Analysis* **17**(1), 97-105.

Harrington, D.P., and T.R. Fleming. (1982). A Class of Rank Test Procedures for Censored Survival Data. *Biometrika* **69**(3), 553-566.

Harter, H.L., and A.H. Moore. (1966). Local-Maximum-Likelihood Estimation of the Parameters of Three-Parameter Lognormal Populations from Complete and Censored Samples. *Journal of the American Statistical Association* **61**, 842-851.

Hashimoto, L.K., and R.R. Trussell. (1983). Evaluating Water Quality Data Near the Detection Limit. Paper presented at the Advanced Technology Conference, American Water Works Association, Las Vegas, Nevada, June 5-9, 1983.

Hattis, D. (1990). Three Candidate "Laws" of Uncertainty Analysis. *Risk Analysis* **10**(1), 11.

Hattis, D., and D.E. Burmaster. (1994). Assessment of Variability and Uncertainty Distributions for Practical Risk Analyses. *Risk Analysis* **14**(5), 713-730.

Hayes, B. (1993). The Wheel of Fortune. *American Scientist* **81**, 114-118.

Helsel, D.R. (1987). Advantages of Nonparametric Procedures for Analysis of Water Quality Data. *Hydrological Sciences Journal* **32**, 179-190.

Helsel, D.R. (1990). Less than Obvious: Statistical Treatment of Data Below the Detection Limit. *Environmental Science and Technology* **24**(12), 1766-1774.

Helsel, D.R. and R.M. Hirsch. (1988). Discussion of Applicability of the t-test for Detecting Trends in Water Quality Variables. *Water Resources Bulletin* **24**(1), 201-204.

Helsel, D.R., and R.J. Gilliom. (1986). Estimation of Distributional Parameters for Censored Trace Level Water Quality Data: 2. Verification and Applications. *Water Resources Research* **22**, 147-155.

Helsel, D.R., and R.M. Hirsch. (1992). **Statistical Methods in Water Resources Research**. Elsevier, New York, NY, 522pp.

Helsel, D.R., and T.A. Cohn. (1988). Estimation of Descriptive Statistics for Multiply Censored Water Quality Data. *Water Resources Research* **24**(12), 1997-2004.

Hettmansperger, T.P. (1984). **Statistical Inference Based on Ranks**. John Wiley and Sons, New York, 323pp.

Heyde, C.C. (1963). On a Property of the Lognormal Distribution. *Journal of the Royal Statistical Society, Series B* **25**, 392-393.

Hill, B.M. (1963). The Three-Parameter Lognormal Distribution and Bayesian Analysis of a Point-Source Epidemic. *Journal of the American Statistical Association* **58**, 72-84.

Hirsch, R.M. (1988). Statistical Methods and Sampling Design for Estimating Step Trends in Surface-Water Quality. *Water Resources Bulletin* **24**(3), 493-503.

Hirsch, R.M. and E.J. Gilroy. (1984). Methods of Fitting a Straight Line to Data: Examples in Water Resources. *Water Resources Bulletin* **20**(5), 705-711.

Hirsch, R.M. and J.R. Slack. (1984). A Nonparametric Trend Test for Seasonal Data with Serial Dependence. *Water Resources Research* **20**(6), 727-732.

Hirsch, R.M., and J.R. Stedinger. (1987). Plotting Positions for Historical Floods and Their Precision. *Water Resources Research* **23**(4), 715-727.

Hirsch, R.M., D.R. Helsel, T.A. Cohn, and E.J. Gilroy. (1993). Statistical Analysis of Hydrologic Data. In: Maidment, D.R., ed. **Handbook of Hydrology**. McGraw-Hill, New York, Chapter 17.

Hirsch, R.M., J.R. Slack, and R.A. Smith. (1982). Techniques of Trend Analysis for Monthly Water Quality Data. *Water Resources Research* **18**(1), 107-121.

Hirsch, R.M., R.B. Alexander, and R.A. Smith. (1991). Selection of Methods for the Detection and Estimation of Trends in Water Quality. *Water Resources Research* **27**(5), 803-813.

Hoaglin, D.C., and D.F. Andrews. (1975). The Reporting of Computation-Based Results in Statistics. *The American Statistician* **29**(3), 122-126.

Hoaglin, D.C., F. Mosteller, and J.W. Tukey, editors. (1983). **Understanding Robust and Exploratory Data Analysis**. John Wiley and Sons, New York.

Hoeffding, W. (1948). A Class of Statistics with Asymptotically Normal Distribution. *Annals of Mathematical Statistics* **19**, 293-325.

Hoffman, F.O., and J.S. Hammonds. (1994). Propagation of Uncertainty in Risk Assessments: The Need to Distinguish Between Uncertainty Due to Lack of Knowledge and Uncertainty Due to Variability. *Risk Analysis* **14**(5), 707-712.

Hollander, M., and D.A. Wolfe. (1973). **Nonparametric Statistical Methods**. John Wiley and Sons, New York, 503pp.

Hoshi, K., J.R. Stedinger, and J. Burges. (1984). Estimation of Log-Normal Quantiles: Monte Carlo Results and First-Order Approximations. *Journal of Hydrology* **71**, 1-30.

Hosking, J.R.M. (1984). Testing Whether the Shape Parameter is Zero in the Generalized Extreme-Value Distribution. *Biometrika* **71**(2), 367-374.

Hosking, J.R.M. (1985). Algorithm AS 215: Maximum-Likelihood Estimation of the Parameters of the Generalized Extreme-Value Distribution. *Applied Statistics* **34**(3), 301-310.

Hosking, J.R.M. (1990). L-Moments: Analysis and Estimation of Distributions Using Linear Combinations of Order Statistics. *Journal of the Royal Statistical Society, Series B* **52**(1), 105-124.

Hosking, J.R.M., and J.R. Wallis (1995). A Comparison of Unbiased and Plotting-Position Estimators of *L* Moments. *Water Resources Research* **31**(8), 2019-2025.

Hosking, J.R.M., J.R. Wallis, and E.F. Wood. (1985). Estimation of the Generalized Extreme-Value Distribution by the Method of Probability-Weighted Moments. *Technometrics* **27**(3), 251-261.

Hubaux, A., and G. Vos. (1970). Decision and Detection Limits for Linear Calibration Curves. *Annals of Chemistry* **42**, 849-855.

Hughes, J.P., and S.P. Millard. (1988). A Tau-Like Test for Trend in the Presence of Multiple Censoring Points. *Water Resources Bulletin* **24**(3), 521-531.

Ibrekk, H., and M.G. Morgan. (1987). Graphical Communication of Uncertain Quantities to Nontechnical People. *Risk Analysis* **7**(4), 519-529.

Iman, R.L., and W.J. Conover. (1980). Small Sample Sensitivity Analysis Techniques for Computer Models, With an Application to Risk Assessment (with Comments). *Communications in Statistics--Volume A, Theory and Methods*, **9**(17), 1749-1874.

Iman, R.L., and W.J. Conover. (1982). A Distribution-Free Approach to Inducing Rank Correlation Among Input Variables. *Communications in Statistics--Volume B, Simulation and Computation*, **11**(3), 311-334.

Iman, R.L., and J.M. Davenport. (1982). Rank Correlation Plots For Use With Correlated Input Variables. *Communications in Statistics--Volume B, Simulation and Computation*, **11**(3), 335-360.

Iman, R.L., and J.C. Helton. (1988). An Investigation of Uncertainty and Sensitivity Analysis Techniques for Computer Models. *Risk Analysis* **8**(1), 71-90.

Iman, R.L. and J.C. Helton. (1991). The Repeatability of Uncertainty and Sensitivity Analyses for Complex Probabilistic Risk Assessments. *Risk Analysis* **11**(4), 591-606.

Isaaks, E.H., and R.M. Srivastava. (1989). **An Introduction to Applied Geostatistics**. Oxford University Press, New York, 561pp.

Israeli, M., and C.B. Nelson. (1992). Distribution and Expected Time of Residence for U.S. Households. *Risk Analysis* **12**(1), 65-72.

Jenkinson, A.F. (1955). The Frequency Distribution of the Annual Maximum (or Minimum) of Meteorological Events. *Quarterly Journal of the Royal Meteorological Society* **81**, 158-171.

Jenkinson, A.F. (1969). Statistics of Extremes. *Technical Note 98*, World Meteorological Office, Geneva.

Johnson, G.D., B.D. Nussbaum, G.P. Patil, and N.P. Ross. (1996). Designing Cost-Effective Environmental Sampling Using Concomitant Information. *Chance* **9**(1), 4-11.

Johnson, N. L., S. Kotz, and A.W. Kemp. (1992). **Univariate Discrete Distributions.** Second Edition. John Wiley and Sons, New York, 565pp.

Johnson, N. L., S. Kotz, and N. Balakrishnan. (1994). **Continuous Univariate Distributions, Volume 1**. Second Edition. John Wiley and Sons, New York, 756pp.

Johnson, N. L., S. Kotz, and N. Balakrishnan. (1995). **Continuous Univariate Distributions, Volume 2**. Second Edition. John Wiley and Sons, New York, 719pp.

Johnson, N.L., and B.L. Welch. (1940). Applications of the Non-Central t-Distribution. *Biometrika* **31**, 362-389.

Johnson, R.A., and D.W. Wichern. (1992). **Applied Multivariate Statistical Analysis**. Third Edition. Prentice Hall, Englewood Cliffs, NJ, 642pp.

Johnson, R.A., S. Verrill, and D.H. Moore. (1987). Two-Sample Rank Tests for Detecting Changes That Occur in a Small Proportion of the Treated Population. *Biometrics* **43**, 641-655.

Journel, A.G., and Huijbregts, C.J. (1978). **Mining Geostatistics**. Academic Press, London.

Judge, G.G., W.E. Griffiths, R.C. Hill, H. Lutkepohl, and T.C. Lee. (1985). Qualitative and Limited Dependent Variable Models: Chapter 18. In **The Theory and Practice of Econometrics**. John Wiley and Sons, New York, 1019 pp.

Kaiser, H. (1965). Zum Problem der Nachweisgrenze. *Fresenius' Z. Anal. Chem.* **209**, 1.

Kalbfleisch, J.D., and R.L. Prentice. (1980). **The Statistical Analysis of Failure Time Data**. John Wiley and Sons, New York, 321pp.

Kaplan, E.L., and P. Meier. (1958). Nonparametric Estimation From Incomplete Observations. *Journal of the American Statistical Association* **53**, 457-481.

Kapteyn, J.C. (1903). **Skew Frequency Curves in Biology and Statistics**. Astronomical Laboratory, Noordhoff, Groningen.

Keith, L.H. (1991). **Environmental Sampling and Analysis: A Practical Guide**. Lewis Publishers, Boca Raton, FL, 143pp.

Kendall, M. G., and A. Stuart. (1991). **The Advanced Theory of Statistics, Volume 2: Inference and Relationship**. Fifth Edition. Oxford University Press, New York, 1323pp.

Kendall, M.G. (1975). **Rank Correlation Methods**. Charles Griffin, London.

Kendall, M.G., and J.D. Gibbons. (1990). **Rank Correlation Methods**. Fifth Edition. Charles Griffin, London, 260pp.

Kennedy, W.J., and J.E. Gentle. (1980). **Statistical Computing**. Marcel Dekker, New York, 591pp.

Kim, P. J., and R.I. Jennrich. (1973). Tables of the Exact Sampling Distribution of the Two Sample Kolmogorov-Smirnov Criterion. In: Harter, H.L., and D.B. Owen, eds. **Selected Tables in Mathematical Statistics, Vol. 1**. American Mathematical Society, Providence, Rhode Island, pp.79-170.

Kodell, R.L., H. Ahn, J.J. Chen, J.A. Springer, C.N. Barton, and R.C. Hertzberg. (1995). Upper Bound Risk Estimates for Mixtures of Carcinogens. *Toxicology* **105**, 199-208.

Kodell, R.L., and J.J. Chen. (1994). Reducing Conservatism in Risk Estimation for Mixtures of Carcinogens. *Risk Analysis* **14**(3), 327-332.

Kolmogorov, A.N. (1933). Sulla determinazione empirica di una legge di distribuzione. *Giornale dell' Istituto Italiano degle Attuari* **4**, 83-91.

Kotz, S., and N.L. Johnson, eds. (1985). **Encyclopedia of Statistics**. John Wiley and Sons, New York.

Kuchenhoff, H., and M. Thamerus. (1996). Extreme Value Analysis of Munich Air Pollution Data. *Environmental and Ecological Statistics* **3**, 127-141.

Kushner, E.J. (1976). On Determining the Statistical Parameters for Pollution Concentrations from a Truncated Data Set. *Atmospheric Environment* **10**, 975-979.

Lambert, D., B. Peterson, and I. Terpenning. (1991). Nondetects, Detection Limits, and the Probability of Detection. *Journal of the American Statistical Association* **86**(414), 266-277.

Lambert, J.H., N.C. Matalas, C.W. Ling, Y.Y. Haimes, and D. Li. (1994). Selection of Probability Distributions in Characterizing Risk of Extreme Events. *Risk Analysis* **14**(5), 731-742.

Land, C.E. (1971). Confidence Intervals for Linear Functions of the Normal Mean and Variance. *The Annals of Mathematical Statistics* **42**(4), 1187-1205.

Land, C.E. (1972). An Evaluation of Approximate Confidence Interval Estimation Methods for Lognormal Means. *Technometrics* **14**(1), 145-158.

Land, C.E. (1973). Standard Confidence Limits for Linear Functions of the Normal Mean and Variance. *Journal of the American Statistical Association* **68**(344), 960-963.

Land, C.E. (1975). Tables of confidence limits for linear functions of the normal mean and variance, in **Selected Tables in Mathematical Statistics, Vol. III**. American Mathematical Society, Providence, RI, pp. 385-419.

Landwehr, J.M., N.C. Matalas, and J.R. Wallis. (1979). Probability Weighted Moments Compared With Some Traditional Techniques in Estimating Gumbel Parameters and Quantiles. *Water Resources Research* **15**(5), 1055-1064.

Latta, R.B. (1981). A Monte Carlo Study of Some Two-Sample Rank Tests with Censored Data. *Journal of the American Statistical Association* **76**(375), 713-719.

Law, A.M., and W.D. Kelton. (1991). **Simulation Modeling & Analysis**. Second Edition. McGraw-Hill, Inc., New York, 759pp.

Lee, E.T. (1980). **Statistical Methods for Survival Data Analysis**. Lifetime Learning Publications, a Division of Wadsworth, Inc., Belmont, California, 557pp.

Lehmann, E.L. (1975). **Nonparametrics: Statistical Methods Based on Ranks**. Holden-Day, Oakland, CA, 457pp.

Lettenmaier, D.P. (1988). Multivariate Non-Parametric Tests for Trend in Water Quality. *Water Resources Bulletin* **24**(3), 505-512.

Likes, J. (1980). Variance of the MVUE for Lognormal Variance. *Technometrics* **22**(2), 253-258.

Looney, S.W., and T.R. Gulledge. (1985). Use of the Correlation Coefficient with Normal Probability Plots. *The American Statistician* **39**(1), 75-79.

Lu, L., and J.R. Stedinger. (1992). Variance of Two- and Three-Parameter GEV/PWM Quantile Estimators: Formulae, Confidence Intervals, and a Comparison. *Journal of Hydrology* **138**, 247-267.

Lundgren, R. (1994). **Risk Communication**. Battelle Press, Columbus, OH, 186pp.

Macleod, A.J. (1989). Remark AS R76: A Remark on Algorithm AS 215: Maximum Likelihood Estimation of the Parameters of the Generalized Extreme-Value Distribution. *Applied Statistics* **38**(1), 198-199.

Maritz, J.S., and A.H. Munro. (1967). On the Use of the Generalized Extreme-Value Distribution in Estimating Extreme Percentiles. *Biometrics* **23**(1), 79-103.

Marsaglia, G. et al. (1973). **Random Number Package: "Super-Duper"**. School of Computer Science, McGill University.

MathSoft, Inc. (1996). **S+SpatialStats User's Manual, Version 1.0**. MathSoft, Inc., Data Analysis Products Division, Seattle, WA.

McBean, E.A., and F.A. Rovers. (1992). Estimation of the Probability of Exceedance of Contaminant Concentrations. *Ground Water Monitoring Review* **Winter**, 115-119.

McKay, M.D., R.J. Beckman., and W.J. Conover. (1979). A Comparison of Three Methods for Selecting Values of Input Variables in the Analysis of Output From a Computer Code. *Technometrics* **21**(2), 239-245.

McKean, J.W., and G.L. Sievers. (1989). Rank Scores Suitable for Analysis of Linear Models Under Asymmetric Error Distributions. *Technometrics* **31**, 207-218.

McKone, T.E. (1994). Uncertainty and Variability in Human Exposures to Soil and Contaminants Through Home-Grown Food: A Monte Carlo Assessment. *Risk Analysis* **14**(4), 449-463.

McKone, T.E., and K.T. Bogen. (1991). Predicting the Uncertainties in Risk Assessment. *Environmental Science and Technology* **25**(10), 1674-1681.

McLeod, A.I., K.W. Hipel, and B.A. Bodo. (1991). Trend Analysis Methodology for Water Quality Time Series. *Environmetrics* **2**(2), 169-200.

McNichols, R.J. and C.B. Davis. (1988). Statistical Issues and Problems in Ground Water Detection Monitoring at Hazardous Waste Facilities. *Ground Water Monitoring Review* **8**(4), 135-150.

Michael, J.R., and W.R. Schucany. (1986). Analysis of Data from Censored Samples. In D'Agostino, R.B., and M.A. Stephens, eds. **Goodness-of Fit Techniques**. Marcel Dekker, New York, 560pp, Chapter 11, 461-496.

Millard, S.P. (1987). Environmental Monitoring, Statistics, and the Law: Room for Improvement (with Comment). *The American Statistician* **41**(4), 249-259.

Millard, S.P. (1987). Proof of Safety vs. Proof of Hazard. *Biometrics* **43**, 719-725.

Millard, S.P. (1996). Estimating a Percent Reduction in Load. *Water Resources Research* **32**(6), 1761-1766.

Millard, S.P., and D.P. Lettenmaier. (1986). Optimal Design of Biological Sampling Programs Using the Analysis of Variance. *Estuarine, Coastal and Shelf Science* **22**, 637-656.

Millard, S.P., J.R. Yearsley, and D.P. Lettenmaier. (1985). Space-Time Correlation and Its Effects on Methods for Detecting Aquatic Ecological Change. *Canadian Journal of Fisheries and Aquatic Science*, **42**(8), 1391-1400. Correction: (1986), **43**, 1680.

Millard, S.P., and S.J. Deverel. (1988). Nonparametric Statistical Methods for Comparing Two Sites Based on Data With Multiple Nondetect Limits. *Water Resources Research*, **24**(12), 2087-2098.

Miller, R.G. (1981). **Simultaneous Statistical Inference**. Springer-Verlag, New York, 299pp.

Miller, R.G. (1981). **Survival Analysis**. John Wiley and Sons, New York, 238pp.

Molak, V., ed. (1997). **Fundamentals of Risk Analysis and Risk Management**. Lewis Publishers, Boca Raton, FL, 472pp.

Moore, D. S. (1986). Tests of Chi-Squared Type. In D'Agostino, R.B., and M.A. Stephens, eds. **Goodness-of Fit Techniques**. Marcel Dekker, New York, pp.63-95.

Morgan, M.G., and M. Henrion. (1990). **Uncertainty: A Guide to Dealing with Uncertainty in Quantitative Risk and Policy Analysis**. Cambridge University Press, New York.

Neely, W.B. (1994). **Introduction to Chemical Exposure and Risk Assessment**. Lewis Publishers, Boca Raton, FL, 190pp.

Nelson, W. (1969). Hazard Plotting for Incomplete Failure Data. *Journal of Quality Technology* **1**(1), 27-52.

Nelson, W. (1972). Theory and Applications of Hazard Plotting for Censored Failure Data. *Technometrics* **14**, 945-966.

Nelson, W. (1982). **Applied Life Data Analysis**. John Wiley and Sons, New York, 634pp.

Nelson, W., and J. Schmee. (1979). Inference for (Log) Normal Life Distributions from Small Singly Censored Samples and BLUEs. *Technometrics* **21**, 43-54.

Nelson, W.R. (1970). Confidence Intervals for the Ratio of Two Poisson Means and Poisson Predictor Intervals. *IEEE Transactions on Reliability* **R-19**(2), 42-49.

Nelson, W.R. (1982). **Applied Life Data Analysis**. John Wiley and Sons, New York, 634pp.

Newman, M.C., and P.M. Dixon. (1990). UNCENSOR: A Program to Estimate Means and Standard Deviations for Data Sets with Below Detection Limit Observations. *American Environmental Laboratory*, **4/90**, 26-30.

Newman, M.C., P.M. Dixon, B.B. Looney, and J.E. Pinder. (1989). Estimating Mean and Variance for Environmental Samples with Below Detection Limit Observations. *Water Resources Bulletin* **25**(4), 905-916.

Odeh, R.E., and D.B. Owen. (1980). **Tables for Normal Tolerance Limits, Sampling Plans, and Screening**. Marcel Dekker, New York, 316pp.

Ott, W.R. (1990). A Physical Explanation of the Lognormality of Pollutant Concentrations. *Journal of the Air and Waste Management Association* **40**, 1378-1383.

Ott, W.R. (1995). **Environmental Statistics and Data Analysis**. Lewis Publishers, Boca Raton, FL, 313pp.

Owen, D.B. (1962). **Handbook of Statistical Tables**. Addison-Wesley, Reading, MA, 580pp.

Owen, W., and T. DeRouen. (1980). Estimation of the Mean for Lognormal Data Containing Zeros and Left-Censored Values, with Applications to the Measurement of Worker Exposure to Air Contaminants. *Biometrics* **36**, 707-719.

Parkin, T.B., J.J. Meisinger, S.T. Chester, J.L. Starr, and J.A. Robinson. (1988). Evaluation of Statistical Estimation Methods for Lognormally Distributed Variables. *Journal of the Soil Science Society of America* **52**, 323-329.

Parkin, T.B., S.T. Chester, and J.A. Robinson. (1990). Calculating Confidence Intervals for the Mean of a Lognormally Distributed Variable. *Journal of the Soil Science Society of America* **54**, 321-326.

Parmar, M.K.B., and D. Machin. (1995). **Survival Analysis: A Practical Approach**. John Wiley and Sons, New York, 255pp.

Patil, G.P., A.K. Sinha, and C. Taillie. (1994). Ranked Set Sampling. In Patil, G.P., and C.R. Rao, eds., **Handbook of Statistics, Vol. 12: Environmental Statistics**. North-Holland, Amsterdam, a division of Elsevier, New York, NY, Chapter 5, 167-200.

Patil, G.P., and C.R. Rao, eds. (1994). **Handbook of Statistics, Vol. 12: Environmental Statistics**. North-Holland, Amsterdam, a division of Elsevier, New York, NY, 927pp.

Patil, G.P., S.D. Gore, and A.K. Sinha. (1994). Environmental Chemistry, Statistical Modeling, and Observational Economy. In Cothern, C.R., and N.P. Ross, eds., **Environmental Statistics, Assessment, and Forecasting**. Lewis Publishers, Boca Raton, FL, Chapter 3, pp.57-97.

Pearson, E.S., and H.O. Hartley, eds. (1970). **Biometrika Tables for Statisticians, Volume 1**. Cambridge Universtiy Press, New York, 270pp.

Persson, T. and H. Rootzen. (1977). Simple and Highly Efficient Estimators for a Type I Censored Normal Sample. *Biometrika* **64**, 123-128.

Pettitt, A.N. (1976). Cramer-von Mises Statistics for Testing Normality with Censored Samples. *Biometrika* **63**(3), 475-481.

Pettitt, A.N., and M.A. Stephens. (1976). Modified Cramer-von Mises Statistics for Censored Data. *Biometrika* **63**(2), 291-298.

Pomeranz, J. (1973). Exact Cumulative Distribution of the Kolmogorov-Smirnov Statistic for Small Samples (Algorithm 487). **Collected Algorithms from CACM**.

Porter, P.S., R.C. Ward, and H.F. Bell. (1988). The Detection Limit. *Environmental Science & Technology* **22**(8), 856-861.

Prentice, R.L. (1978). Linear Rank Tests with Right Censored Data. *Biometrika* **65**, 167-179.

Prentice, R.L. (1985). Linear Rank Tests. In Kotz, S., and N.L. Johnson, eds. **Encyclopedia of Statistical Science**. John Wiley and Sons, New York. Volume 5, pp.51-58.

Prentice, R.L., and P. Marek. (1979). A Qualitative Discrepancy Between Censored Data Rank Tests. *Biometrics* **35**, 861-867.

Prescott, P., and A.T. Walden. (1980). Maximum Likelihood Estimation of the Parameters of the Generalized Extreme-Value Distribution. *Biometrika* **67**(3), 723-724.

Prescott, P., and A.T. Walden. (1983). Maximum Likelihood Estimation of the Three-Parameter Generalized Extreme-Value Distribution from Censored Samples. *Journal of Statistical Computing and Simulation* **16**, 241-250.

Press, W.H., S.A. Teukolsky, W.T.Vetterling, and B.P. Flannery. (1992). **Numerical Recipes in FORTRAN: The Art of Scientific Computing**. Second Edition. Cambridge University Press, New York, 963pp.

Ripley, B.D. (1981). **Spatial Statistics**. John Wiley and Sons, New York.

Ripley, B.D. (1987). **Stochastic Simulation.** John Wiley and Sons, New York.

Rocke, D.M., and S. Lorenzato. (1995). A Two-Component Model for Measurement Error in Analytical Chemistry. *Technometrics* **37**(2), 176-184.

Rodricks, J.V. (1992). **Calculated Risks: The Toxicity and Human Health Risks of Chemicals in our Environment**. Cambridge University Press, New York, 256pp.

Rogers, J. (1992). Assessing Attainment of Ground Water Cleanup Standards Using Modified Sequential t-Tests. *Environmetrics* **3**(3), 335-359.

Rosebury, A.M., and D.E. Burmaster. (1992). Lognormal Distributions for Water Intake by Children and Adults. *Risk Analysis* **12**(1), 99-104.

Rowe, W.D. (1994). Understanding Uncertainty. *Risk Analysis* 14(5), 743-750.

Royston, J.P. (1982). Algorithm AS 177. Expected Normal Order Statistics (Exact and Approximate). *Applied Statistics* **31**, 161-165.

Royston, J.P. (1992a). Approximating the Shapiro-Wilk *W*-Test for Non-Normality. *Statistics and Computing* **2**, 117-119.

Royston, J.P. (1992b). Estimation, Reference Ranges and Goodness of Fit for the Three-Parameter Log-Normal Distribution. *Statistics in Medicine* **11**, 897-912.

Royston, J.P. (1992c). A Pocket-Calculator Algorithm for the Shapiro-Francia Test of Non-Normality: An Application to Medicine. *Statistics in Medicine* **12**, 181-184.

Royston, P. (1993). A Toolkit for Testing for Non-Normality in Complete and Censored Samples. *The Statistician* **42**, 37-43.

Rubinstein, R.Y. (1981). **Simulation and the Monte Carlo Method**. John Wiley and Sons, New York, 278pp.

Ruffle, B., D.E. Burmaster, P.D. Anderson, and H.D. Gordon. (1994). Lognormal Distributions for Fish Consumption by the General U.S. Population. *Risk Analysis* **14**(4), 395-404.

Saltelli, A. and J. Marivoet. (1990). Non-Parametric Statistics in Sensitivity Analysis for Model Output: A Comparison of Selected Techniques. *Reliability Engineering and System Safety* **28**, 229-253.

Sarhan, A.E. and B.G. Greenberg. (1956). Estimation of Location and Scale Parameters by Order Statistics for Singly and Doubly Censored Samples, Part I, The Normal Distribution Up to Samples of Size 10. *Annals of Mathematical Statistics* **27**, 427-457.

Saw, J.G. (1961a). Estimation of the Normal Population Parameters Given a Type I Censored Sample. *Biometrika* **48**, 367-377.

Saw, J.G. (1961b). The Bias of the Maximum Likelihood Estimators of Location and Scale Parameters Given a Type II Censored Normal Sample. *Biometrika* **48**, 448-451.

Scheffe, H. (1959). **The Analysis of Variance**. John Wiley and Sons, New York, 477pp.

Scheuer, E.M., and D.S. Stoller. (1962). On the Generation of Normal Random Vectors. *Technometrics* **4**(2), 278-281.

Schmee, J., D.Gladstein, and W. Nelson. (1985). Confidence Limits for Parameters of a Normal Distribution from Singly Censored Samples, Using Maximum Likelihood. *Technometrics* **27**(2) 119-128.

Schneider, H. (1986). **Truncated and Censored Samples from Normal Populations**. Marcel Dekker, New York, New York, 273pp.

Schneider, H., and L. Weisfeld. (1986). Inference Based on Type II Censored Samples. *Biometrics* **42**, 531-536.

Seiler, F.A. (1987). Error Propagation for Large Errors. *Risk Analysis* **7**(4), 509-518.

Seiler, F.A., and J.L. Alvarez. (1996). On the Selection of Distributions for Stochastic Variables. *Risk Analysis* **16**(1), 5-18.

Shapiro, S.S., and C.W. Brian. (1981). A Review of Distributional Testing Procedures and Development of a Censored Sample Distributional Test. In Taillie, C., G.P. Patil, and B.A. Baldessari, eds. **Statistical Distributions in Scientific Work**, Volume 5 -- Inferential Problems and Properties, 1-24.

Shapiro, S.S., and M.B. Wilk. (1965). An Analysis of Variance Test for Normality (Complete Samples). *Biometrika* **52**, 591-611.

Shapiro, S.S., and R.S. Francia. (1972). An Approximate Analysis of Variance Test for Normality. *Journal of the American Statistical Association* **67**(337), 215-219.

Shea, B.L., and A.J. Scallon. (1988). Remark AS R72. A Remark on Algorithm AS 128. Approximating the Covariance Matrix of Normal Order Statistics. *Applied Statistics* **37**, 151-155.

Sheskin, D.J. (1997). **Handbook of Parametric and Nonparametric Statistical Procedures**. CRC Press, Boca Raton, FL, 719pp.

Shlyakhter, A.I. (1994). An Improved Framework for Uncertainty Analysis: Accounting for Unsuspected Errors. *Risk Analysis* **14**(4), 441-447.

Shumway, R.H., A.S. Azari, and P. Johnson. (1989). Estimating Mean Concentrations Under Transformations for Environmental Data With Detection Limits. *Technometrics* **31**(3), 347-356.

Size, W.B., ed. (1987). **Use and Abuse of Statistical Methods in the Earth Sciences**. Oxford University Press, New York, 169pp.

Slob, W. (1994). Uncertainty Analysis in Multiplicative Models. *Risk Analysis* **14**(4), 571-576.

Smirnov, N.V. (1939). Estimate of Deviation Between Empirical Distribution Functions in Two Independent Samples. *Bulletin Moscow University* **2**(2), 3-16.

Smirnov, N.V. (1948). Table for Estimating the Goodness of Fit of Empirical Distributions. *Annals of Mathematical Statistics* **19**, 279-281.

Smith, A.E., P.B. Ryan, and J.S. Evans. (1992). The Effect of Neglecting Correlations When Propagating Uncertainty and Estimating the Population Distribution of Risk. *Risk Analysis* **12**(4), 467-474.

Smith, E.P. (1994). Biological Monitoring: Statistical Issues and Models. In Patil, G.P., and C.R. Rao, eds., **Handbook of Statistics, Vol. 12: Environmental Statistics**. North-Holland, Amsterdam, a division of Elsevier, New York, NY, Chapter 8, 243-261.

Smith, E.P., and K. Rose. (1991). Trend Detection in the Presence of Covariates: Stagewise Versus Multiple Regression. *Environmetrics* **2**(2), 153-168.

Smith, R.L. (1985). Maximum Likelihood Estimation in a Class of Nonregular Cases. *Biometrika* **72**(1), 67-90.

Smith, R.L. (1994). Use of Monte Carlo Simulation for Human Exposure Assessment at a Superfund Site. *Risk Analysis* **14**(4), 433-439.

Smith, R.M., and L.J. Bain. (1976). Correlation Type Goodness-of-Fit Statistics with Censored Sampling. *Communications in Statistics A - Theory and Methods* **5**(2), 119-132.

Sokal, R.F., and F.J. Rolfe. (1981). **Biometry: The Principals and Practice of Statistics in Biological Research**. Second Edition. W.H. Freeman and Company, San Francisco, 859pp.

Springer, M.D. (1979). **The Algebra of Random Variables**. John Wiley and Sons, New York, 470pp.

Stedinger, J. (1983). Confidence Intervals for Design Events. *Journal of Hydraulic Engineering* **109**(1), 13-27.

Stedinger, J.R. (1980). Fitting Lognormal Distributions to Hydrologic Data. *Water Resources Research* **16**(3), 481-490.

Stedinger, J.R., R.M. Vogel, and E. Foufoula-Georgiou. (1993). Frequency Analysis of Extreme Events. In: Maidment, D.R., ed. **Handbook of Hydrology**. McGraw-Hill, New York, Chapter 18.

Stein, M. (1987). Large Sample Properties of Simulations Using Latin Hypercube Sampling. *Technometrics* **29**(2), 143-151.

Stephens, M. A. (1970). Use of the Kolmogorov-Smirnov, Cramer-von Mises and Related Statistics Without Extensive Tables. *Journal of the Royal Statistical Society, Series B*, **32**, 115-122.

Stephens, M.A. (1986a). Tests Based on EDF Statistics. In D'Agostino, R. B., and M.A. Stevens, eds. Goodness-of-Fit Techniques. Marcel Dekker, New York, Chapter 4, pp.97-193.

Stephens, M.A. (1986b). Tests Based on Regression and Correlation. In D'Agostino, R. B., and M.A. Stevens, eds. Goodness-of-Fit Techniques. Marcel Dekker, New York, Chapter 5, pp.195-233.

Suter, G.W. (1993). **Ecological Risk Assessment**. Lewis Publishers, Boca Raton, FL, 538pp.

Taylor, J.K. (1990). **Statistical Techniques for Data Analysis**. Lewis Publishers, Boca Raton, FL, 200pp.

Thompson, K.M., and D.E. Burmaster. (1991). Parametric Distributions for Soil Ingestion by Children. *Risk Analysis* **11**(2), 339-342.

Thompson, K.M., D.E. Burmaster, and E.A.C. Crouch. (1992). Monte Carlo Techniques for Quantitative Uncertainty Analysis in Public Health Risk Assessments. *Risk Analysis* **12**(1), 53-63.

Tiago de Oliveira, J. (1963). Decision Results for the Parameters of the Extreme Value (Gumbel) Distribution Based on the Mean and Standard Deviation. *Trabajos de Estadistica* **14**, 61-81.

Tiago de Oliveira, J. (1983). Gumbel Distribution. In Kotz, S., and N. Johnson, eds. **Encyclopedia of Statistical Sciences**. John Wiley and Sons, New York. Volume 3, pp.552-558.

Travis, C.C., and M.L. Land. (1990). Estimating the Mean of Data Sets with Nondetectable Values. *Environmental Science and Technology* **24**, 961-962.

Tsui, K-L., N.P. Jewell, and C.F.J. Wu. (1988). A Nonparametric Approach to the Truncated Regression Problem. *Journal of the American Statistical Association* **83**(403), 785-792.

USEPA. (1983). Standard for Remedial Actions at Inactive Uranium Processing Sites; Final Rule (40 CFR Part 192). *Federal Register* **48**(3), 590-604.

USEPA. (1987a). **Data Quality Objectives for Remedial Response Activities, Development Process**. EPA/540/G-87/003. U.S. Environmental Protection Agency, Washington, D.C.

USEPA. (1987b). **Data Quality Objectives for Remedial Response Activities, Example Scenario RI/FS Activities at a Site with Contaminated Soils and Ground Water**. EPA/540/G-87/004. U.S. Environmental Protection Agency, Washington, D.C.

USEPA. (1987c). List (Phase 1) of Hazardous Constituents for Ground-Water Monitoring; Final Rule. *Federal Register* **52**(131), 25942-25953 (July 9, 1987).

USEPA. (1988). Statistical Methods for Evaluating Ground-Water Monitoring from Hazardous Waste Facilities: Final Rule. *Federal Register* **53**, 39, 720-31.

USEPA. (1989a). **Methods for Evaluating the Attainment of Cleanup Standards, Volume 1: Soils and Solid Media**. EPA/230-02-89-042. Office of Policy, Planning, and Evaluation, U.S. Environmental Protection Agency, Washington, D.C.

USEPA. (1989b). **Statistical Analysis of Ground-Water Monitoring Data at RCRA Facilities, Interim Final Guidance**. EPA/530-SW-89-026. Office of Solid Waste, U.S. Environmental Protection Agency, Washington, D.C.

USEPA. (1989c). **Risk Assessment Guidance for Superfund, Volume 1: Human Health Evaluation Manual (Part A), Interim Final Guidance**. EPA/540/1-89/002. Office of Emergency and Remedial Response, U.S. Environmental Protection Agency, Washington, D.C.

USEPA. (1990). **Test Methods for Evaluating Solid Waste, 3rd Edition, Proposed Update I**. Office of Solid Waste and Emergency Response, U.S. Environmental Protection Agency, Washington, D.C.

USEPA. (1991a). **Data Quality Objectives Process for Planning Environmental Data Collection Activities, Draft**. Quality Assurance Management Staff, Office of Research and Development, U.S. Environmental Protection Agency, Washington, D.C.

USEPA. (1991b). **Risk Assessment Guidance for Superfund, Volume 1: Human Health Evaluation Manual, Supplemental Guidance/Standard Default Exposure Factors, Interim Final**. OSWER Directive 9285.6-03. Office of Emergency and Remedial Response, U.S. Environmental Protection Agency, Washington, D.C.

USEPA. (1991c). Solid Waste Disposal Facility Criteria: Final Rule. *Federal Register* **56**, 50978-51119.

USEPA. (1992a). **Characterizing Heterogeneous Wastes: Methods and Recommendations**. EPA/600/R-92/033. U.S. Environmental Protection Agency, Washington, D.C.

USEPA. (1992b). **Methods for Evaluating the Attainment of Cleanup Standards, Volume 2: Groundwater**. EPA/230-R-92-014. Office of Policy, Planning, and Evaluation, U.S. Environmental Protection Agency, Washington, D.C.

USEPA. (1992c). **Statistical Analysis of Ground-Water Monitoring Data at RCRA Facilities: Addendum to Interim Final Guidance**. Office of Solid Waste, U.S. Environmental Protection Agency, Washington, D.C. Currently available as part of: *Statistical Training Course for Ground-Water Monitoring Data Analysis*, EPA/530-R-93-003, which may be obtained through the RCRA Docket (202/260-9327).

USEPA. (1992d). **Supplemental Guidance to RAGS: Calculating the Concentration Term**. Publication 9285.7-081, May 1992. Intermittent Bulletin, Volume 1, Number 1. Office of Emergency and Remedial Response, Hazardous Site Evaluation Division, OS-230. Office of Solid Waste and Emergency Response, U.S. Environmental Protection Agency, Washington, D.C.

USEPA. (1994a). **Guidance for the Data Quality Objectives Process. EPA QA/G-4**. EPA/600/R-96/055. Office of Research and Development, U.S. Environmental Protection Agency, Washington, D.C., September, 1994.

USEPA. (1994b). **Statistical Methods for Evaluating the Attainment of Cleanup Standards, Volume 3: Reference-Based Standards for Soils and Solid Media.** EPA/230-R-94-004. Office of Policy, Planning, and Evaluation, U.S. Environmental Protection Agency, Washington, D.C.

USEPA. (1994c). **Use of Monte Carlo Simulation in Risk Assessments**. EPA/903-F-94-001. Hazardous Waste Management Division, U.S. Environmental Protection Agency, Region III, Philadelphia, PA.

USEPA. (1995a). **EPA Observational Economy Series, Volume 1: Composite Sampling**. EPA/230-R-95-005. Office of Policy, Planning, and Evaluation, U.S. Environmental Protection Agency, Washington, D.C.

USEPA. (1995b). **EPA Observational Economy Series, Volume 2: Ranked Set Sampling**. EPA/230-R-95-006. Office of Policy, Planning, and Evaluation, U.S. Environmental Protection Agency, Washington, D.C.

USEPA. (1995c). **EPA Risk Characterization Policy and Guidance**. Memorandum from Carol M. Browner, Administrator, U.S. Environmental Protection Agency, March 21, 1995.

USEPA. (1996). **Guidance for Data Quality Assessment: Practical Methods for Data Analysis. EPA QA/G-9, QA96 Version**. EPA/600/R-96/084. Office of Research and Development, U.S. Environmental Protection Agency, Washington, D.C., July, 1996.

USEPA. (1997). **Policy for Use of Monte Carlo Analysis in Risk Assessment, With Attachment (Draft)**. Memorandum from William P. Wood, January 29, 1997. U.S. Environmental Protection Agency, Washington, D.C.

van Belle, G., and J.P. Hughes. (1984). Nonparametric Tests for Trend in Water Quality. *Water Resources Research* **20**(1), 127-136.

Venables, W.N., and B.D. Ripley. (1997). **Modern Applied Statistics with S-PLUS**. Second Edition. Springer-Verlag, New York, 548pp.

Verrill, S., and R.A. Johnson. (1987). The Asymptotic Equivalence of Some Modified Shapiro-Wilk Statistics -- Complete and Censored Sample Cases. *The Annals of Statistics* **15**(1), 413-419.

Verrill, S., and R.A. Johnson. (1988). Tables and Large-Sample Distribution Theory for Censored-Data Correlation Statistics for Testing Normality. *Journal of the American Statistical Association* **83**, 1192-1197.

Vogel, R.M. (1986). The Probability Plot Correlation Coefficient Test for the Normal, Lognormal, and Gumbel Distributional Hypotheses. *Water Resources Research* **22**(4), 587-590. (Correction, *Water Resources Research* **23**(10), 2013, 1987.)

Vogel, R.M., and D.E. McMartin. (1991). Probability Plot Goodness-of-Fit and Skewness Estimation Procedures for the Pearson Type 3 Distribution. *Water Resources Research* **27**(12), 3149-3158.

Vogel, R.M., and N.M. Fennessey. (1993). L Moment Diagrams Should Replace Product Moment Diagrams. *Water Resources Research* **29(6)**, 1745-1752.

Vose, D. (1996). **Quantitative Risk Analysis: A Guide to Monte Carlo Simulation Modelling**. John Wiley and Sons, New York, 328pp.

Wald, A., and J. Wolfowitz. (1946). Tolerance Limits for a Normal Distribution. *Annals of Mathematical Statistics* **17**, 208-215.

Webster, R., and M.A. Oliver. (1990). **Statistical Methods in Soil and Land Resource Survey**. Oxford University Press, New York, 316pp.

Weinstein, N.D., P.M. Sandman, and W.K. Hallman. (1994). Testing a Visual Display to Explain Small Probabilities. *Risk Analysis* **14**(6), 895-896.

Weisberg, S. (1985). **Applied Linear Regression**. Second Edition. John Wiley and Sons, New York, 324pp.

Weisberg, S., and C. Bingham. (1975). An Approximate Analysis of Variance Test for Non-Normality Suitable for Machine Calculation. *Technometrics* **17**(1), 133-134.

Wicksell, S.D. (1917). On Logarithmic Correlation with an Application to the Distribution of Ages at First Marriage. *Medd. Lunds. Astr. Obs.* **84**, 1-21.

Wilks, S.S. (1941). Determination of Sample Sizes for Setting Tolerance Limits. *Annals of Mathematical Statistics* **12**, 91-96.

WSDOE. (1992). **Statistical Guidance for Ecology Site Managers**. Washington State Department of Ecology, Olympia, WA.

WSDOE. (1993). **Statistical Guidance for Ecology Site Managers, Supplement S-6: Analyzing Site or Background Data with Below-Detection Limit or Below-PQL Values (Censored Data Sets)**. Washington State Department of Ecology, Olympia, WA.

Zacks, S. (1970). Uniformly Most Accurate Upper Tolerance Limits for Monotone Likelihood Ratio Families of Discrete Distributions. *Journal of the American Statistical Association* **65**, 307-316.

Zar, J.H. (1984). **Biostatistical Analysis.** Second Edition. Prentice-Hall, Englewood Cliffs, NJ, 718pp.

Zar, J.H. (1996). **Biostatistical Analysis.** Third Edition. Prentice-Hall, Englewood Cliffs, NJ.

APPENDIX B:

License Agreement

Purchase or operation of ENVIRONMENTALSTATS for S-PLUS indicates your acceptance of the following terms and conditions. If you have any questions or do not agree to the License Agreement, please contact PSI immediately.

1. **LICENSE.** Probability, Statistics & Information (PSI), grants you a personal, non-transferable and non-exclusive right ("License") to use and store ENVIRONMENTALSTATS for S-PLUS and any future temporary fixes, programming temporary fixes, updates, and upgrades that are received separately by you (collectively, the "Software") on any single designated computer or workstation at a single location. It is your duty to employ a reasonable mechanism to ensure that at any one time the number of concurrent Software users on a multi-user system never exceeds the number of users on the License you have lawfully acquired. However, if you have obtained multiple Licenses of the Software, then you may have as many copies of the Software in use as you have Licenses for that system.
2. **MODIFICATION.** You may modify the Software and/or merge it into another program for your own use on the Licensed computer or workstation. Any portion of the Software merged into another program is the exclusive property of PSI and will be subject to the terms and conditions of this Agreement. You must reproduce and include all copyright notices on any copy, modification, or portion of the Software that is merged into another program.
3. **COPYRIGHT.** The Software is licensed, not sold, to you. No title to the intellectual property in ENVIRONMENTALSTATS for S PLUS is transferred to you by this License. The Software is protected by the copyright laws of the United States and various international treaties. You may not copy the Software, except that, for each License you obtain, you may make one copy of the Software solely for backup purposes. All authorized copies and/or modifications that you make shall contain the following notice: "©1995 Steven P. Millard, All Rights Reserved." Any copy of the Software is subject to this agreement. You may not copy the written materials accompanying the Software.
4. **TRANSFER and USE RESTRICTIONS.**
    a) You may not sell, rent, license, lend, or otherwise transfer the Software to any other party including a subsidiary except as expressly authorized by PSI in writing, except that you may transfer on a permanent basis all of your license rights to the Software if you do not keep any copies of the Software or accompanying materials and the recipient agrees to the terms of this License. cf. to 10.
    b) You acknowledge that the Software and any updates thereto contain confidential and trade secret material, and as such you or your agents may not reverse engineer, decompile, disassemble, or otherwise attempt to discover the source code to the Software.

5. **TERM.**
   a) This license is effective until terminated. You may terminate it at any time by destroying all copies of the Software in your possession. It will also terminate, without the necessary or written notice from PSI, if you do not comply with any term or condition of this Agreement or if you fail to pay the appropriate fees. Upon such termination, you must destroy all originals and copies of the Software in your possession.
   b) PSI warrants that PSI has the right to license this software.
   c) You agree upon termination to notify PSI in writing and to destroy the Software and all copies, modifications, and merged portions in any form, as well as all related software documentation.
   d) In the event of termination, PSI has no obligation to refund to you any fees paid under this Agreement.
6. **LIMITED WARRANTY.**
   a) For ninety (90) days from the date of initial receipt, PSI warrants that the media (for example, diskette) on which the Software is contained will be free from defects in material and workmanship. This warranty does not cover damage caused by improper use or neglect.
   b) PSI does not warrant the contents of the Software. IT IS FURNISHED "AS IS" AND WITHOUT WARRANTY AS TO THE PERFORMANCE OR RESULTS YOU MAY OBTAIN USING IT. It may not be error free. The entire risk as to the results and performance of the Software is assumed by you.
   c) PSI's entire liability and your exclusive remedy for failure of the Software to comply with this limited warranty shall be, at PSI's sole option, to either (i) repair or replace the software or (ii) return a fair portion of the price paid for the software. To receive warranty service during the 90-day warranty period, you must return the Software (postage paid) with a sales receipt and a description of the problem to the place where you obtained it. Any replacement software will be warranted for the remainder of the original 90-day warranty period or thirty (30) days from the date you receive the replacement, whichever is longer. This Limited Warranty applies only to the original recipient of the software and is void if the failure of the software has resulted from accident, abuse, or misuse.
7. **DISCLAIMER OF WARRANTY AND LIMITATION OF REMEDIES.** You understand and agree as follows:
   a) THE LIMITED WARRANTY IN THIS AGREEMENT REPLACES ALL WARRANTIES, EXPRESS OR IMPLIED, INCLUDING, BUT NOT LIMITED TO, ANY WARRANTIES OF MERCHANTABILITY, FITNESS FOR A PARTICULAR PURPOSE, AND NONINFRINGEMENT. NO ORAL OR WRITTEN INFORMATION OR ADVICE GIVEN BY PSI, ITS EMPLOYEES, DISTRIBUTORS, DEALERS, OR AGENTS SHALL INCREASE THE SCOPE OF THE ABOVE WARRANTIES OR CREATE ANY NEW WARRANTIES. PSI AND ITS SUPPLIERS DISCLAIM AND EXCLUDE ALL OTHER IMPLIED OR EXPRESS WARRANTIES.
   b) IN NO EVENT WILL PSI OR OUR SUPPLIERS HAVE ANY LIABILITY FOR ANY INDIRECT, SPECIAL, INCIDENTAL, CONSEQUENTIAL DAMAGES (INCLUDING WITHOUT LIMITATION ANY LOSS OF PROFITS OR DATA, BUSINESS INTERRUPTION OR OTHER PECUNIARY LOSS) ARISING OUT OF THE USE OR INABILITY TO USE THE SOFTWARE, OR THE ACCOMPANYING MATERIALS, OR PERFORMANCE OF OR FAILURE TO PERFORM ANY RELATED SERVICES, EVEN IF PSI OR ITS SUPPLIERS HAVE KNOWLEDGE OF THE POTENTIAL LOSS OR DAMAGE. IN NO EVENT SHALL THE AGGREGATED LIABILITY OF PSI AND ITS SUPPLIERS (WHETHER IN CONTRACT, WARRANTY, TORT (INCLUDING NEGLIGENCE), PRODUCT LIABILITY, OR OTHER THEORY) (I) ARISING OUT OF THE USE OR INABILITY TO USE THE SOFTWARE EXCEED THE LICENSE FEE

PAID BY YOU FOR THE SOFTWARE; OR (II) ARISING OUT OF THE PERFORMANCE OF OR FAILURE TO PERFORM ANY RELATED SEVICES EXCEED THE PRICE YOU PAID FOR SUCH SERVICES DURING THE YEAR IN WHICH THE CLAIM AROSE. FURTHER, PSI WILL NOT BE LIABLE FOR ANY DELAY OR FAILURE TO PERFORM ANY RELATED SERVICES OR ANY OF ITS OBLIGATIONS UNDER THIS LICENSE AS A RESULT OF ANY CAUSES OR CONDITIONS BEYOND PSI'S RESONABLE CONTROL.

c) This limited warranty gives you specific legal rights, and you may also have other rights that vary from state to state in the United States. Some states do not allow the exclusion or limitation of implied warranties, incidental damages, or consequential damages, so the above limitation or exclusion may not apply to you.

8. **GOVERNMENT LICENSEE - RESTRICTED RIGHTS.** The Software and accompanying written materials are provided with restricted rights. Use, duplication, or disclosure is subject to restrictions set forth in subparagraph(c)(1)(ii) of the Rights in Technical Data and Computer Software clause at 48 C.F.R. § 252.227-7013 or in subparagraphs (c)(1) and (2) of the Commercial Computer Software-Restricted Rights clause at 48 C.F.R. §52.227-19, as applicable. The contractor/manufacturer is PSI at the address described below.
9. **DUAL MEDIA.** If you acquire a single License and the Software is supplied to you in both 3.5-inch and 5.25-inch disks, then you may use only the disks appropriate for your single designated computer or workstation. You shall not use the other disks on any other computer or workstation, or rent, sell, lease, or otherwise transfer them to anyone else except as permitted in this Agreement.
10. **EXPORT.** IN YOUR USE OF the Software, YOU MUST COMPLY WITH THE U.S. EXPORT ADMINISTRATION ACT, REGULATIONS OF THE U.S. DEPARTMENT OF COMMERCE, AND OTHER EXPORT CONTROLS OF THE UNITED STATES ("EXPORT LAWS"), AS THEY MAY BE AMENDED, AS WELL AS THE IMPORT LAWS OF THE FOREIGN NATION TO WHICH YOU INTEND TO TRANSFER. YOU AGREE AND CERTIFY THAT NEITHER the Software NOR ANY PRODUCTS DERIVED OR GENERATED THEREFROM WILL BE ACQUIRED, SHIPPED, TRANSFERRED, OR REEXPORTED, DIRECTLY OR INDIRECTLY, INTO ANY COUNTRY PROHIBITED BY THE EXPORT LAWS OR WILL BE USED FOR ANY PURPOSE PROHIBITED BY THE SAME.
11. **MISCELLANEOUS.**
    a) This Agreement will be governed by the laws of the State of Washington without reference to its choice of law principles. The parties agree that jurisdiction and venue for any dispute arising hereunder shall be in the Superior Court of King County, Washington. This Agreement is made at Seattle, Washington.
    b) This Agreement shall inure to the benefit of PSI, its successors, administrators, heirs, and assigns.
    c) This Agreement constitutes the full and complete understanding between PSI and you with respect to all matters related to the subject hereof. The Agreement may only be modified by a written amendment signed by PSI. No provision appearing on any form originated by you (other than terms contained therein identifying the products and quantities ordered) that is different from or in addition to the provisions of this License (whether or not it would materially alter this License) shall be applicable unless it is expressly agreed to in writing by PSI. Statements or representations made by employees, distributors, resellers, or any other third party do not constitute warranties by PSI, do not bind PSI, and should not be relied upon by you.
    d) If any portion or provision of this Agreement is deemed to be invalid or unenforceable under any applicable statute or rule of law, it is to that extent deemed to be omitted. This Agreement shall continue to be effective in all other respects.
    e) The waiver or failure of either party to exercise in any respect any right provided for in this Agreement shall not be deemed a waiver of any further right under this Agreement.

APPENDIX C :

Index

This appendix provides an index to Chapters 0-10. Please note that all functions are listed under the heading Functions, and all data sets are listed under the heading Datasets.

`.Distribution.frame`, 117, 165

A

Accuracy, 173
ACL. *See* Alternate concentration limit
Alternate concentration limit, 185
Analysis of variance, 216
    nonparametric one-way, 216
    nonparametric two-way, 216
ANOVA. *See* Analysis of variance
Autocorrelation plot, 93
Average. *See* Mean

B

Bar chart, 65, 66
Beta distribution, 118, 124, 166
Bias, 173
Binomial coefficient, 153
Binomial distribution, 118, 124, 153, 166
    computing cdf, 137
    computing pdf, 135, 153
    computing quantiles, 140
    confidence interval for p, 183
    estimating parameters, 183
    generating quantities from, 155
    maximum likelihood estimator, 173
    mean and variance, 155
    method of moments estimator, 170
    minimum variance unbiased estimator, 174
    plotting cdf, 137
Binomial test, 216
Bivariate data, 96
Bootstrapping, 272
Boxplot, 65, 75, 76, 77
Brushing, 96, 103

C

Cauchy distribution, 124
CDF. *See* Cumulative distribution function
Censored data, 69, 243
    data qualifiers, 248
    detection limit, 247
    empirical cdf plot, 244, 249
    estimation, 258
        Lognormal distribution, 244, 259
        Normal distribution, 244, 258
        Poisson distribution, 244
    examples of, 246
    goodness-of-fit test, 244
        plotting results, 263
    left vs. right vs. double, 245
    linear rank test, 244, 266
    Q-Q plot, 244, 253
    single vs. multiple, 245
    terminology, 245
    truncated vs. censored, 245
    type I vs. type II, 246
    vs. zero-modified distribution, 255
Central Limit Theorem, 149
Chi distribution, 118, 124
Chi-square distribution, 118, 124
Chi-square test

goodness-of-fit, 221
two-way table, 217
Cloud plot, 96, 103
Coefficient of variation, 142, 149
Conditioning plots. *See* Multi-panel conditioning plots
Confidence coefficient, 175
prediction interval, 197
Confidence interval, 175
Binomial distribution, 47, 183
for quantile, 185
general formula, 176
Land's method, 181
Lognormal distribution, 181
Normal distribution, 45, 176, 177
related to hypothesis test, 221
sample size for, 44
Confidence level, 175
prediction interval, 197
tolerance interval, 205
Continuous random variable, 129
Contour plot, 96, 104
Coplots. *See* Multi-panel conditioning plots
Correlation, 93
estimates of, 234
test for, 217
Coverage, 205
Cuachy distribution, 118
Cumulative distribution function, 127, 135, 136
computing, 136
plotting, 137
Customizing. *See* ENVIRONMENTALSTATS for S-PLUS
CV. *See* Coefficient of variation

D

Data qualifiers, 40, 248
Datasets
`air`, 94, 97, 101, 107, 111
`environmental`, 94
`epa.89b.aldicarb1.df`, 45, 51, 178
`epa.89b.aldicarb2.df`, 206
`epa.89b.cadmium.df`, 47, 56
`epa.89b.lead.df`, 66, 186
`epa.89b.loglead.df`, 53
`epa.89b.sulfate.df`, 258
`epa.92c.arsenic2.df`, 69, 203
`epa.92c.benzen1.df`, 190
`epa.92c.benzene1.df`, 87, 202, 209
`epa.92c.benzene2.df`, 200
`epa.92c.chrysene.df`, 188, 207
`epa.92c.copper2.df`, 192, 210
`epa.92c.nickel1.df`, 222
`epa.92c.zinc.df`, 255, 262
`epa.92d.chromium.vec`, 181
`epa.94b.tccb.df`, 9, 62, 129, 229, 233
`epa.98b.cadmium.df`, 183
`helsel.cohn.88.silver.df`, 249, 253
`millard.deverel.88.df`, 251, 266
Delta distribution. *See* Zero-modified distribution:Lognormal
Density plot, 65, 74
Design. *See* Sampling design
Detection limit, 69, 183, 247
Discrete random variable, 129
Dot chart, 65, 66
Duplicate points, 88

E

ecdf plot. *See* Empirical cdf plot
Empirical cdf plot, 65, 78, 80
with censored data, 244, 249
Empirical cumulative distribution function, 127
Empirical distribution, 124
Environmental statistics, 23
ENVIRONMENTALSTATS for S-PLUS, 24
customizing, 31
UNIX, 35
Windows, 32
help, 6, 28
font sizes, 29
UNIX, 7, 28, 30
using, 30
Windows, 6, 28
installing, 2
UNIX, 4
Windows, 2
intended users, 25
removing, 31
starting, 6, 27
system requirements, 1
technical support, 7
user pre-requisites, 1
Estimation, 165
Binomial distribution, 183
estimating functions, 165
Lognormal distribution, 181
with censored data, 244, 259
maximum likelihood, 171
Binomial distribution, 173
Normal distribution, 173
mean squared error, 175
method of moments, 168
Binomial distribution, 170
Normal distribution, 170
methods of, 168
minimum variance unbiased, 173
Binomial distribution, 174
Normal distribution, 174
nonparametric, 167, 191
Normal distribution, 177
with censored data, 244, 258
of distribution parameters, 165, 168
of quantiles, 167, 185
Lognormal distribution, 188
nonparametric, 191
Normal distribution, 186

Poisson distribution, 190
Poisson distribution
with censored data, 244
with censored data, 258
Euler's constant, 160
Expected value, 141
Exponential distribution, 119, 124, 166
Exposure
computing, 270
Exposure assessment. *See* Risk assessment
Expression
evaluating, 270
Extreme value distribution, 119, 124, 158, 166
Generalized, 161
generating quantities from, 161
goodness-of-fit test, 221
mode, mean, and variance, 160
probability density function, 158

F

F distribution, 119, 124
False negative rate, 220
False positive rate, 220
Fisher's exact test, 217
ForeHelp, 28
Functions
`.First`, 28, 31
?, 28
`acf`, 94, 95
`aov`, 216
`aov.n`, 50, 53
`aov.power`, 50, 53
`barchart`, 65, 66
`barplot`, 65, 66
`binom.test`, 216
`boxplot`, 12, 54, 65, 75, 77
`brush`, 96, 103
`bwplot`, 65, 75, 77
`cdf.compare`, 12, 65, 78, 127
`cdf.compare.censored`, 244, 249, 251, 252
`cdfplot`, 127, 137, 307
`chisq.gof`, 216, 221
`chisq.test`, 217
`choose`, 153
`ci.binom.half.width`, 44, 47
`ci.binom.n`, 44, 47
`ci.norm.half.width`, 44, 45
`ci.norm.n`, 44, 45, 46
`cloud`, 96, 103
`co.intervals`, 111
`contour`, 96, 104, 106
`contourplot`, 96, 104, 106
`coplot`, 96, 111
`cor.test`, 217
`dbinom`, 135, 155
`density`, 65
`densityplot`, 65, 74
`devd`, 161
`dlnorm`, 150
`dlnorm.alt`, 135, 150
`dnorm`, 147
`dotchart`, 65
`dotplot`, 65
`dpois`, 158
`ebeta`, 166
`ebinom`, 166, 183, 184
`ecdfplot`, 12, 65, 78, 127
`ecdfplot.censored`, 244, 249
`eevd`, 166
`eexp`, 166
`egamma`, 166
`egeom`, 166
`egevd`, 166
`ehyper`, 166
`elnorm`, 16, 166, 181
`elnorm.alt`, 17, 166, 181
`elnorm.alt.multiply.censored`, 244, 259
`elnorm.alt.singly.censored`, 244
`elnorm.multiply.censored`, 244, 259
`elnorm.singly.censored`, 244
`elnorm3`, 166
`elogis`, 166
`enbinom`, 166
`enorm`, 166, 177
`enorm.multiply.censored`, 244
`enorm.singly.censored`, 244, 258
`epois`, 166
`epois.multiply.censored`, 244
`epois.singly.censored`, 244
`eqlnorm`, 19, 167, 186, 188
`eqnorm`, 167, 186
`eqnpar`, 167, 186, 191
`eqpois`, 186, 190
`equal.count`, 112
`eunif`, 166
`eval.expr`, 270, 305, 315
`eweibull`, 166
`expand.grid`, 105
`ezmlnorm`, 167
`ezmlnorm.alt`, 167
`ezmnorm`, 167
`fact`, 153
`fisher.test`, 217
`for`, 336
`friedman.test`, 216
`full.summary`, 9, 63
`help`, 6, 28
`hist`, 10, 65, 72
`histogram`, 65, 72
`image`, 96, 107
`kendall.trend.test`, 217, 235, 237
`kruskal.test`, 216
`ks.gof`, 216, 221
`kurtosis`, 64
`l.moment`, 64
`lag.plot`, 94
`levelplot`, 96, 107

lm, 217
location.m, 61
loess, 104, 105
mad, 61
mantelhaen.test, 217
mcnemar.test, 217
mean, 61
median, 61
module, 6, 27, 31
multicomp, 217
pairs, 96, 101
pbinom, 137, 155
pdfplot, 127, 136, 307
persp, 96, 109
pevd, 161
pie, 65, 69
piechart, 65, 69
plnorm, 150
plnorm.alt, 137, 150
plot, 96
plot.ci.binom.design, 44, 47
plot.ci.norm.design, 44, 45
plot.gof.censored.summary, 263
plot.gof.summary, 18, 225
plot.prop.test.design, 50, 56
plot.t.test.design, 50, 51, 220
pnorm, 147
ppcc.evd.gof, 216, 221
ppcc.norm.gof, 216, 221
ppcc.norm.singly.censored.gof, 244, 261
ppois, 158
pred.int.lnorm, 196, 199
pred.int.norm, 196, 199
pred.int.npar, 196, 203
pred.int.npar.n, 196, 203
pred.int.pois, 196, 202
predict, 105
prop.test, 183, 217
prop.test.mdd, 50, 55
prop.test.n, 50, 55
prop.test.power, 50, 55
pw.moment, 64
qbinom, 140, 155
qevd, 161
qlnorm, 150
qlnorm.alt, 140, 150
qnorm, 147
qpois, 158
qq, 65, 81
qqmath, 65, 81
qqnorm, 65, 81
qqplot, 14, 65, 81, 83, 85, 86, 92, 127
qqplot.censored, 244, 253, 255
qqplot.gestalt, 89, 127
quantile, 61, 185
quantile.test, 19, 216, 233
quantile.test.p.value, 233
rbinom, 128, 155
revd, 161
rlnorm, 150
rlnorm.alt, 144, 150
rmvnorm, 270, 310
rnorm, 147, 310
rpois, 158
runif, 276
sapply, 62, 336
scale.a, 61
scale.tau, 61
scatter.smooth, 96, 99
search, 27
seasonal.kendall.trend.test, 217, 235, 239
set.seed, 90, 144, 276
sf.gof, 216, 221
sf.singly.censored.gof, 244, 261
simulate.mv.matrix, 270, 283, 309
simulate.vector, 270, 283, 309
skew, 64
spin, 96, 103
splom, 96, 101
stripplot, 65, 71
summary, 63
sw.gof, 17, 216, 221, 224, 227
sw.singly.censored.gof, 244, 261
t.test, 177, 216
t.test.n, 50, 51
t.test.power, 50, 51
t.test.scaled.mdd, 50, 51
tol.int.lnorm, 197, 206
tol.int.norm, 197, 206
tol.int.norm.K, 197
tol.int.npar, 197, 210
tol.int.npar.n, 197, 210
tol.int.pois, 197, 209
trellis.device, 65
ts.plot, 94
two.sample.linear.rank.test, 19, 216, 229, 244
two.sample.linear.rank.test.censored, 266
var, 61
var.test, 216
wilcox.test, 216, 229
wireframe, 96, 109
xyplot, 96, 100, 112

G

Gamma distribution, 120, 125, 166
Gaussian distribution. *See* Normal distribution
Generalized extreme value distribution, 119, 125, 161, 166
Geometric distribution, 120, 125, 166
Gestalt. *See* Q-Q plot
Goodness-of-fit test, 216, 221
  plotting results of, 224
    with censored data, 263
  with censored data, 244, 261, 262
Graphics devices, 65

H

Help. *See* ENVIRONMENTALSTATS for S-PLUS
Histogram, 65, 72, 76
HTML, 7, 28
Hypergeometric distribution, 120, 125, 166
Hypothesis test, 50, 93, 218
  analysis of variance, 216
  binomial, 216
  chi-square for two-way table, 217
  correlation, 217
  Fisher's exact for two-way table, 217
  goodness-of-fit, 216, 221
    chi-square, 221
    Kolmogorov-Smirnov, 221
    probability plot correlation coefficient, 221
    Shapiro-Francia, 221
    Shapiro-Wilk, 221
    with censored data, 244, 262
  linear rank, 216, 229
    with censored data, 244, 266
  linear regression, 217
  Mantel-Haenszel for 3-D table, 217
  McNemar's test for paired count data, 217
  Mood's median, 229, 231
  multiple comparisons, 217
  nonparametric one-way analysis of variance, 216
  nonparametric test for trend, 217, 234
  nonparametric two-sample, 229
  nonparametric two-way analysis of variance, 216
  parametric vs. nonparametric, 219
  power, 219
  proportions, 217
  p-value, 220
  quantile, 216, 233
  related to confidence interval, 221
  signed rank sum, 216
  significance level, 219
  Student's t-test, 216
  two-sample variance, 216
  two-sided vs. one-sided, 218
  type I error, 219
  type II error, 219
  Wilcoxon rank sum, 216, 229, 231

I

Image plot, 96, 107
Installing. *See* ENVIRONMENTALSTATS for S-PLUS

J

Judgment sampling. *See* Sampling

K

Kendall's tau, 234
Kendall's trend test, 217, 235
  with seasons, 217, 235
Kolmogorov-Smirnov goodness-of-fit test, 221
Kurtosis, 64, 142, 143

L

Land's method, 181
Latin Hypercube sampling, 270, 278
  algorithm for, 279
  ENVIRONMENTALSTATS for S-PLUS functions for, 283
  properties, 282
  replicated, 293
  vs. simple random sampling, 278, 283
Law of Large Numbers, 132
Level plot, 96, 107
LHS. *See* Latin Hypercube sampling
`license.EnvironmentalStats`, 8
Linear rank test, 216, 229
  with censored data, 244, 266
Linear regression, 217
L-Moments, 64
Locally most powerful test, 229
Loess, 104
Loess smooth, 98, 100, 111
Logistic distribution, 120, 125, 166
Lognormal distribution, 120, 125, 147, 166, 167
  alternative parameterization, 149, 166
  coefficient of variation, 149
  computing cdf, 137
  computing pdf, 135
  computing quantiles, 140
  confidence interval for mean, 181
  estimating parameters, 181
    with censored data, 244, 259
  estimating quantiles, 188
  generating quantities from, 150
  Land's method, 181
  median, 149
  mixture, 121, 125, 161
  plotting cdf, 137
  prediction interval for, 196, 199
  probability density function, 149
  random numbers from, 144
  three-parameter, 121, 125, 151, 166
  tolerance interval for, 197, 206
  truncated, 121, 125
  zero-modified, 123, 124, 126, 162

M

Mantel-Haenszel test, 217
Masked functions, 27
Maximum concentration limit, 51, 178, 185
Maximum likelihood estimator, 171
  Binomial distribution, 173
  Normal distribution, 173
MCL. *See* Maximum concentration limit
McNemar's test, 217
Mean, 141, 168
  sample, 61, 169, 173

unbiased estimator, 173
Mean squared error, 175
Median, 140
sample, 61
Method of moments estimator, 168
Binomial distribution, 170
Normal distribution, 170
Minimum variance unbiased estimator, 173
Binomial distribution, 174
Normal distribution, 174
Mixture distribution, 161
in risk assessment, 294, 330
Lognormal, 121, 125, 161
Normal, 122, 126, 161
MLE. *See* Maximum likelihood estimator
MME. *See* Method of moments estimator
Mode, 140
Moment, 168
central, 169
Monte Carlo simulation, 271
in risk assessment, 309
in sensitivity analysis, 287, 291
in uncertainty analysis, 287, 293
input parameters, 271
related to bootstrapping, 272
sum of two normal random variables, 273
Mood's median test, 229, 231
MSE. *See* Mean squared error
Multi-panel conditioning plots, 96, 111
Multiple comparisons, 195, 217
Multivariate data, 96
MVUE. *See* Minimum variance unbiased estimator

N

Negative binomial distribution, 121, 126, 166
Non-central distribution
Beta, 118
Chi, 119
F, 119
t, 123
Nonparametric estimation, 191
of quantiles, 191
Nonparametric prediction interval, 196, 203
Nonparametric test. *See* Hypothesis test:nonparametric
Nonparametric tolerance interval, 197, 210
Normal distribution, 122, 126, 144, 166, 167
confidence interval for mean, 176, 177
estimating parameters, 177
with censored data, 244, 258
estimating quantiles, 186
generating quantities from, 147
maximum likelihood estimators, 173
mean and standard deviation, 146
method of moments estimators, 170
minimum variance unbiased estimators, 174
mixture, 122, 126, 161
multivariate, 270
prediction interval for, 196, 199
probability density function, 144
sum of two random variables, 273
tolerance interval for, 197, 206
truncated, 122, 126
zero-modified, 124, 126, 162, 263
Z-transformation, 146

O

Outlier, 76
Outside value, 76

P

Pareto distribution, 122, 126
PDF. *See* Probability density function
Pearson product moment correlation, 234
Percentile, 139, 185
related to quantile, 139
Perspective plot. *See* Wireframe plot
Pie chart, 65, 69
Plotting
autocorrelation, 93
bar chart, 66
bivariate data, 96
boxplot, 75
brushing a scatterplot matrix, 103
cloud plot, 103
contour plot, 103
cumulative distribution function, 137
density plot, 74
dot chart, 66
empirical cdf, 12, 78
with censored data, 244, 249
goodness-of-fit results, 18
goodness-of-fit test
with censored data, 263
histogram, 72
image plot, 103
loess smooth, 98
multi-panel conditioning plots, 111
multivariate data, 96
pie chart, 69
power and sample size, 52, 57
probability density function, 136
Q-Q plot, 80, 87, 92
Gestalt, 89
with censored data, 253
quantile plot, 78
results of goodness-of-fit test, 224
sample size for confidence intervals, 46, 49
scatterplot, 96
scatterplot matrix, 101
strip plot, 71
Tukey mean-difference plot, 85
wireframe plot, 103
Plotting functions, 65
PMF. *See* Probability mass function
Poisson distribution, 122, 126, 156, 166, 167

estimating parameters
with censored data, 244
estimating quantiles, 190
generating quantities from, 158
mean and variance, 158
prediction interval for, 196, 202
probability density function, 156
tolerance interval for, 197, 209
Population, 23, 39, 117
Power, 42, 43, 50, 195, 219
Binomial distribution, 55
Normal distribution, 51, 53
Precision, 173
Prediction interval, 195, 197
confidence coefficient for, 197
confidence level for, 197
Lognormal distribution, 196, 199
nonparametric, 196, 203
Normal distribution, 196, 199
Poisson distribution, 196, 202
related to tolerance interval, 205
Probabilistic risk assessment. *See* Risk assessment
Probability, 24
Probability density function, 74, 117, 127, 132
Binomial distribution, 153
computing, 135
Extreme value distribution, 158
Lognormal distribution, 149
Normal distribution, 144
plotting, 136
Poisson distribution, 156
Probability distribution, 117, 130
describing, 140
Probability mass function, 132
Probability plot. *See* Q-Q plot
Probability plot correlation coefficient goodness-of-fit test, 221
Probability-weighted moments, 64
Proportions test, 217
P-value, 220

Q

QA. *See* Quality Assurance
QC. *See* Quality Control
Q-Q plot, 65, 78, 80, 92, 127
Gestalt, 89, 91
Lognormal distribution, 83
Normal distribution, 81, 89
other distributions, 87
Poisson distribution, 87, 91
Tukey mean-difference, 85, 90
with censored data, 244, 253
with fitted line, 83, 85
zero-modified normal distribution, 255
Quality Assurance, 42
Quality Control, 42
Quantile, 139, 185
computing, 139
confidence interval for, 185
confidence interval for related to tolerance interval, 206
estimating, 185
Lognormal distribution, 188
nonparametrically, 191
Normal distribution, 186
Poisson distribution, 190
Quantile plot. *See* Empirical cdf plot
Quantile test, 216, 233
Quantile-Quantile plot. *See* Q-Q plot

R

Random numbers, 276
from arbitrary distribution, 277
from Uniform distribution, 276
generating, 143, 270, 276, 309
in S-PLUS, 276
inverse transformation method, 277
Latin Hypercube sampling, 270, 278
multivariate, 270
period of generator, 276
seed, 276
Random sampling. *See* Sampling
Random variable, 127
continuous, 129
discrete, 129
Regression. *See* Linear regression
Removing. *See* ENVIRONMENTALSTATS for S-PLUS
Risk, 296
computing, 270
point estimate, 301
Risk assessment, 269, 296
cancer potency factor, 296, 299
cancer slope factor, 296
characterizing distribution of exposure and risk, 315
characterizing input distributions, 307, 311
conservative estimate of risk, 302
dose, 296
dose-response curve, 296
functions to describe exposure and risk, 305
guidelines for conducting, 303
guidelines for presenting results of, 304
limitations of, 295
model for, 298
model for benzene in soil, 300
model for PCE in groundwater, 299
Monte Carlo simulation, 309
point estimate of risk, 270, 301
probabilistic risk assessment, 270, 302
RME (reasonable maximum exposure), 301
sensitivity analysis, 322
steps of, 297
uncertainty analysis, 325
using ENVIRONMENTALSTATS for S-PLUS, 305
using Latin Hypercube sampling, 310
using mixture distributions, 330
using simple random sampling, 310
Risk characterization, 297

Risk communication, 298
Risk management, 298

S

Sample, 23, 39
Sample mean, 169, 173
Sample size, 42, 43, 50
  Binomial distribution, 55
  nonparametric prediction interval, 196, 204
  nonparametric tolerance interval, 197, 212
  Normal distribution, 51, 53
Sample standard deviation, 170, 174
Sample variance, 170, 174
Sampling, 41
  judgment, 41, 43
  Latin Hypercube, 270, 278, 279, 282, 283
  probability, 41
  random, 41
  simple random, 41, 270
    vs. Latin Hypercube, 278, 283
Sampling design, 39, 41
Scatterplot, 96
Scatterplot matrix, 96, 101
  interactive, 96
Secon-order random variable, 294
Sensitivity analysis, 287
  example of, 322
  important parameter, 287
  methods of, 289
  sensitive parameter, 287
  using Monte Carlo simulation, 291
Serial correlation, 93
Shapiro-Francia goodness-of-fit test, 221
Shapiro-Wilk goodness-of-fit test, 221
  with censored data, 262
Signed rank sum test, 216
Significane level, 219
Simple random sampling, 41, 270
  vs. Latin Hypercube sampling, 278, 283
Skew, 64, 142
Spearman's rho, 234
SRS. *See* Simple random sampling
Stable distribution, 126
Standard deviation, 141, 169
  sample, 170, 174
Starting. *See* ENVIRONMENTALSTATS for S-PLUS
Statistics, 24
Strip plot, 65, 71
Student's t distribution. *See* t distribution
Student's t-test, 216
Summary statistics, 61
System requirements. *See* ENVIRONMENTALSTATS for S-PLUS

T

t distribution, 122, 126
Technical support. *See* ENVIRONMENTALSTATS for S-PLUS
Three-dimensional cloud plot. *See* Cloud plot
Three-parameter lognormal distribution, 121, 125, 151, 166
Time series, 93
Tolerance interval, 186, 195, 205
  confidence level for, 205
  coverage, 205
  Lognormal distribution, 197, 206
  nonparametric, 197, 210
  Normal distribution, 197, 206
  Poisson distribution, 197, 209
  related to confidence interval for quantile, 206
  related to prediction interval, 205
  β-content, 205
  β-expectation, 205
Trellis graphics, 65, 96, 111
Trend
  test for, 234
Triangular distribution, 123, 126
Truncated distribution
  Lognormal, 121, 125
  Normal, 122, 126
Tukey mean-difference plot. *See* Q-Q plot
Tukey, John, 75
Tutorial, 9
Type I error, 43, 195, 219
Type II error, 43, 219

U

Unbiased, 173
Unbiased estimator, 173
  mean, 173
  variance, 173
Uncertainty, 289
  vs. uncertainty analysis, 289
  vs. variability, 289
Uncertainty analysis, 287
  example of, 325
  methods of, 293
  vs. uncertainty, 289
  with Monte Carlo simulation, 293
Uniform distribution, 123, 126, 166
  random numbers from, 276

V

Variability, 173, 289
  vs. uncertainty, 289
Variance, 141, 169
  sample, 61, 170, 174
  two-sample variance test, 216
  unbiased estimator, 173
`version`, 2, 8
`version.EnvironmentalStats`, 8

W

Weibull distribution, 123, 126, 166
Wilcoxon distribution, 123, 126

Wilcoxon rank sum test, 216, 229, 231
Wireframe plot, 96, 109

Z

Zero-modified distribution, 162, 255
    Lognormal, 123, 124, 126, 162, 167
    Normal, 124, 126, 162, 167, 263
    Q-Q plot for, 255
Z-transformation, 146

ESSENTIAL S-PLUS RESOURCES!

NEW SECOND EDITION!

W. N. VENABLES, University of Adelaide, Australia and **B. D. RIPLEY,** University of Oxford, England

MODERN APPLIED STATISTICS WITH S-PLUS

This second edition is intended for users of S-PLUS 3.4, 4.0, or later. It covers the recent developments in graphics and new statistical functionality. The material on S-PLUS programming has been re-written to explain the full story behind the object-oriented programming features. The authors have written several software libraries which enhance S-PLUS; these and all the datasets used are available on the Internet in versions for Windows and UNIX. There are also on-line complements covering advanced material, further exercises and new features of S-PLUS as they are introduced.

Contents: Introduction • The S Language • Graphical Output • Programming in S • Distributions and Data Summaries • Linear Statistical Models • Generalized Linear Models • Robust Statistics • Non-linear Models • Random and Mixed Effects • Modern Regression • Survival Analysis • Multivariate Analysis • Tree-based Methods • Time Series • Spatial Statistics • Classification • Appendices: A. Datasets and Software • B. Common S-PLUS Functions • C. Using S-PLUS Libraries

1997/545 PP., 144 ILLUS./HARDCOVER/ISBN 0-387-98214-0
STATISTICS AND COMPUTING

ANDREAS KRAUSE, GeneData AG, Basel, and **MELVIN OLSON,** Ares Serono, Geneva, Switzerland

THE BASICS OF S AND S-PLUS

Unlike the S-PLUS manuals, this book is not comprehensive, but instead introduces the most important ideas of S-PLUS through the use of many examples. Each chapter also includes a collection of exercises. This book is well-suited for self-study and as a textbook.

Contents: Introduction • System Design • A First Session • A Second Session • Graphics • Exploring Data • Statistical Modeling • Programming • Input and Output • Useful Hints and Techniques • Special Topics • References

1997/242 PP., 34 ILLUS./SOFTCOVER
ISBN 0-387-94985-2
STATISTICS AND COMPUTING

Springer

JOHN M. CHAMBERS, Bell Laboratories, Lucent Technologies, Murray Hill, NJ

PROGRAMMING WITH DATA

A Guide to the S Language

Here is a thorough and authoritative guide to the latest version of the S language and to its programming environment, the premier software platform for computing with data. *Programming With Data* describes a new and greatly extended version of S, and is written by the chief designer of the language. The book is a guide to the complete programming process, starting from simple, interactive use and continuing through ambitious software projects. S is designed for computing with data—for any project in which organizing, visualizing, summarizing, or modeling data is a central concern. Its focus is on the needs of the programmer/user, and its goal is "to turn ideas into software, quickly and faithfully." S is a functional, object-based language with a huge library of functions for all aspects of computing with data. Its long and enthusiastic use in statistics and applied fields has also led to many valuable libraries of user-written functions. The new version of S provides a powerful class/method structure, new techniques to deal with large objects, extended interfaces to other languages and files, object-based documentation compatible with HTML, and powerful new interactive programming techniques. This version of S underlies the S-PLUS system, versions 5.0 and higher.

1998/480 PP./SOFTCOVER
ISBN 0-387-98503-4

STEPHEN P. KALUZNY, SILVIA C.VEGA, TAMRE P. CARDOSO and **ALICE A SHELLY,** all of MathSoft, Inc., Seattle Division

S+SPATIALSTATS

User's Manual for Windows® and UNIX®

S+SPATIALSTATS is the first comprehensive, object-oriented package for the analysis of spatial data. *S+SPATIALSTATS* was created specifically for the exploration and modeling of spatially correlated data. It can be used to analyze data arising in all areas where data is sampled spatially. This user's manual provides the documentation for the *S+SPATIALSTATS* module.

Contents: Introduction to Spatial Data and S+SPATIALSTATS • Getting Started with S+SPATIALSTATS • Visualizing Spatial Data • Analyzing Geostatistical Data • Analyzing Lattice Data • Analyzing Spatial Point Patterns • and S+SPATIALSTATS and GIS • Appendices: A: Installing S+SPATIALSTATS • B. S+SPATIALSTATS Index by Category • C. Sample Data Sets • D. Function Reference

1998/384 PP., 122 ILLUS./SOFTCOVER
ISBN 0-387-98226-4

ANDREW BRUCE and **HONG-YE GAO,** both of MathSoft, Inc., Seattle Division

APPLIED WAVELET ANALYSIS WITH S-PLUS

This book introduces applied wavelet analysis through the S-Plus software system. Using a visual data analysis approach, wavelet concepts are explained in a way that is intuitive and easy to understand. In addition to wavelets, a range of related signal processing techniques are covered.

1996/338 PP., 192 ILLUS./SOFTCOVER
ISBN 0-387-94714-0
STATISTICS AND COMPUTING

To Order or For Information:

In North America:
Call: 1-800-SPRINGER • **Fax:** (201)-348-4505 • **Write:** Springer-Verlag New York, Inc., 175 Fifth Avenue, New York, NY 10010 • **Visit** our Website: http://www.springer-ny.com • **Email:** orders@springer-ny.com

Outside North America:
Call: +49 (30) 827 870 • **Fax:** +49 (30) 827 87 301 • **Write:** Springer-Verlag, P.O. Box 3113 40, D-10643 Berlin, Germany • **Visit** our Website: http://www.springer.de • **Email:** orders@springer.de

3/98 **Promotion Number H227**